Summer Wildflowers of the Northeast

Summer Wildflowers of the Northeast

A NATURAL HISTORY

Carol Gracie

WITH A FOREWORD BY ROBERT NACZI

PRINCETON UNIVERSITY PRESS · PRINCETON AND OXFORD

For my husband, Scott Mori, with thanks for his encouragement and his patient company on many of the field trips for this project

Copyright © 2020 by Princeton University Press
Requests for permission to reproduce material from this work
should be sent to permissions@press.princeton.edu
Published by Princeton University Press
41 William Street, Princeton, New Jersey 08540
6 Oxford Street, Woodstock, Oxfordshire OX20 1TR
press.princeton.edu

All Rights Reserved

Library of Congress Cataloging-in-Publication Data

Names: Gracie, Carol, author.
Title: Summer wildflowers of the Northeast : a natural history / Carol Gracie.
Description: Princeton, New Jersey : Princeton University Press, [2020] | Includes bibliographical references and index.
Identifiers: LCCN 2019033704 (print) | LCCN 2019033705 (ebook) | ISBN 9780691199344 (paperback) | ISBN 9780691203300 (ebook)
Subjects: LCSH: Wild flowers—Northeastern States.
Classification: LCC QK117 .G682 2020 (print) | LCC QK117 (ebook) | DDC 582.130974—dc23
LC record available at https://lccn.loc.gov/2019033704
LC ebook record available at https://lccn.loc.gov/2019033705

British Library Cataloging-in-Publication Data is available

Editorial: Robert Kirk and Abigail Johnson
Production Editorial: Mark Bellis
Text Design: D & N Publishing, Wiltshire, UK
Cover Design: Lorraine Doneker
Production: Steven Sears
Publicity: Matthew Taylor and Julia Hall
Copyeditor: Lucinda Treadwell
Cover Credit: Carol Gracie

All photos by the author unless otherwise credited

This book has been composed in Minion Pro
Printed on acid-free paper. ∞
Printed in Singapore
1 3 5 7 9 10 8 6 4 2

Contents

Foreword by Robert Naczi	vi
Preface	viii
Acknowledgments	xi

Species Accounts

Alpine Wildflowers	3
American Cranberry	17
American Ginseng	27
American Lotus	35
Asters	43
Beechdrops	51
Blackberry-lily	57
Bog Orchids	65
Broad-leaved Helleborine	75
Buckbean	83
Bunchberry	89
Cardinal Flower	97
Chicory	107
Common Milkweed	117
Common Mullein	141
Evening-primrose	153
Fringed Gentian	159
Fringed Orchids	169
Goldenrods	179
Grass-of-Parnassus	193
Indian Pipe	201
Jewelweed	209
Jimsonweed	221
Lilies	229
Partridge-berry	241
Passion-flowers	247
Pipsissewa and Related Species	259
Prickly Pear	265
Purple Pitcher Plant	277
Queen Anne's Lace	287
Showy Lady-slipper	299
Swamp Rose-mallow	305
Wild Leek	317
Wild Lupine	323
Yellow Pond-lily	331
Glossary	336
References	345
Index	365

Foreword

Carol Gracie has done it again. She gave wildflower enthusiasts reason to rejoice with publication of her *Spring Wildflowers of the Northeast: A Natural History*. Now, with *Summer Wildflowers of the Northeast: A Natural History*, we benefit anew from Gracie's exquisite talent for presenting the beauty, biology, and ethnobotany of wild plants in a most engaging way.

The objects of *Summer Wildflowers* are those plants blooming from early June onward in the northeastern United States and adjacent Canada. Gracie seamlessly integrates diverse facets about these plants—history, geography, habitats, human uses, morphology, classification, pollination, conservation, and more. Truly, this book has something for everyone, whether beginner or expert, hiker or gardener, entomologist or etymologist. Though many of the species are familiar and beloved (e.g. Cardinal Flower, Common Milkweed, Goldenrods, Purple Pitcher Plant), we learn new facts and find new reasons to love these plants in the pages of *Summer Wildflowers*.

Most of the species featured are geographically widespread and ecologically important. They occur in a wide variety of habitats, including meadows, roadsides, forests, lakes, marshes, bogs, and mountaintops. Many are native to the Northeast, but a few are Eurasian species introduced so long ago they are fully part of the Northeast's flora (e.g. Chicory, Common Mullein, Queen Anne's Lace). Gracie's knack for choosing fascinating species, richly illustrating the essays with her stunning photographs, and communicating a wealth of botanical information in an understandable way, even when the topic is complex, make the book tremendously alluring. Too, by sharing her experiences and sense of humor, Gracie makes her subjects immediately approachable. I found myself chuckling knowingly when I read her remark about Jewelweed, "Much to the delight of children (and a fair number of adults), the fully ripe fruits explode at the slightest touch." With such personal details, even unfamiliar plants quickly become botanical personalities we want to meet.

It is most fitting that I first met Gracie on a field trip. The occasion was the joint field meeting of the Northeastern Section of the Botanical Society of America, the Philadelphia Botanical Club, and the Torrey Botanical Society. Such a meeting happens only once each year and is attended by a crowd of enthusiasts who visit diverse plant communities, share their love of plants in the field, and learn from each other in a jovial and supportive manner. Within minutes of meeting Gracie, I was impressed with how she embodied the best of the field botanists' community. Extensive expertise, passion for plants, and imparting of knowledge in a generous way characterized Gracie then, and continue to be her guiding principles, as evidenced in *Summer Wildflowers*.

Through *Summer Wildflowers*, Gracie attempts to foster an appreciation for summer's plants, to "encourage you to become a keen observer of wildflowers." Her unique background and abilities enable her to succeed admirably. Gracie's decades of teaching classes, leading field trips, and delivering popular lectures brighten her prose and permit her to masterfully communicate all matters botanical. Over the course of more than 40 years, she has taken tens of thousands of botanical photographs, many of which she has used in five other books she has authored and illustrated. Indeed, the beautiful photographs Gracie saved for *Summer Wildflowers* enliven both the book and the profiled species. Most of all, her passion for wildflowers provides the real spark of this book. Evidence for this passion includes herculean efforts to investigate diverse aspects of life cycles, pollination, and fruit dispersal, including literally tens of thousands of miles of travel to observe species in the field and to capture superb photos. On several occasions, I have witnessed Gracie's dogged quest to learn as much as possible about wildflowers, especially from firsthand experience in the field. These quests serve as inspirations to me, and now readers will benefit from her efforts, too.

Beyond the beautiful photography and clear prose, *Summer Wildflowers* includes many welcome characteristics. Gracie's clear explanations inform, demystify, and entertain, clearly attesting to her talents as an educator. Witness her gift for explaining the derivations of both scientific and common names. These etymologies serve as vivid mnemonics for each of the featured species. Gracie is an entertaining scholar, too. For example, her accounts of the history of introduction of non-native species are well researched and

Foreword

intriguing. We learn how Thomas Jefferson obtained Blackberry-lily, from seeds via a horticulturist he respected so much that Jefferson appointed him to grow living plants from the Lewis and Clark expedition. Gracie's bibliography numbers in the hundreds, thorough documentation that provides ample opportunity for further reading for each one of her chapters. Original observations provide Gracie the opportunity to describe unique aspects of her subjects, relate fascinating stories, and weave a holistic account of the biology of each species.

Summer Wildflowers is a celebration of beauty and complexity in the natural world, and our wonder and fascination with it. It is a story about relationships—yes, about a flower and its pollinators, but ultimately, it is about the relationship Gracie hopes we cultivate with plants. I wish this book had been available when I was first learning wildflowers but am thrilled it is available now.

Robert Naczi
Arthur J. Cronquist Curator of
North American Botany
New York Botanical Garden

Preface

The wildflowers of spring are beloved by all who enjoy nature. The color they bring to the landscape after winter's long period of browns, grays, and white makes the heart sing and spirits rise, drawing wildflower lovers into the woods each spring to seek them out. However, by the time summer arrives, we seem to have become accustomed to nature's multi-hued palette, and sadly, the wildflowers of summer tend to get less of our attention. Yet the natural history of summer-blooming flowers is every bit as interesting as that of spring wildflowers, and in some cases more so because of the greater diversity of insects present during the summer months. The interrelationships between flowers, the insects that visit them, the predators and parasites of those insects, and the animals that disperse their seeds can be even more complex than those involving spring wildflowers.

The essays of this book are intended to encourage you to become a keen observer of wildflowers. Just as serious birders are interested in observing what a bird is eating or doing rather than just ticking off a life list of birds seen, wildflower observers can learn much by taking the time to look carefully at the flowers for insects or other creatures on or near them, and then trying to determine what they are doing and whether it appears to be harmful or beneficial to the plant. Careful observation of wildflowers can open doors into other aspects of natural history, including the lives of caterpillars, butterflies, beetles, and other insects. Such serendipity fosters an appreciation of the intricacy of nature, leading to a broader appreciation of natural history in general. When possible, I like to compare and contrast the life histories of related species I have encountered in other parts of the world with those familiar to us in the Northeast.

For the purposes of this book, I am defining summer more broadly than simply the dates between the summer solstice and the autumn equinox. In my earlier book, *Spring Wildflowers of the Northeast: A Natural History*, I included flowers that bloom in April and May, with one outlier—skunk cabbage—which may be found in bloom as early as mid-February and is considered by many the first wildflower of spring. The flowers included in the present book are those found in bloom from early June through the end of September, again with an outlier—the gentians, many of which do not flower until October.

As in *Spring Wildflowers*, in this book on summer wildflowers I discuss a number of species that inhabit the northeastern quarter of the United States and adjacent Canada from Nova Scotia to Virginia and west to Minnesota and northern Missouri. From a practical standpoint, the majority of the species in this book can be found throughout that range, with a few very interesting species that are more geographically restricted. For the sake of diversity, I have included species from different habitats: alpine (e.g., diapensia and alpine azalea), aquatic (e.g., American lotus), arid rocky or sandy soil (e.g., prickly pear), coniferous forest (e.g., Indian pipe), swamps and bogs (e.g., pitcher plant and several orchids), and fields and roadsides (most of the rest). Blooming times are given primarily for the central portion of the range (New York, New Jersey, Connecticut), yet some species may bloom earlier in the southern part of the range and later further north or higher in elevation.

Both common and scientific names are used. This has the advantage of avoiding confusion, which arises when multiple common names are used for the same species in different parts of its range. The common names in this book are those that are most widely used.

We live in an exciting era when new tools, including DNA sequencing, have resulted in insights into plant relationships, which have sometimes caused plant classification to be reevaluated and changed from long-standing perceptions. When this has occurred with species included in the text, explanations for the changes are offered. This book follows the classification system provided by the Angiosperm Phylogeny Group (http://www.mobot.org/MOBOT/research/APweb/), adopting the currently accepted family names and the genera included in each family. The system is based on both morphological characters and molecular data. This new system may cause readers to be taken aback when they discover that some well-known favorite plants are now placed in different families or genera. For example, in the Scrophulariaceae (figwort family), most northeastern members have been placed into other families, primarily the Plantaginaceae (plantain family) and the Orobanchaceae (broomrape family), with only a few northeastern genera remaining in what were always called "the scrophs." Similarly, the Liliaceae (lily family) have been segregated into several more narrowly defined families. Molecular studies at the genus and species levels have also resulted in changes to what had been previously well-known accepted names; e.g., all our native "asters," formerly in the genus *Aster*, now are placed into six different genera. If a species account has already been published in the ongoing, monumental multivolume work on the flora of the United States and Canada, *Flora of North America North of Mexico*, I have used that generic placement of the species. Otherwise, I have followed the scientific names accepted by TROPICOS, an online resource for nomenclature of New World plants provided by the Missouri Botanical Garden (http://www.tropicos.org).

To supplement my own observations on wildflower natural history, I have relied on research published in scientific journals and books, plus additional information available on the internet (especially through Google Scholar). Botanists labor for months or years to tease out some of the fascinating life histories of plants and their relationships to their environment, but seldom is this interesting information ever distilled for the general reader. I hope to bring the results of their research to a wider audience.

This book is written for all who share an interest in wildflowers. Plants are arranged alphabetically by common name (but always with the scientific name and family given as well) in order to allow even the beginning wildflower lover to easily locate a species treated in the book. I have done my best to avoid the use of scientific jargon, but when I found it necessary to use a term that might not be familiar to the reader, I have defined it either within the text or in the glossary at the end of the book. I have not included formal citations or footnotes within the text since the book is intended for a general audience; however, references for each species are provided under the relevant species divisions in the reference section at the end of the book. The reader can refer to the original sources for further information.

I have chosen to focus on the life histories of 35 of the hundreds of summer-blooming wildflower species found in the northeastern United States, but in many cases, I have included information about other closely related species, bringing the total number of species discussed to somewhere closer to 50. There are plants selected because of their beauty; others are included because they have compelling natural history stories, historical or medicinal uses, or interesting interactions with insects. Both common and less well-known species are covered. In addition to our native species, you may be surprised to find plants that have been introduced from afar if they are commonly found naturalized in our area; some are even considered "weeds"! The ubiquity of some of the introduced species, such as Queen Anne's lace and chicory, may lead novices to assume they are native species, and I often receive questions about them when speaking about plants or leading field trips. These plants, too, have interesting life histories worthy of telling. For the most part, our native summer wildflowers, as well as those introduced from other parts of the world, are found in open sunny areas, but there are enough forest dwellers treated here (e.g., beechdrops, Indian pipe, and some milkweeds) to tempt the reader to enjoy summer wildflower walks within our cool, shady woodlands.

Medicinal uses and reported edibility are mentioned in many of the descriptions, but in no way do I endorse any such use; in fact, I strongly discourage it because mistakes are so often made in plant identification. Certainly, many of our local plants are edible or have medicinal properties, but they have rarely been thoroughly studied or scientifically tested for efficacy or negative side effects. In many cases, reports of medicinal properties are based on the use of plants by Native Americans before and during Colonial times. Native peoples, of necessity, experimented with a vast variety of plants in their attempts to treat disease, sometimes with disastrous results. If the patient recovered, the good outcome would then be attributed to the plant(s) administered, when, in fact, the "cure" might well have been unrelated to the treatment. The standardization and certification requirements necessary for FDA approval of true medications do not apply to herbal remedies sold in today's market, and thus, such remedies might vary in strength or even actual ingredients.

You will find a 10× magnification lens extremely useful for observing the fine details of floral structure; it does for flowers what binoculars do for birds. With such a lens, and a good field guide, you are on your way to discovering a new and fascinating world. I urge you to follow your interests, as curiosity can lead to interesting discoveries in the study of nature. Do remember that in nature there is always variation: you may

Preface

encounter a normally purple flower in its white form, or one with more or fewer floral parts than is "normal."

A short explanation of the naming of plants and the various levels of plant classification should help to alleviate any fear of using Latin-based names, as well as provide an understanding of the relationships among species, genera, and families of plants. In 1753, Carl (Carolus) Linnaeus, the founder of the system we use to name plants and other living organisms, published his monumental work, *Species Plantarum*, in which he provided names for all the species of plants then known to European botanists. Linnaeus sought to simplify and standardize the naming of plants by giving each species a unique combination of two names (termed a binomial) to designate the genus (plural genera) plus its modifier, the species (used for both singular and plural); the species name (also known as the specific epithet) never stands alone; rather it is always used in combination with the genus name—hence the binomial or scientific name. The name (or a standardized abbreviation) of the botanist who first described and published the species is appended to the scientific name of the plant (e.g., *Matelea graciea* Morillo was first named and described by [Gilberto] Morillo). Individual species sharing many characteristics in common were grouped into larger categories called genera, and the genera, in turn, were grouped into families, based on similar characteristics. Today, family names have been standardized to always end in -aceae.

Botanists may choose to give genus or species names to a plant using a descriptive term (e.g., a one-flowered plant might be given the specific epithet *uniflora*: *Monotropa uniflora*), or a geographic locality (e.g., a plant originally collected in New England might be named *novae-angliae*: *Sympyotrichum novae-angliae*). A genus or species may also be given a name that commemorates a person, perhaps the first one to collect the plant or someone of importance (e.g., *Kalmia latifolia* [mountain laurel] was named by Linnaeus to honor Peter Kalm, a student of Linnaeus who collected many new species in North America [*latifolia* indicates that the species has broad leaves]). It is not permitted for the person who describes and names a plant to name it for him- or herself, even if this person was the first to discover the plant. Binomial scientific names are always italicized, and the names of genera are always capitalized, while the specific epithet is always written in lower case, even if based on a proper noun. A species may be further subdivided into subspecies, varieties, or forms. Although the binomial system of nomenclature was meant to provide a single definitive name recognizable worldwide for each species, there is sometimes disagreement among botanical taxonomists (those who classify plants), resulting in alternative names for some species. A strict code (the International Code of Botanical Nomenclature) governs the naming of plants, and decisions on new or contested nomenclature are made by taxonomists at an international meeting convened every five years. When writing a scientific name (e.g., *Monotropa uniflora*), it is accepted practice that after a plant has been referred to by its full scientific name (genus + species) in a written work, any subsequent mention of the plant in that work can generally be written with the genus abbreviated by only the first letter followed by a period (e.g., *M. uniflora*).

Knowing a bit about the Latin or Greek roots of words makes learning the scientific name of a plant easier since the name may in fact describe the plant or something about it. Thus, scientific names become more meaningful and easier to remember.

I'm certain that as you become more familiar with the natural history of wildflowers, you will want to become an advocate for their protection and conservation. Many of our once common wildflowers are now rare for many reasons. You can help to preserve our wildflower heritage for future generations by becoming involved in, and supporting, local and regional conservation organizations. The Native Plant Trust (formerly The New England Wildflower Society), based in Framingham, Massachusetts, does much to promote wildflower conservation throughout New England and is a reputable source of native wildflower species for cultivation; its lovely Garden in the Woods is a showplace for native wildflowers throughout spring and summer. Other regional centers known for their native wildflower displays are Bowman's Hill Wildflower Preserve in New Hope, Pennsylvania, the Mt. Cuba Center near Wilmington, Delaware, and The New York Botanical Garden's Native Plant Garden. Local organizations whose mission it is to promote and protect native species include garden clubs, nature centers, land trusts, and the Native Plant Center in my home county of Westchester, New York. In addition, national conservation organizations such as the National Audubon Society, The Nature Conservancy (and local chapters of both), and the Lady Bird Johnson Center are advocates for our local flora and fauna.

Acknowledgments

In writing *Summer Wildflowers*, I was fortunate to be able to draw upon the knowledge and generosity of many friends and acquaintances who provided assistance during various stages of producing this book. First and foremost are those who read and offered their insightful comments on the entire manuscript as it was in progress: my husband, Scott Mori; Garrett Crow; and Ed Hecklau, all of whom helped improve the text both in content and in form. My dear late friend Ginnie Weinland still serves as an inspiration; her broad knowledge and quick mind, still sharp at 102, contributed much in terms of editing to the earlier stages of the book. I'm sorry she will not get to see the final product. Others, Rob Naczi and Eric Lamont, read and commented on accounts of plants within their particular realms of expertise. To all the above I offer my deep gratitude for their time and helpful comments. Of course, I assume full responsibility for any remaining errors.

I'm also indebted to those who alerted me to places and blooming times of plants I wanted to include in the book, many of them taking me (by car, foot [sometimes in waders], canoe, and off-road vehicle) to the plants—or inviting me onto their property to photograph them: Carmine Bellotti, Nancy Brogden, Zaac Chavez, Bob Duncan, Jackie Donnelly, Evelyn Greene, Ed Hecklau, Taro Ietaka, Tait Johannsen, Scott LaFleur, Paul Lewis, Eric Lind, Jay Luzzi, Tim Motley, Jody Payne, Michael Penzinger, Helle Raheem, Kate Reeves, Barb and Pete Rzasa, Sam Saulys, and Bonnie Vicki.

My thanks also go to people who alerted me to research papers that were important for me to read in preparation for writing about some of the included species: Robbin Moran, Rob Naczi, Garrett Crow, Peter Bernhardt, and Skip Blanchard.

My greatest pleasure was derived from the fieldwork done in preparation for writing about the plants in this book. Having the companionship of many people who also enjoyed "botanizing" was one of the most rewarding parts of this endeavor. Thanks for happy times and good memories to: ALFASAC (my walking group), Carmine Bellotti, Jackie Donnelly, Bob Duncan, Polly Goodwin, Evelyn Greene, Nancy Harrison, Ed Hecklau, Taro Ietaka, Francesca Jones, Paul and Jean Lewis, Charlie Roberto, Barb and Pete Rzasa, Sam Saulys, Phyllis Tortora, and Bonnie Vicki—and, of course, my husband, Scott.

Many people aided me in identifying the insects, fungi, or foreign plants in my photos or provided explanations to my questions about taxonomic changes and genetics; I appreciate their contributions, which allowed me to add information to both my text and photos: Amy Berkov, Rick Cech, Bob Gibbons, Sabine Huhndorf, Rich Kelly, Gary Lincoff, Amy Litt, Roz Lowen, Rob Naczi, Guy Nesom, Emily Peyton, John Pruski, Charlie Staines, David Wagner, and the experts who volunteer their expertise on BugGuide.com, among them: John Ascher, Bill Dean, Charlie Eisenman, and Molly Jacobson.

The kind hospitality of friends made it possible for me to visit places more than a daytrip from home or to get to places not accessible by car by borrowing a canoe to reach an otherwise inaccessible bog or providing an exciting ride through the forest on an off-road vehicle to reach a secluded patch of ginseng. Many thanks to Carmine and Dottie Bellotti, Jackie and Denis Donnelly, Ed and El Hecklau, and Phyllis Tortora.

The staff at The LuEsther T. Mertz Library at The New York Botanical Garden were of great help in facilitating access to their vast collection of botanical resources. I would like to especially acknowledge Susan Fraser, Marie Long, Don Wheeler, and Susan Lynch.

The scientists whose papers and books I read as part of my research for this book deserve a large dose of gratitude. The work they do is often time-consuming and tedious but results in new, and often exciting, contributions to the fields of botany and natural history. These results are generally published in scientific journals that are not easily accessible to the general public or written in terms that are too technical for the average reader. I hope I have conveyed some of their findings in a manner that allows the lay person to enjoy the intricacies of the fascinating world centered around wildflowers.

Lastly, I am grateful to the staff at Princeton University Press (PUP) for their encouragement and expertise: Robert Kirk, Publisher (Field Guides and Natural

Acknowledgments

History), for believing in this project and exercising patience until I had the time to complete it; Kristen Zodrow for expediting the prepublication review process (and to the two anonymous reviewers who recommended that the book be published); Laurie Schlesinger for preparing promotional material; Lorraine Donneker for the striking cover design; Namrita Price-Goodfellow for her excellent abilities in designing the layout of the book; Mark Bellis for overseeing the production process; and to all the other PUP staff who have been a pleasure to work with at various points during the publication process. Special thanks go to Lucinda Treadwell, whose exacting copyediting skills were quick to spot misplaced hyphens and commas and whose fine-tuned nuances in grammatical usage did much to improve this manuscript. I assume responsibility for any remaining errors.

Summer Wildflowers of the Northeast

Various Families

Fig. 1. Alpine azalea (*Kalmia* [formerly *Loiseleuria*] *procumbens*) (pink flowers), diapensia (*Diapensia lapponica*) (white flowers), and Lapland rosebay (*Rhododendron lapponicum*) (magenta flowers) cascade over a rocky slope on the upper reaches of Mt. Washington in New Hampshire.

Alpine Wildflowers

Fig. 2. A carpet of white-flowered alpine bluets (*Houstonia caerulea* var. *faxonorum*) covers a hillside along a cairn-marked trail looking southwest across the Presidential Range in northern New Hampshire.

Representative Species

Alpine Bluets: *Houstonia caerulea* L. (Rubiaceae); Alpine Azalea: *Kalmia* (formerly *Loiseleuria*) *procumbens* (L.) Desv. (Ericaceae); Diapensia: *Diapensia lapponica* L. (Diapensiaceae); Lapland Rosebay: *Rhododendron lapponicum* (L.) Wahlenb. (Ericaceae); Moss Plant: *Harrimanella hypnoides* (L.) Corville (Ericaceae); Mountain Avens: *Geum peckii* Pursh (Rosaceae); Mountain Cranberry: *Vaccinium vitis-idaea* L. subsp. *minus* (Lodd., G. Lodd and W. Lodd) Hultén (Ericaceae); Mountain Heath: *Phyllodoce caerulea* (L.) Bab. (Ericaceae); Mountain Sandwort: *Minuartia groenlandica* (Retz.) Ostenf. (Caryophyllaceae); Robbins' Cinquefoil: *Potentilla robbinsiana* (Lehm.) Oakes ex Rydb. (Rosaceae).

Alpine wildflowers are a hardy group. The environment where they live is consistently cold and windswept, and they must complete their reproductive cycle during a short growing season, when the mountain summits are free of snow and warm enough for pollinators to be active. Despite the small stature of the plants, many of them are prostrate woody subshrubs of great age whose growth is infinitesimal during the brief summers. Since the plants are typically small and low to the ground, the flowers appear comparatively large and showy (fig. 1).

Habitat: In general, alpine flowers inhabit the land above tree line, with some species restricted to special areas within that zone (e.g., snowbanks, krummholz, or sedge-rush meadows).

Range: In the Northeast, only on mountains with harsh climatic conditions that limit tree growth; these include only the highest mountains of New York's Adirondacks and the tallest summits in New Hampshire, Vermont, and Maine, with New Hampshire's White Mountains having the largest alpine area.

Various Families

One need not travel to Iceland or the far northern reaches of Canada to see true tundra vegetation. The summits of the tallest mountains of New England and New York, those that reach above tree line (the point above which trees become stunted and gnarled, forming a krummholz (meaning twisted wood in German), provide a glimpse of this alpine wonderland. The largest expanse of such habitat is found on the summits of the Presidential Range and a few other mountaintops in New Hampshire's White Mountains and encompasses a total of more than seven and a half square miles. It is one of my favorite places—one that has lured me back many times over the past four decades, particularly in June when the alpine flora is spectacularly in bloom. The feeling of being where I can see forever and have masses of flowers at my feet (fig. 2) has been equaled only by times spent in the Alps or in the Andean páramo at twice the altitude.

Some background information is necessary to explain why these special types of plants, known as alpine tundra vegetation, occur so far south and at such a low elevation. Both geology and climate have played a role. Briefly, the Appalachians were once as high as the tallest mountains on earth—more than 20,000 feet. Glacial events repeatedly covered even the tallest mountains as ice sheets slowly advanced and then receded when temperatures warmed. Cirques, tarns, and other glacially generated features were formed. Each time the climate cooled, many plants migrated south ahead of the advancing glaciers. As the last continental ice sheet retreated, 11,000–12,000 years ago, and temperatures warmed, plants gradually began to migrate back northward. Eventually, those plants most tolerant of the coldest climatic conditions reached the northern continental limits of possible plant life, but then some of those cold-adapted plants that had migrated to the upper reaches of the mountains became stranded there as temperatures continued to warm. These same plants were replaced on the lower slopes by forest tree species as the trees migrated northward from their more southerly refugia. Above the krummholz tree line, the alpine species were able to survive, as temperatures there were more like those of their Arctic origins. Thus, these stranded bits of alpine tundra vegetation remain far to the south of similar vegetation in the Arctic. It gives one pause to think of what may become of our New England alpine plants as our climate continues to warm and tree species are able to migrate further up the mountain slopes. There is no more "up" in the White Mountains to which the alpine species could migrate.

The Appalachian Mountain Club has been compiling data on climate and bloom time throughout the Northeast. While it is too early to draw conclusions about the effect of climate change on the plants of our northeastern mountains, it has been noted that the plants of alpine regions have shown an advance in blooming time of only one or two days. This change differs from that of many plants of lower elevations, some of which have begun to bloom two weeks or more in advance of earlier average dates. The data gathered to date have shown a clear pattern between the number of degree-days necessary to trigger bloom in *Diapensia* as compared with mountain cranberry (*Vaccinium vitis-idaea*), the latter requiring twice as much accumulated heat as *Diapensia* before blooming. Thus, the blooming time of *Diapensia* would more likely be rapidly affected by an increase in temperature.

The tallest peak in the Northeast is Mt. Washington in New Hampshire's White Mountains. At 6288 feet in elevation, it is a midget by western standards, where foothills in the southern Rocky Mountains may be higher than Mt. Washington's summit, and tree line is not reached until at least 8000 feet. However, the summits of the tallest Presidential mountains are subject to such extreme weather conditions that they are similar in climate and habitat to regions much higher in altitude or many degrees of latitude to the north. "The Whites" lie at the confluence of prevailing westerly winds with frigid Arctic air typically pushing down from the northwest. This unfortunate conjunction triggers strong, bitterly cold winds throughout the year. Until the 1990s, the highest wind ever recorded on earth, other than in a tornado, occurred on Mt. Washington in 1934: 231 mph. It was not surpassed until 1996, when a gust of 253 mph was measured during a Category 4 cyclone off the coast of Australia, and since then, by a 239 mph gust recorded in the 2003 Atlantic Hurricane Isabel, and a gust of approximately 253 mph during Typhoon Haiyan that devastated the Philippines in 2013. However, the Mt. Washington record still holds for a top surface wind not associated with a hurricane, typhoon, or other extreme weather event. One hundred mph winds are recorded for every month of the year on Mt. Washington, and 75 mph winds occur on two to four days a month each summer. With

Alpine Wildflowers

a mean summer temperature at the summit of only 48°F (8.9°C), hikers must take the summit conditions into account when planning to hike from the base, where conditions may be up to 30°F warmer. Many deaths have occurred in the alpine region of the Presidential Range, often due to poor preparation by hikers and/or rapidly changing weather conditions. On many days the alpine zone is completely "socked in" by low cloud cover and visibility is limited to only a few feet; it is easy to lose sight of the trail ahead, and hypothermia can set in even in summer. Yet, nothing compares with being on Mt. Washington on a perfectly clear day when the sun warms the land, and views extend as far as the Atlantic Ocean to the east and the Adirondack Mountains to the west. Given that more than 75% of summer days are overcast, if not actually producing precipitation in some form, it is indeed a lucky occurrence to enjoy that view.

On more than one occasion in the White Mountains, I have met a Canadian ecology professor who had brought his students *south* to study tundra vegetation on Mt. Washington. And no wonder—two-thirds of the plants composing the flora of Mt. Washington's upper slopes are Arctic plants commonly found hundreds of miles to the north. In fact, the vegetation above tree line in the Presidentials is closer in floristic composition to that of the Arctic tundra of far northeastern Canada or Greenland than it is to that only 1000 feet below at the trailhead. Aside from elevation and cold temperatures, the beginning of tree line is dependent on exposure to strong, turbulent montane winds, and in places krummholz begins at 4500 feet (or even lower on some exposed ridges in the Presidentials).

In addition to a limited alpine zone in the Franconia Ridge section of the White Mountains, there are smaller areas of alpine vegetation, totaling around 5.5 square miles, scattered on summits in three other states: New York (e.g., Mt. Marcy and Whiteface), Vermont (e.g., Mt. Mansfield and Camel's Hump), and Maine (e.g., Mt. Katahdin, the northern terminus of the Appalachian Trail). One doesn't have to hike up to the alpine zone to see these fascinating plants; in addition to the Mt. Washington Auto Road and the Cog Railway that takes people up the opposite side of Mt. Washington, both Whiteface Mountain in New York's Adirondacks and Vermont's Mt. Mansfield have auto roads to their summits.

However, by hiking up, the observant hiker will enjoy seeing the changes in vegetation as the elevation increases, starting in mixed deciduous/white pine/hemlock forest, climbing through mixed evergreen/red spruce/white birch forest, on up through balsam fir/white birch forest, and emerging into the world of wind-stunted trees: the krummholz. Ultimately one reaches the land above the trees—a land of lichen-covered boulders interspersed with varying amounts of Lilliputian vegetation. Within this land above tree line are found several microhabitats, each with its own community of plants. I will focus on the vegetation of Mt. Washington and its environs, the region I know best. I have selected for discussion here a number of my favorite plants, representative of just a few of the 10–12 specialized microhabitats. Left out of this discussion are the more numerous (but less showy) sedges, rushes, mosses, and lichens—all very important components of the special flora that makes up the alpine tundra.

Alpine Heath and Diapensia Communities

A dominant vegetation type in the tundra region is the alpine heath/diapensia community. This suite of flora inhabits one of the harshest areas of the alpine zone, an area blown almost free of snow by fierce winds. As the community name indicates, many of the plants belong to the heath family (Ericaceae), the family of blueberries, bilberries, cranberries, and rhododendrons. The alpine heath community shares some species with the cushion-tussock (or diapensia) community, and they intermingle in part. Plants that survive in this community are perennial plants that characteristically hug the ground, growing only an inch or two in height and forming dense creeping mats or tightly knit rounded clumps known as cushions that conserve warmth and present a hard surface to the fiercely desiccating winds that can cause the plant to lose precious moisture. The microenvironment at ground level may be 10°F–15°F warmer than the air only a few inches above. Most cushion plants have thick, tough evergreen leaves or leaves covered with scales or hairs, both adaptations that aid in preventing desiccation. Grasses, sedges, rushes, and alpine willows are important members of the alpine flora here. The prime area of the mountain to see alpine heath and diapensia communities is a locality known as the Alpine Garden, a more or

Various Families

Fig. 3. A dense, rounded cushion of diapensia edges onto a rock covered in yellow lichen (*Rhizocarpon geographicum*) in Mt. Washington's Alpine Garden.

less level plateau on the southeast slope of Mt. Washington. During mid-June of the short alpine summer this natural "garden" becomes a spectacularly colorful patchwork of pinks and whites sprinkled with yellows.

Diapensia (*Diapensia lapponica*) is a common plant in the alpine heath community but is not itself a member of the heath family (Ericaceae). Diapensia belongs to the Diapensiaceae, a small family of only five genera with a total of 14 species. Long considered a circumpolar species with two subspecies (subsp. *lapponica* and subsp. *obovata*), the recent treatment of *Diapensia* in *Flora of North America* now treats these two former subspecies of *D. lapponica* as distinct species: *Diapensia lapponica*, an amphi-Atlantic species occurring in eastern North America and western Europe, and *D. obovata*, found in western North America and Asia. The range of both species extends southward along mountain tops in alpine environments: *D. lapponica* reaches the southern limit of its range in the Adirondacks, and *D. obovata* extends its range as far south as British Columbia in North America and to Japan and Korea in Asia. Although the European range of *D. lapponica* expands across the northern Scandinavian countries and Russia, in 1951, much to the excitement of British plant lovers, *D. lapponica* was discovered in Glenfinnan, Scotland, at the head of Loch Shiel. It is still known from only that one locality in all of Great Britain.

Diapensia is a classic cushion plant, rounded in outline, or forming somewhat flat mats low to the ground (fig. 3). Tightly compressed cushion plants like diapensia provide an important ecological function in the tundra environment by providing protection to less hardy plants growing close to it. These more delicate plants might not survive otherwise, and it is in this way that cushion plants contribute to greater diversity in the alpine ecosystem.

Individual plants of diapensia can live to be hundreds of years old, adding only a few leaves each year. The small, shiny, evergreen leaves closely resemble those of alpine azalea. Living leaves on the perimeter of the plant carry out photosynthesis for the plant, while years of dead leaves fill the interior of diapensia's cushion shape. As with wintergreen plants (see section on *Rhododendron lapponicum*), evergreen plants, with their green leaves ready to absorb the first rays of sunlight in early spring, have an extended period in which to make and store carbohydrates. As the days shorten in autumn, the leaves accumulate anthocyanin, a red pigment that causes them to become a deep burgundy color through the winter. A new and unique leaf venation pattern was discovered and described from the leaves of *Diapensia* and the closely related *Pyxidanthera* in the mid-1970s. The veins form a pattern much like that created by the branching architecture of a weeping willow.

The flowers of diapensia are large, compared with its tiny leaves. The creamy white flowers are tubular with five rounded lobes, between which are five distinctively flattened stamens (fig. 4). The stamens have

Alpine Wildflowers

Fig. 4. The flattened filaments and incurved anthers of *Diapensia lapponica* can be seen in this close-up of the flower growing amidst the small pink flowers of alpine azalea.

Fig. 5. In these three flowers of diapensia, the stamens are lengthening, becoming longer than the style, as the male, pollen-shedding phase of the flowers begins. An old capsular fruit from the previous year can be seen in the lower left of the image.

incurved anthers, a feature also found in other members of the family including *Galax*, *Pyxidanthera*, and *Shortia*. The flowers are protogynous: that is, the female reproductive parts (stigmas) mature before the anthers (male reproductive parts) of the stamens, minimizing self-pollination and helping to ensure cross-pollination. The three-lobed stigma is elevated above the stamens when the flower opens, but the anthers surpass the stigmas as the filaments lengthen during the subsequent two days; the horizontally oriented anthers then split open to release their pollen (fig. 5). Flies are the principal pollinators, as is true with most white flowers in the tundra environment. Infrequently, when there is a slight overlap in female and male maturity, self-pollination may occur. Most flowers appear very early in the season, placing them in the so-called pollen-risker category. If, in a given year, pollination is minimal due to a shortage of pollinating insects early in the season, the production of pollen will have been "wasted," and seed set will be correspondingly low in that year. A lesser bloom period occurs about a month later—perhaps serving as a backup reproductive attempt. Strong winds can easily blow the corollas from the flowers, but the pistils remain upright and continue to mature. The resultant capsular fruits do not split open to spill their tiny seeds until October or later when they are shaken from the capsules and dispersed by wind. The opposite strategy is that of the later-flowering plants, which are classified as "seed-riskers." They avoid the chance of early spring blizzards but take a chance that their seeds might not have time to mature before the autumn snows arrive. Life is precarious in the Arctic-alpine zone.

Many of the plants that inhabit alpine areas (which have a small but widely variable number of days free of snow) propagate by a process called "layering." Layering is the development of new plants from stems that have produced roots when in contact with the soil. When the stems detach from the parent plant to form new plants vegetatively, they are called "layers." Aging such clonal plants is therefore difficult.

Various Families

Diapensia, however, does not reproduce by layering, but instead grows outward from a central taproot, making it is easier to calculate the age of a plant by measuring its diameter. From the central point, the plants increase equally in diameter by a more or less known amount each year (0.6 mm per year in one study) unless rocks or other obstacles interfere. The plants do not produce flowers until an average age of 18 years, and then they are not reliably in flower every year until they are approximately two decades older, but then continue to flower even into their declining years. Plants in a study area in Swedish Lapland have been aged at nearly 700 years, having begun life in the early fourteenth century.

Diapensia relies strictly on reproduction from seed, a method with a low rate of success. Although seedlings *are* produced, they frequently succumb during their first several years to late snowstorms early in the growing season. Furthermore, plants already reproductive are more vulnerable early in the season because their new leaves do not "harden off" until midsummer. In addition, their flowers are vulnerable to frost, resulting in low or no seed set for that season. Because there is a trend toward an earlier beginning to the growing season due to climate change, diapensia may suffer greater losses (both of young plants and seed production) as a result of early spring blizzards; however, this unfortunate circumstance allows diapensia to serve as an indicator species in monitoring the impact of climate change on the Arctic environment. Indeed, a study conducted in Sweden in the 1990s showed a 20%

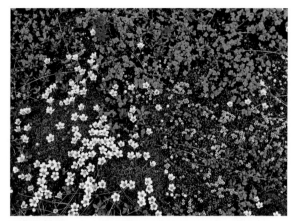

Fig. 7. Diapensia (white flowers), alpine azalea (pink flowers), Lapland rosebay (magenta/purple flowers), and the light green leaves of bog bilberry (*Vaccinium uliginosum*) intermingle in the Alpine Garden, a natural "garden" on the southwest slope of Mt. Washington.

dieback in *Diapensia lapponica* populations (including plants aged at 700 years) in just five years, a circumstance attributed to the warming climate.

A rare form of diapensia occurs in the White Mountains—one in which the flower has been doubled such that it looks like a miniature white rose (affectionately called "double-di") (fig. 6). Doubled flowers are the result of a genetic mutation in which a gene responsible for the reproductive structures is missing or damaged. I have seen these doubled flowers in the same area (probably on the same plant) over a period of years. They are quite lovely, but since the extra petals are a result of the loss of the reproductive parts, pollen cannot be produced, the plant cannot reproduce sexually, and no seeds are formed. When diapensia intermingles with other flowering alpine plants, the effect is particularly beautiful (fig. 7).

Lapland rosebay (*Rhododendron lapponicum*) grows intermixed with other heaths such as alpine azalea and species of dwarf blueberries, mountain cranberry, and bilberries. In this extreme environment, this miniature *Rhododendron* reaches a height of only a few inches (fig. 8)—yet the lovely, 1 inch, purple flowers are readily recognizable as those of a rhododendron. In a study conducted on Arctic-alpine plants in Canada, nectar production in Lapland rosebay was determined to be among the highest and most sucrose-rich of the plants studied. This plant is thus highly attractive to insects. The leaves are leathery, covered with scales,

Fig. 6. A rare double-flowered form of diapensia in which a genetic mutation has occurred, causing the stamens to be converted into additional petals.

Alpine Wildflowers

Fig. 8. The relatively large purple-pink flowers of Lapland rosebay, a miniature species of rhododendron, put on an eye-catching show in early June. *Rhododendron lapponicum* is a wintergreen species; the leaves are produced in summer and remain on the plant through winter and into the early part of the following summer, when they fall as new leaves are produced.

and curled under at the edges, all aiding in water conservation. Leaves and flowers are poisonous due to the presence of grayanotoxin (also known as andromedotoxin), a diterpene compound common in many plants of the Ericaceae. Although the amount of this toxin (as found in honey from other species of *Rhododendron*) is usually not enough to cause illness, larger amounts can cause perspiration, vomiting, dizziness, low blood pressure, and more serious symptoms, but rarely death. "Mortal paralysis," caused by the toxin from *Rhododendron ponticum*, played a role in the 2009 film, *Sherlock Holmes* (although the actual effects of rhododendron poisoning are not accurately portrayed in the film).

Rhododendron lapponicum has an almost complete circumarctic/circumboreal distribution, occurring in the Western Hemisphere from the Arctic southward to Alaska and the alpine regions of the mountains of the Northeast, with a single odd outlier population disjunct on the non-alpine sandstone cliffs in the Dells of Wisconsin (a relic of the Ice Age). It is also found in Arctic-alpine regions of Eurasia (including Lapland, as reflected in its scientific name) but is strangely absent in the subarctic portions of Russia. *Rhododendron lapponicum,* like many other Arctic-alpine plants, is an example of a wintergreen plant, one that produces its leaves in one year and holds them throughout the winter. This enables the leaves to continue to mature during the following season before they are shed after the production of new leaves. This adaptation to the short growing season allows the previous year's leaves to begin photosynthesis early in the year; the increased period of photosynthesis results in a gain of carbon, particularly during early summer when days are longest. The carbon produced by last year's leaves is particularly important for the growth of the current year's reproductive branches. Those branches that will produce flowers during the following year form their flower buds by midsummer of the current season. About one-third of the plants in the less extreme regions of Arctic-alpine tundra benefit from this wintergreen strategy.

Despite its common name, **alpine azalea** (*Kalmia procumbens*) is not an azalea, a term commonly applied to some groups in the genus *Rhododendron*. Nevertheless, it *is* a member of the plant family Ericaceae. The species ranges into some of the highest, most exposed areas of the slopes of Mt. Washington and other alpine ranges worldwide, as well as into true Arctic regions. A cushion and mat–forming plant, its leaves are leathery with a white hairy undersurface and recurved edges. The tiny plants may live for 50 to 60 years, growing only negligibly (if at all) each year. Unlike many other alpines, alpine azalea has a sizable rootstock that goes deep into the rocky soil and makes up 70% of the plant's biomass.

The flowers of *Kalmia procumbens* are small, pink bells with five pointed lobes that face upward (fig. 9). Each individual bell is held within five narrow, dark

Fig. 9. The tiny, upright pink flowers of alpine azalea (not a true azalea) attract insects with a colorful display made by the number of flowers. The principal pollinators are small bumblebees.

Various Families

red sepals. As with most pink-to-purple flowers in the alpine zone, the principal pollinators are bees. The stigmas mature first, followed by stamen maturation—promoting cross-pollination. The ovaries develop into upright, round, brown capsules that split open to release their seeds. Like many alpine plants, the main form of propagation is vegetative, by layering.

Members of the genus *Vaccinium* found in the alpine region include *V. vitis-idaea*, known as **mountain cranberry** (or partridgeberry in Newfoundland); *V. oxycoccus* (small cranberry); *V. cespitosum* (dwarf bilberry); *V. angustifolium* (lowbush blueberry, which is also found at lower elevations); and *V. uliginosum* (bog bilberry, with deciduous leaves—the most common heath shrub in the lower portion of the alpine zone on Mt. Washington). I have a particular fondness for mountain cranberry. Its small, red fruits succeed the tiny, pendulous pink flowers. Both flower and fruit contrast attractively with the glossy, deep green leaves, which have a single groove running the length of the blade (fig. 10). The small, but sturdy, flowers have four lobes and are borne at the tips of the branches. Both bees and wasps have been recorded visiting the flowers, with small bumblebees being the more likely pollinators. While visiting a flower, they employ a special technique of rapidly vibrating their indirect flight muscles. The resulting vibration causes pollen to be dislodged from the anthers, a process known as buzz-pollination. Their leathery leaves with rolled margins are an adaptation for conserving moisture in the windswept environment. As with all cranberries, the red fruits are tart, but tasty (fig. 11), and, like the fruits of small cranberry (*V. oxycoccus*), those that linger under winter's snow have a sweeter taste the following summer. There are five to eight small seeds, which are dispersed by birds, particularly the ground-dwelling spruce grouse that visit the tundra vegetation zone in summer and fall. Mountain cranberry is a woody, prostrate, creeping plant, only an inch or two tall. It reproduces primarily by spreading subterranean rhizomes and by adventitious rootlets on its aboveground stems. It is a dominant subshrub, growing in several different communities within the alpine zone. The subspecies found in the Northeast is subsp. *minus*. Like several other plants mentioned here, it has a circumboreal distribution. Another subspecies, subsp. *vitis-idaea*, is known as lingonberry in the colder, peaty/forested regions of Europe. The flowers are white to slightly pink (fig. 12). The red fruits are wild-collected and made into both jam and a sauce that is served with meat. There is interest in cultivating the species commercially in both Europe and Newfoundland, where wild "partridgeberry" is already a favorite fruit for making jams and pies.

Fig. 10. (Left) Mountain cranberry (*Vaccinium vitis-idaea* subsp. *minus*) has the glossiest leaves of the ground-hugging plants in the alpine region. They are tough and leathery, as are the corollas of the four-lobed flowers, a feature that aids in preventing damage and desiccation in the windy environment. **Fig. 11.** (Right) The small, deep red berries of mountain cranberry are tart, like those of the familiar cranberry (*Vaccinium macrocarpon*) enjoyed as part of Thanksgiving dinner. If mountain cranberry fruits are not eaten by birds, such as the spruce grouse, before the snow arrives, they remain on the plant over winter and are sweeter the following summer.

Alpine Wildflowers

Fig. 12. (Left) A different subspecies of mountain cranberry (*Vaccinium vitis-idaea* subsp. *vitis-idaea*) is known as lingonberry in Europe. Its flowers are white, tinged with pink, rather than the deep pink to red color of the flowers of our local plants. **Fig. 13.** (Right) Mountain sandwort (*Minuartia groenlandica*) is one of the few wildflowers above tree line that has a long blooming season and great tolerance of disturbance. It often grows right along the rocky trails used by hikers.

A plant found in more than one community is the **mountain sandwort** (*Minuartia groenlandica*), a white-flowered plant that is visited, and often pollinated by, many different species of flies. Bumblebees visit as well, their hairy bodies making them the most effective pollinators. Insects visit for both pollen and nectar, and mountain sandwort, being one of the most common plants on the mountain, serves as an important resource for them. The plants have a long flowering period, beginning in late June and extending into August, but reaching a peak in July. Its ability to live in the most disturbed areas allows mountain sandwort to grow adjacent to—and sometimes in the middle of—the rocky trails. During times of low visibility, the white flowers help the hiker to follow the trails above tree line. True to its specific epithet, mountain sandwort grows in Greenland, as well as in the Canadian Maritimes and parts of northern Canada. In the United States, it follows the mountains from New England south to the Carolinas. Although often found growing in shallow soils on rock outcrops at high elevations (fig. 13), the common name "sandwort," meaning "sand-plant," recalls its former genus name, *Arenaria* (Latin for "sand-loving"). Oddly, there is a confirmed report of this species growing on a mountain in Santa Catarina, a state in southern Brazil.

Surprisingly, in this harsh environment there are two rare butterflies that inhabit the tundra zone and, in fact, are found only there. The White Mountain Arctic (*Oenesis melissa semidea*) is a subspecies of the Melissa Arctic and is endemic to the alpine zone of the Presidential Range. It is reported to visit mountain sandwort to drink the droplets of nectar produced at the base of each stamen. This cryptically colored, dark gray-brown butterfly, like many of the plant species found here, was stranded on the mountain tops when warming temperatures eliminated its host plant, Bigelow sedge (*Carex bigelowii*), and its preferred nectar sources from the vegetation in the intervening, lower area further north. It thus could migrate no further. The taxonomic placement of a second possibly endemic butterfly, a subspecies of the White Mountain Fritillary, is less certain. For now, it is being treated as *Boloria titania montinus*, but additional study is needed to determine whether that name is correct. It emerges later than the White Mountain Arctic, has white spots on its purple or red tinged wings, and nectars on alpine goldenrod (*Solidago cutleri*) and a few species from other genera.

Snowbank Communities

Two types of snowbank communities are found in White Mountain alpine regions; the first develops at the edge of the krummholz, where short, but often very old, wind-stunted trees shelter the plants; protection is also found in areas sheltered by large boulders on the east/southeast side of Mt. Washington and other mountains above tree line. A long-lasting snowy blanket covers plants for part of the year, protecting them from fierce spring storms, and they have the benefit of sunny warmth on favorable summer days. These protected snowbank communities are the most species-rich in the alpine zone. In addition to boreal species not found elsewhere in these mountains, snowbank communities harbor plants more typically found at lower, forested elevations, including gold thread (*Coptis trifolia* subsp. *groenlandica* [formerly *C. groenlandica*]), blue bead lily (*Clintonia borealis*),

Various Families

bunchberry (*Cornus canadensis*), Labrador tea (*Rhododendron groenlandicum* [formerly *Ledum groenlandicum*]), false hellebore (*Veratrum viride*), and Canada mayflower (*Maianthemum canadense*). All are smaller and bloom later than their lower elevation counterparts. At my home elevation of 540 feet, false hellebore blooms in mid to late May, yet I have seen it in July at the edge of the area of melting snow in Tuckerman's Ravine, with its leaves still not expanded.

Two plants found only in this community are **mountain heath** (aka mountain heather) (*Phyllodoce caerulea*) and moss plant (*Harrimanella hypnoides*), both members of the heath family. Mountain heather is quite rare here because it is limited to the exposed edges of snowbank habitat on the mountain. It is a plant found in both Arctic and boreal environments in Western Europe, Asia, and North America. The leaves are needlelike, resembling those of another widespread inhabitant of the alpine region, crowberry (*Empetrum nigrum*), a species in a genus recently moved from its own family (Empetraceae) into the Ericaceae. When examined closely, the fine serrations on the leaves of mountain heath differentiate them from those of crowberry. Mountain heath flowers are bright magenta-purple, elongated urns that divide into five tiny lobes that frame the small opening of the flower (fig. 14); the corollas fall quickly after flowering. Mountain heath's scientific name seems entirely inappropriate: "Phyllodoce" is the name of one of the Greek sea nymphs (certainly misplaced in this high-elevation environment), and "caerulea," meaning blue, is hardly descriptive of the flower color.

Research on pollination of mountain heath in three widely separated parts of its overall range yields differing results. In Japan it was found to be dependent on cross-pollination by bumblebees. In Great Britain, bees, butterflies, and beetles visited the flowers, although pollination by insects was thought to be low, with the flowers relying for the most part on self-pollination. In the northeastern United States, the flowers are thought to be self-pollinating. Since mountain heath flowers produce an ample amount of nectar to satisfy insects, it would seem that, in order to confirm the means of pollination further observations should be undertaken.

When not in bloom **moss plant** may easily be mistaken for a moss. Like mountain heath, it is limited to sheltered, damp, rocky areas and therefore not often seen. Five deep red sepals clasp its bell-shaped, white flowers—a pretty combination (fig. 15). When I first encountered this plant, more than 30 years ago, it was still generally considered a member of the genus *Cassiope*, named for Cassiopeia, the vain queen of Greek mythology—a name also applied to the W-shaped constellation in the northern sky. Despite the negative association with the queen's vanity, *Cassiope* seemed a prettier name, more suitable for such a delicate plant than the present name of *Harrimanella*. That name was given to the plant in 1901 to honor Edward Henry Harriman, a nineteenth-century railroad baron. It was Harriman who organized an important scientific expedition to Alaska in 1899, and who funded John Muir's travels and writings during that period. Thus, perhaps Harriman was more deserving of being commemorated than the conceited queen. The change in

Fig. 14. Mountain heath (*Phyllodoce caerulea*) grows in sheltered areas where it has the protection of longer lasting snow cover.

Fig. 15. Moss plant (*Harrimanella hypnoides*), as its common name indicates, can be mistaken for a moss when not in flower. Like mountain heath, it is an inhabitant of protective snowbank communities.

Alpine Wildflowers

Fig. 16. One of the largest flowers in the alpine region of the White Mountains' Presidential Range is that of mountain avens (*Geum peckii*). The shape of the flower (a curved, bowl shape) helps to hold heat and attracts insects that need to warm up to continue to fly. Here, a small syrphid fly heads for a warm respite.

Fig. 17. A dark-colored fly warms itself in the solar-collecting flower of mountain avens, a near-endemic of the White Mountains. Aside from the Presidential Range, it is found only in a limited area of Nova Scotia.

Fig. 18. (Right) *Geum peckii* can be identified easily even when not in flower by its large, rounded terminal leaflet. Closer inspection is needed to spot the smaller leaflets beneath the large one.

the scientific name, though, is the result of more careful comparison of all the plants assigned to the genus *Cassiope*, and the realization that morphological differences noted in two species warranted the establishment of a new genus, *Harrimanella*, while the other species were retained in the genus *Cassiope*. No matter how well known or poetic the older genus placement for moss plant, we must accept the new one with its valid reasons for existence (differences in leaf arrangement, flower position on the plant, corolla shape, stigma type, types of hairs, etc.). The species epithet of moss plant is highly appropriate: *hypnoides*, meaning like *Hypnum*, a genus of mosses.

The second snowbank community inhabits depressions or small ravines formed within other communities, such as the alpine-heath and diapensia communities. There one can find colonies of alpine bluets (*Houstonia caerulea*), bright yellow-flowered mountain avens (*Geum peckii*), and other plants, all of which benefit from a sheltering habitat where snow accumulates and thus affords protection from the desiccating winds. These plants and others are also found in the mossy strips along streams and springs.

Mountain avens, a member of the rose family, was discovered on an early scientific expedition (1804) by a Harvard professor, William D. Peck, and later named for him. This plant has one of the largest flowers found above tree line, up to a full 1.5 inches across. As such, the bright yellow flowers serve as solar collectors and provide a site where insects can warm themselves for continued flight (figs. 16–17). The large compound leaves are also conspicuous for their size. One has to look carefully to see the smaller leaflets beneath the large, rounded terminal lobe (fig. 18). The leaves,

Various Families

Fig. 19. Several of the species above tree line have leaves that turn a deep red or burgundy, as seen in this autumn photo of a fruiting alpine bearberry (*Arctostaphylos alpina*), another member of the Ericaceae.

Fig. 20. An early frost decorates the leaves of mountain avens before they have even completed their fall color change.

like those of diapensia and bearberry (*Arctostaphylos alpina*), a member of the Ericaceae (fig. 19), turn red or burgundy as temperatures drop in September and frost often coats the leaves in early morning (fig. 20). Mountain avens was once thought to be restricted to the alpine areas of New Hampshire, but the 1948 discovery of a population of mountain avens on Brier Island, off the coast of Nova Scotia, and at five other sites on mainland Nova Scotia, deprived mountain avens of its endemic status. Endemic means occurring *solely* in one specific geographic area (which may vary in size depending on the plant being discussed).

Alpine bluets are much like those of the same species seen at lower elevations that have blue petals with a yellow center, but in the Presidential Range, the flowers are white with a yellow center (fig. 21). Research on the pollination of *Houstonia caerulea* at lower elevations showed bombyliid (bee) flies to be the principal pollinators, but I could find no information about pollination of the white-flowered variation in the alpine area. The white color form of alpine bluets, formerly named var. *faxonorum*, was thought endemic to the White Mountains, but, in fact, the white-flowered *Houstonia* also grows on two islands off Newfoundland, and a taxonomic decision to submerge var. *faxonorum* into *Houstonia caerulea* as only a color variant of the species disqualified it as an endemic variety.

The only true endemic in the alpine zone of New Hampshire's Presidential Range (and at one locality in the Franconia Range), is a diminutive member of the rose family known as **Robbins' cinquefoil** (*Potentilla robbinsiana*). When not in flower, Robbins' cinquefoil is an easy-to-miss 2.5–5 cm (1–2 in.) rosette of deeply toothed leaves. Flowers appear in June at the perimeter of the plants; they are small, yellow, and five-petaled but generally numerous enough to catch one's attention at that time of year (fig. 22). The plant was discovered in the 1820s, not long after completion of the Crawford Path, the first trail built to the top of Mt. Washington. The flowers of Robbins' cinquefoil are either self-pollinated or reproduce asexually by the production of seed without fertilization (termed agamospermy, from the Greek *agamos*, meaning "unmarried," and *sperma*, meaning "seed"). Dispersal by wind carries the seeds only ca. 8 inches from the plant, so even the larger, main population, south of the summit of Mt. Washington, has remained confined to a section known as Monroe Flats, an area of small, loose rocks, covering less than an acre that is perhaps the most windswept spot in the whole alpine zone. The plant was quickly recognized as rare, and as such became an object of desire for plant collectors eager to add a rare specimen to their herbarium, or for rock gardeners who wanted to grow the plants in their gardens. Those plants surely died a quick death when removed from their very specific habitat requirements. An estimated half the plants in the Monroe Flats population were removed by collectors, almost 75% of them in a nine-year period around the turn of the twentieth century.

Alpine Wildflowers

The species is named for James W. Robbins, a Massachusetts physician/botanist, but despite reports that Robbins was the person who first discovered the plant, there is no evidence that he had even seen the plant before it was described and named by William Oakes. There is controversy regarding who actually first found the plant in 1824, but the first documented collection was made by Thomas Nuttall, a British botanist and ornithologist who lived and worked in the United States during the early nineteenth century. His collection, originally labeled as a different species of *Potentilla*, is in the Hooker herbarium at Kew Gardens in Richmond, England. In fact, many of the early collections of *Potentilla robbinsiana* were made for European herbaria—or sold to them later. One collection is even in an herbarium in South Africa. Only 20% of the known collections of *Potentilla robbinsiana* are in New England herbaria.

Uncertainty about the proper taxonomic classification of *Potentilla robbinsiana* still exists, but because of its geographic isolation and other criteria, it is currently considered a separate species, endemic to the White Mountains. Robbins' cinquefoil is a plant that was placed on the Endangered Species List in 1980, after its population had been severely reduced both by overcollecting and by unintentional trampling by hikers. Individual plants may live for several decades if undisturbed, but they are shallow-rooted and easily dislodged from the soil by careless hikers. The well-used Crawford Path crossed directly through the main population of *Potentilla*, south of the summit of Mt. Washington. In 1983, in order to protect the species, the trail through the main population was rerouted, signage was posted asking hikers to stay on the trails, and educational programs were instituted to explain the need for protecting the fragile plant life above tree line. Those efforts, combined with off-site propagation of wild-collected seeds for reintroduction, made it possible for Robbins' cinquefoil to rebound to the degree that it was removed from the national endangered species list in 2002, but it still remains on the list of protected plants in New Hampshire. Eighteen miles to the west, a much smaller population, differing in both habit and habitat, was discovered in 1940, more than 100 years after the initial discovery. A robust population of plants is now thriving in the original two sites as well as in two other suitable areas in the White Mountains. Thanks to the collaborative efforts of The Appalachian Mountain Club, The New England Wildflower Society, the Fish and Wildlife Service, and the United States Forest Service, *Potentilla robbinsiana* ranks as one of the major success stories of the Endangered Species Act. In 2014, Robbins' cinquefoil was named one of the 10 most important results of the Act, the only plant to earn that distinction.

Fig. 21. (Left) Alpine bluets are white, rather than blue, in the alpine zone of the Presidentials. The yellow center serves as a nectar guide for insects. **Fig. 22.** (Right) The tiny rosettes of Robbins' cinquefoil (*Potentilla robbinsiana*) are barely larger than a quarter in diameter. In June, five-petaled yellow flowers are produced around the edge of the plant. (Photo by Scott Mori).

Heath Family

Fig. 23. The early colonists thought that the flowers of cranberry (*Vaccinium macrocarpon*) resembled the head and neck of a crane (the pedicel resembling the long, curved neck; the corolla representing the head; and the long, exserted stamens appearing like the bill of the bird. They, thus, called the plant crane-berry, which over time became shortened to cranberry.

American Cranberry

Fig. 24. Cranberry vines grow in acidic wetlands. Here, they are flowering in a cultivated cranberry bog in June.

American Cranberry

Vaccinium macrocarpon Ait.
Heath Family (Ericaceae)

One probably doesn't think first of flowers when hearing the word "cranberry," yet it is the flower of this species that is the basis for its English common name. The early settlers thought the flowers of this berry-producing plant resembled the curved neck (the pedicel), head (the corolla), and long bill (the stamens) of a crane, hence: "crane-berry" (fig. 23). Other common names include "bearberry," because bears were observed to eat the fruits; "fenberry," for the boggy (but, unlike fens, acidic) wetlands where they grow; "sassamenesh," the Wampanoag (an Algonquian language) word meaning "very sour berry"; and "ibimi," the name used by members of the Lenni-Lenape tribe, meaning "bitter berry."

Habitat: Bogs, swamps, wet shores, interdunal swales along the Atlantic shore, and headlands with acidic soil.

Range: Northeastern United States from Maine west to Minnesota and northern Illinois and Indiana, and south to North Carolina and Tennessee (but not in Kentucky); introduced in Washington and Oregon. In eastern Canada from Ontario, Quebec, New Brunswick, Newfoundland (but not Labrador), Nova Scotia, and Prince Edward Island, plus the French archipelago of St. Pierre and Miquelon off the coast of Newfoundland; introduced in western Canada in British Columbia. American cranberry is also introduced in Europe.

Cranberry is a member of the large and diverse heath family, Ericaceae, members of which include such plants as azaleas and rhododendron (*Rhododendron*), heather (*Erica*), blueberries (*Vaccinium* spp.), and heath (*Calluna*), plus tropical American epiphytic shrubs such as *Macleania*. Recently the families Monotropaceae, Pyrolaceae, and Empetraceae have been merged into this family as well. *Vaccinium*, the genus of cranberry, is a large genus, numbering about 500 species worldwide, with 25 of those growing in North America.

Cranberry plants grow as trailing subshrubs in acidic wetlands (preferably with a pH of 4–5.5) (fig. 24) and produce flowers in early summer (June); the fruit is ready to harvest in late September through October (fig. 25). Cranberry plants are mycorrhizal, benefiting from a symbiotic relationship with a fungus that

Heath Family

infiltrates the roots of the plants and enables the roots to more easily absorb nitrogen from the soil. In return, the fungus receives carbohydrates made by the plant.

The leaves of American cranberry (*Vaccinium macrocarpon*) are evergreen, small and narrowly elliptic (5–15 mm long) with blunt, rounded tips and entire (smooth) margins that are somewhat incurved. The undersides of the leaves are lighter in color than the top surface (see fig. 23). Cranberry flowers are produced on long, curved, slightly pubescent pedicels that arise from the wiry stems of the cranberry plant at the base of the current season's (or sometimes previous season's) growth. Each pedicel has a pair of small green, leaflike bracts (bracteoles) located above the middle of the pedicel (these can be seen above and behind the flower in fig. 26 and in fig. 23). Flowers have a green tubular calyx with four short lobes and four pink or white petals that reflex strongly backward when the flower opens. The eight stamens each have anthers ending with long tubules opening by a pair of pores. The anthers are connivant (tightly appressed to each other) and protrude from the center of the flower, giving the appearance of a bird's long beak (see fig. 26). The narrow style, tipped with a tiny stigma, emerges from the stamens when the flower is in the female phase. Cranberry flowers are protandrous; that is, their male parts mature first, followed by receptivity of the stigmas.

In a hectare of cranberries, an average of 33% of the flowers set fruit, the equivalent of 370 barrels (100 lbs.) of berries. However, a good year may yield almost twice that amount, so it is important to maximize pollination success in any way possible.

Although American cranberry is native to North America and its pollinators (primarily bumblebees) coevolved with the flowers long before the arrival of Europeans, when cranberry farming became an important commercial enterprise, hives of non-native honeybees were brought to the cranberry ponds when the plants were in flower, with the hope of increasing productivity (fig. 27). However, honeybees, overall, are inefficient pollinators, which results in lower fertilization of the ovules and, thus, fewer and smaller berries. Honeybees forage for nectar both by "legitimate" and "illegitimate" methods. The illegitimate method involves probing the base of the flower from behind to reach the nectary, resulting in lack of contact with the reproductive structures. Only rarely do honeybees forage for pollen by hanging under the flower and "drumming" the anthers with their front legs, thus causing the pollen to sprinkle out of the pores.

Since cranberry flowers have anthers that open by a pair of pores at the tip of the anther, vibration is the most effective means of dislodging the pollen. Our native bumblebees are masters of "buzz-pollination," a technique in which they grasp the flower with their legs and decouple their indirect flight mechanism, thereby preventing movement of the wings. They then vibrate the indirect flight muscles in their thorax to vibrate the pollen from the pore, becoming covered with it. Most of the pollen is groomed off by the bumblebees, who mix it with saliva and pack it into the pollen baskets on their legs to take back to the nest. But some pollen lands on a "safe spot" on the thorax, one that cannot be reached by the bee; it is this spot that comes into contact with the stigma of a flower on another plant, resulting in pollination. Cranberries are not self-fertile. Some growers are now hiring bumblebee keepers to bring hives of these more efficient bumblebee pollinators to their fields.

Fig. 25. (Left) Ripe cranberries ready for harvesting in Nantucket, Massachusetts. **Fig. 26.** (Right) On the pedicel of each cranberry flower are two green leaflike bracteoles between the middle of the pedicel and the flower. They can be seen here arising behind the flower. The green bracteoles distinguish the American (large) cranberry (*Vaccinium macrocarpon*) from its smaller relative, small cranberry (*V. oxycoccos*), which has hairlike reddish bracteoles situated lower on its pedicel.

American Cranberry

Fig. 27. Honeybee hives have been brought to this cranberry bog in the hope of increasing pollination of the flowers and, thus, the yield of the crop. Honeybees, however, are not efficient pollinators of cranberry.

Potentially, certain species of syrphid (flower) flies could be good pollinators of cranberry flowers. Many syrphids visit the flowers, but only a few species carry sizable amounts of pollen. Syrphid fly visits to cranberry flowers tend to increase if the fields surrounding the cranberry bogs are not mowed and have a diversity of flowers to attract insects. Obviously, avoiding mowing could result in greater visitation to cranberry flowers by these potentially important pollinators.

The first colonists to arrive in what is now New England may have been introduced to cranberries by Native Americans living in that region. Cranberries were an important resource for native peoples, most importantly as in a preparation called pemmican, a mixture of dried cranberries, dried deer meat, and tallow that was pounded together and dried. The tallow fat and the acidity of the cranberries served as preservatives, allowing the mixture to be kept for a long time. It was carried by hunters and traders heading off on long excursions and provided them with a source of protein, vitamins, and energy. To this day, the Wampanoag tribe on Martha's Vineyard celebrates Cranberry Day on the first Tuesday of October by first harvesting the berries and afterward enjoying a potluck with tribal members and all others who wish to participate.

Other Native American uses included the production of a red dye (made by boiling the berries) used for dying blankets. The Cree used the dye to color porcupine quills for clothing ornamentation. And the Chippewa used cranberries as a bait to trap snowshoe hares. Various native groups used preparations of the berries for medicinal purposes: treating wounds and fever, "purifying" the blood, and treating urinary tract infections. There seems to be some validity to this latter use; cranberry juice is still recommended to aid in the prevention of urinary tract infections today. It is thought that a component in cranberry juice (as well as blueberry juice) inhibits the ability of bacteria to attach to the lining of the bladder and urethra. This same property also seems to inhibit the adhesion of dental plaque to the teeth. Sailors learned early on that the berries (which contain vitamin C) prevented scurvy and began taking barrels of them on sea voyages. They were as effective in staving off this disease as the limes used by British sailors.

The early colonists used the berries to make a tart sauce to serve with roasted game birds or as an ingredient in a stuffing for the birds. Colonial fur traders adopted the Indian practice of making pemmican to fuel themselves on long journeys, and soon the fruit became economically important enough that it actually resulted in skirmishes between Native Americans and fur traders.

In 1622, British settlers brought honeybees with them to the colonies, and the ready availability of honey as a sweetener opened up new culinary possibilities. Cranberries could now be used to make pies, tarts, and a sweetened relish called cranberry sauce. A recipe for cranberry sauce appeared in a Pilgrim cookbook as early as 1663. Cranberry's association with Thanksgiving began long ago, apocryphally as a food contributed by the Native Americans to the first Thanksgiving feast, which is thought to have taken place in the fall of 1621. If true, the cranberries would not have been in the form of a sweet sauce such as that we eat today.

Heath Family

American cranberry is one of few commercially grown fruits native to North America, the others being blueberries (both highbush and lowbush blueberries) and Concord grapes. Common fruits that we might consider "ours," like apples from Washington and oranges from Florida and California, have long been cultivated here but had their origins elsewhere (apples from Central Asia and oranges presumably from tropical or subtropical regions of Asia, especially Indonesia and the Philippines). Oranges, as we know them, are no longer found as a wild-growing species.

Cranberries were first grown as a commercial crop in the United States during the 1830s–1840s on Cape Cod in Dennis, Massachusetts. But prior to that, in 1808, Sir Joseph Banks, noted British botanist, naturalist, explorer, and president of the Royal Society, already had grown cranberries from American stock in Britain. Cranberry farming expanded in North America, spreading to New Jersey, Wisconsin, and other northern regions having the proper environment for growing the crop. More than 100 cultivars of *Vaccinium macrocarpon* have been developed with the aim of producing fruit that doesn't bruise easily, resists disease, or is sweeter; four of these account for most of the cranberries grown today. One of these cultivars, Early Black (which is deep red in color and may be somewhat pear-shaped) (fig. 28), is now the most widely grown cranberry in North America.

The establishment of a productive cranberry field requires a substantial initial investment in land and equipment, plus hard work, and patience. The bogs are dug in areas having a high water table and acidic soils (fig. 29). Around the margins of the bogs, the excavated soil is used to build dikes (fig. 30), which are high enough to contain water when the fields are flooded and wide enough to allow use of the vehicles needed to maintain the fields. Irrigation and a pumping system are installed, a layer of sand added to the bed, and the vines are planted. The first good harvest will not be ready for five to seven years. However, once established, a cranberry bog may remain productive for more than 150 years. Pruning the vines is beneficial to the plants, and the cuttings can be used to start new beds. Many cranberry farms are multigenerational, with the younger generations often going to college but ultimately returning to the farm with new ideas for more efficient management of the business and ways to apply new technologies, such as using satellite and aerial imagery in conjunction with GPS surveys to locate and monitor "fairy rings," an indication of fungal disease.

Until 1930, cranberries were sold fresh, with most families making their own cranberry sauce or relish from the berries, generally by adding prodigious amounts of sugar (and sometimes orange rind or nuts), and then usually only for Thanksgiving and Christmas meals. Our family's cranberry sauce was

Fig. 28. (Left) A cranberry field in New Jersey with ripe fruit of the Early Black variety of cranberry. **Fig. 29.** (Right) The waters adjacent to a New Jersey bog are tea-colored due to the acidity of the water that leaches tannins from the surrounding vegetation as well as to naturally occurring iron in many of the streams.

Fig. 30. Cranberry bogs are dug in areas having a high water table and acidic soil. The excavated soil is used to build dikes surrounding the bogs that are used for water management.

American Cranberry

Fig. 31. An old wooden cranberry scoop with metal tines once represented a major advance over hand-picking during cranberry harvest as it could be scooped through a wider area of vines at one time.

Fig. 32. Fresh cranberries were once stored in low wooden boxes with slatted bottoms to allow for air circulation (a box is on the left; to the right is an old-fashioned sorter). Today, plastic boxes are more commonly used.

always made by my grandmother, who grew up in cranberry territory on Prince Edward Island.

Depending on the use of the fruit, cranberries are harvested in two different ways. Originally, Native Americans and the first colonial cranberry farmers harvested by hand—an extremely labor-intensive process that required the employment of large numbers of seasonal "pickers." Picking cranberries was hard work, and pickers often had to wrap their fingers with linen cloth to protect them from injury as they stripped the vines of their fruit. As farms grew larger, wooden cranberry hand scoops (fig. 31) were developed that allowed for more rapid harvesting at lower cost. To prevent the berries on the bottom from being crushed, the fresh berries were stored in shallow wooden boxes (fig. 32) that had openings between the wooden slats to allow for air circulation, helping to prevent mold from developing. Unlike many fruits, cranberries are harvested only when they are ripe.

An improvement over the scoop was the long-handled wooden cranberry rake—similar to the scoop, but larger, allowing for more efficient dislodgement of the berries. Cranberry fields were flooded, and the berries were raked from the vines by swinging the rake back and forth through the vines, thus "raking the flood." Since the berries have an internal air pocket (fig. 33), they float to the top of the water, where they can be collected easily.

Today, "dry harvesting" of cranberries represents only 5% of the annual crop and takes place mainly

Fig. 33. A cranberry cut in half to show the four-chambered air pocket surrounding the seeds. The air pocket causes the berries to float after they have been knocked from the vines, facilitating their harvesting.

Fig. 34. Dry harvesting cranberries in an unflooded field requires more manpower and time, but it results in fruit that is less damaged and can, thus, be sold as fresh berries. Only 5% of harvesting is done using this method.

Heath Family

in New Jersey. Dry harvesting of cranberries is more labor-intensive but yields a more unblemished crop that can be sold as raw fruit. Small lawn mower–style, walk-behind harvesters (fig. 34) are guided through the cranberry fields to knock the berries from the vines into plastic boxes, which are left in place in the fields once they are full. Other workers then load the boxes onto trucks to be taken directly to the processor for cleaning and packaging. I have found that cranberries purchased fresh can be frozen for use throughout the year with no deterioration in quality.

By the 1940s, the majority of harvested cranberries were frozen at the receiving stations and sent to factories to be processed into canned whole-berry or strained cranberry sauce. Mechanical harvesters, called water reels or beaters (fig. 35), were invented in the 1950s; these are driven through the flooded fields to remove the berries from the cranberry plants with much greater speed. Boards or foam booms are then used to corral the floating berries into a corner of the field (fig. 36) where they are moved by conveyer belts or pumps into trucks to be taken to the processing plants. This is a colorful sight that can be witnessed each autumn in many of the states where cranberries are grown. I have seen cranberries grown and/or harvested in Massachusetts, Wisconsin, and New Jersey, but I was surprised to encounter a cranberry bog in the center of Manhattan. For several years a traveling exhibit, "Bogs Across America," sponsored by Ocean Spray, has made a stop at Rockefeller Center during harvest season, setting up a 1500 square foot "bog" filled with floating fresh berries (along with several wader-wearing cranberry growers to talk to the public about the farming of cranberries) (fig. 37). Living cranberry vines are planted in a narrow bed surrounding the water-filled "bog" to illustrate how the plants look when growing.

Harvested cranberries are cleaned of any remnants of leaves and stems before they are sorted and tested for quality. Sorting is done by passing the berries over a device with openings (see fig. 32), through which the berries that are not fully developed will pass. The method of testing to determine whether the berries are of high quality is called the bounce test. It supposedly was developed after a New Jersey cranberry grower spilled his berries down a flight of stairs. He noticed that only the freshest of the cranberries bounced down the steps to the bottom, leaving behind any that were soft, or otherwise imperfect, closer to the top. Thus, the bounce test was developed. Initially, this was a rather primitive method (fig. 38) that required the berries to bounce over a 10 cm (4 in.) board on their way to the sorter. Today, this is done using a mechanical device; those not passing the test after seven tries are discarded.

Most cranberries are now harvested using the "wet" method, described above, and are destined for the production of juice, sauce, and other cranberry products. Contrary to common belief, commercial cranberries are not grown in water. Although irrigated as needed during the growing season, the fields are flooded only at harvest time (fig. 39) and during the winter months when the ice protects the plants from extreme cold.

Fig. 35. Wet-harvesting of cranberries today is done with large, efficient, ride-on machines called water reels or beaters that can knock the berries from large swaths of the field with one pass.

Fig. 36. Once the berries are knocked from the vines and are floating on the surface of the water, workers gradually corral them into one corner of the bog using boards or foam booms.

American Cranberry

Fig. 37. Cranberry growers wading in a "bog" of cranberries in New York's Rockefeller Center as part of a traveling exhibit by Ocean Spray to familiarize the public with how cranberries are grown and processed. Living cranberry vines surround the bog.

Fig. 38. An old-fashioned bounce board that was once used to test cranberries for quality at Whitesbog in New Jersey. Only those berries that would bounce over a 4-inch barrier were sold as fresh berries; the rest were discarded.

Fig. 39. A cranberry field in Nantucket flooded for harvesting. The berries are being knocked from the plants as the beaters are driven through the field.

Fig. 40. A sign at a Wisconsin cranberry farm that is part of the Ocean Spray co-op.

Farmers have learned that spreading a layer of sand on the ice every few years (which is then deposited evenly on the beds when the ice melts) stimulates new vine growth once the ice melts and deposits the sand on the beds.

Cranberries are an important commercial crop in the states of Massachusetts, New Jersey, Oregon, Washington, Minnesota, and Wisconsin, with Wisconsin being the most important producer, providing more than half the cranberries sold in the country; Massachusetts is in second place. The estimated worth of the U.S. cranberry crop today is close to $300 million. Cranberry cultivation is done on a large scale with individual farmers joining cooperatives that set the price of the crop and share the costs of marketing.

The largest among these co-ops is Ocean Spray, established in 1930 and now encompassing approximately 700 farms in four states (fig. 40), plus others in Canada. When founded, the Ocean Spray co-op accounted for more than 90% of the cranberry market, and today still represents 65% of the market (an agricultural cooperative exemption prevents antitrust laws from breaking up such co-op organizations). The price per barrel has swung widely due to supply and demand. A Federal Marketing Order has helped to stabilize the market in years when crop yields have been unusually large or small.

Cranberries are also grown in many provinces of Canada and to a lesser degree as an introduced crop in southern Argentina and Chile. Canadian production has increased exponentially over the last decade.

Heath Family

Cranberry farmers, like other farmers, are at the mercy of extreme weather events, losses due to insects and disease, and, particular to cranberry, the native parasitic vine known as dodder (*Cuscuta gronovii*), which uses *Vaccinium macrocarpon* as one of its host plants. Infestation of a cranberry field by dodder may result in a 50% loss in crop productivity. Controlled flooding for 24–48-hour periods can kill some, but not all, of the dodder.

One of the industry's gravest problems occurred as a result of inadequate oversight of pest management. In the fall of 1959, it was discovered that some of that year's cranberry crop was contaminated with traces of a carcinogenic herbicide. The market collapsed, many people eliminated cranberry sauce from their traditional holiday menus, and the industry learned that it had to be more careful in their use of herbicides and pesticides—and that it was foolish to rely on just one product (cranberry sauce) for its revenue. The period of the 1960s was one of intensive product development that resulted in the now popular cranberry drinks, dried fruit, and other products.

Today, the cranberry business is booming, due in part to claims of cranberry's antioxidant compounds benefiting the cardiovascular and immune systems, and the possibility that cranberries may also have anticancer properties. An explosion of products took place in the final few decades of the twentieth century, which introduced cranberry juice drinks (usually mixed with other kinds of juices such as apple, grape, or blueberry; white cranberry juice is made from mature berries that have not yet turned red); dried cranberries as a snack food and for cooking; cranberry compounds in pill form to be taken as a dietary supplement, and even cranberry wine. All these products make use of the frozen berries, which account for 95% of the total cranberry harvest; the remaining 5% of berries are sold fresh to the consumer.

Three species in our area that we call cranberry have tough, leathery, evergreen leaves, but a fourth, a more southerly species that extends into the southern edge of our northeastern region, has deciduous leaves. The treatment of Ericaceae in *Flora of North America* divides *Vaccinium* into 10 sections, with American (large) cranberry (*V. macrocarpon*) and small cranberry (*V. oxycoccos*, which was once in its own genus, *Oxycoccos*) the only two members composing the section Oxycoccus. Small cranberry (fig. 41) resembles a diminutive *V. macrocarpon*, but its leaves are often strongly in-rolled along the margins and have a pointed apex; the undersides of the leaves are whitish. The pedicels have small, scalelike, reddish bracteoles situated lower on the pedicel than the larger (almost leaflike) ones of *V. macrocarpon*. These tiny plants grow on sphagnum hummocks in bogs and fens. They can usually be distinguished by size from *V. macrocarpon*, but in the case of depauperate individuals of *V. macrocarpon*, the bracts are critical to determine the species. The range of small cranberry extends beyond that of American cranberry to Alaska, Idaho, and California in the United States, and includes nearly all of Canada with the exception of the Canadian Arctic Archipelago; it also grows in northern Europe, northern Asia, and Greenland.

The other species referred to as a cranberry throughout the northern part of our region is known

Fig. 41. A plant of small cranberry (*Vaccinium oxycoccos*) grows half-buried in a deep mat of *Sphagnum* moss in an Adirondack bog. The flower of one pedicel has been broken off, but each of the two pedicels has a tiny red bracteole attached to its left side (a remnant of the opposite bracteole is visible as just a slight bump on the right side of the stalk with the flower). The three-parted leaf in front belongs to buckbean (*Menyanthes trifoliata*), another bog-dweller discussed in this book.

American Cranberry

Fig. 42. (Left) A white-flowered plant of mountain cranberry (*Vaccinium vitis-idaea* subsp. *vitis-idaea*) in the Dolomite mountains in Italy.

Fig. 43. (Right) Deeply pink flowers of mountain cranberry (*Vaccinium vitis-idaea* subsp. *minus*) in the White Mountains of New Hampshire. Note the bell-like corollas, unlike those of *Vaccinium macrocarpon*.

as mountain cranberry (*V. vitis-idaea*) and typically grows in the alpine zone of our northeastern mountains. It is a creeping, evergreen shrub having shiny, roundish-elliptic leaves with a distinctive indented midvein and with black dots on the lower leaf surface; it also has bell-like flowers that are more like those of a blueberry or bilberry than other species we call cranberry. It belongs to different section of *Vaccinium*: sect. Vitis-idaea. The flowers range from white in the European subsp. *vitis-idaea* (fig. 42) to deep pink in subsp. *minus* found in the Northeast (fig. 43), and the fruit is bright red (fig. 44). The plant is also known as partridgeberry in Newfoundland and lingonberry in Europe, where the fruits are enjoyed in much the same way we enjoy American cranberries. Mountain cranberry is native throughout New England (but is rare in Massachusetts and has not been collected in Connecticut since the late 1880s), Michigan, Wisconsin, Minnesota, and Alaska in the States, and in all provinces and territories in Canada, plus Greenland. It is also native to northern Europe and in higher mountain ranges to the south in Europe.

A fourth cranberry, southern mountain cranberry (*Vaccinium erythrocarpum*), belongs to yet another section of *Vaccinium*, sect. Oxycoccoides, and just barely extends north into the southern part of our region. It inhabits the southern Appalachian Mountains from West Virginia to northeastern Georgia (another subspecies of this entity is native to China, Japan, and Korea). Southern mountain cranberry is a deciduous shrub, growing to about 1.5 m tall, with flowers similar to, but more delicate than, those of American and small cranberry.

Not only humans enjoy the fruits of these cranberry species; our American cranberries are also eaten by various songbirds. If any mountain cranberries remain through winter, they become sweeter and are quite palatable to people as well. Another benefit to wildlife is that the cranberry bogs of New Jersey serve as a winter refuge for tundra (whistling) swans, which fly from northwestern Canada and Alaska in search of food and open water. The older, unused bogs, which are kept flooded in winter, are the ones most visited.

And do not be fooled by the name high-bush cranberry, which was bestowed upon a shrub, *Viburnum trilobum* (formerly known as *V. opulus* var. *americanum*), from an entirely different family, the Caprifoliaceae, because of its clusters of bright red fruits.

Fig. 44. Mountain cranberry in fruit on Mt. Washington in New Hampshire. The fruits are a favorite summer and autumn food of spruce grouse.

Ginseng Family

Fig. 45. American ginseng (*Panax quinquefolius*) in the foreground and maidenhair fern (*Adiantum pedatum*) in the background. These species are often found growing together on the slopes of rich forests in the Northeast.

American Ginseng

Fig. 46. A mature plant of ginseng usually has three leaves, each with five leaflets arranged in a palmate manner (here, one leaf has only four leaflets). The inconspicuous, ball-like inflorescence is located in the center of the image. It is atop a stalk (the peduncle) separate from the leaf stalks but arising from the same point at the apex of the plant's stem.

American Ginseng
Panax quinquefolius L.
Ginseng Family (Araliaceae)

American ginseng is increasingly difficult to find in the wild due to the overharvesting of plants for the herbal remedy trade. Middlemen ship the roots to China, usually via Hong Kong, where they command a high price for their purported therapeutic properties. The genus name, *Panax*, is from the Greek *pan*, meaning "all," and *akos*, meaning "cure," thus a cure-all or panacea. Linnaeus gave this name to the genus based on the reputation of the Asian species, *Panax ginseng*, that was—and still is—used by the Chinese in the treatment of all manner of ills.

Habitat: Cool microclimates in rich woods, most commonly on north or northeast-facing slopes where winter's protective blanket of snow remains longest; however, in the most northern part of its range, ginseng grows best on south- or southwest-facing slopes. Ginseng inhabits deciduous forest, with the tree species varying according to geographic locality. These may include oak, sugar maple, beech, hickory, and ash. Certain wildflowers are also indicators of potentially hospitable habitat for ginseng, among them: jack-in-the-pulpit, mayapple, blue cohosh, and wild leek; maidenhair and rattlesnake ferns also grow in association with ginseng (fig. 45). Good air circulation, moist soils, shade, and ample calcium sulfate (in the form of gypsum) are important factors in prime ginseng habitat.

Range: The eastern half of North America, extending from Ontario and Quebec (but not the Maritimes), south to isolated pockets in South Carolina and northern Georgia, and west to Minnesota, with scattered localities in South Dakota, Nebraska, Kansas, and Oklahoma. The highest concentration of ginseng habitat occurs from New York and Pennsylvania, west to the easternmost states of the Midwest, and south through the southern mountains, with the Appalachian region of Tennessee, Kentucky, North Carolina, and West Virginia considered to provide the most productive forests for this species. Nevertheless, ginseng is considered rare to uncommon throughout its range.

Ginseng Family

American ginseng (*Panax quinquefolius*) is an easy plant to recognize even when not in flower or fruit. Its long-stalked, palmately compound leaves are arranged in whorls of three (varying from one in young plants to five in long-established plants), making ginseng difficult to confuse with other plants within its range. Each leaf stalk is attached to the apex of the plant's solitary stem and generally bears three larger leaflets (6–15 cm) and two smaller ones, all growing from the same point. The leaflets are stalked and have noticeably serrate margins (fig. 46). In ginseng parlance, the leaves are referred to as "prongs" and the leaflets as "leaves." When in flower, the inflorescence, 1–12 cm in diameter, grows from the same point as the leaf stalks and is topped with a ball-like umbel of small, five-lobed greenish flowers with white anthers in the center (fig. 47). The flowers are perfect, having both male and female organs, and are reported to have an odor similar to vanilla, which likely is the attractant that entices pollinators to the insignificant flowers. Nectar secreted by the nectary at the flower's base is the reward. Ginseng flowers are self-compatible; thus, they can be pollinated with pollen from the same flower or from another flower in the same inflorescence. Outcrossing occurs through the visits of small generalist pollinators such as syrphid flies and halictid bees. Once the ovules have been fertilized, development of the ovary into a mature fruit takes about two and a half months (see immature fruits in fig. 47). The mature fruits are bright red berrylike drupes.

While seeds of ginseng fruits have generally been considered gravity dispersed, and thus fall not far from the parent plant, a study of possible dispersal by songbirds, published in 2014, demonstrated the likelihood of longer-range dispersal by birds that are beginning to migrate at the time the fruits are mature. Birds are known to seek out juicy, nutrient-rich fruits to fuel their long journey, and they are particularly attracted to red fruits, making the hypothesis of bird dispersal feasible. Motion sensor cameras captured songbirds of several species eating the fruits, with the most frequently observed being three species of thrush (wood, hermit, and Swainson's). Captive feeding experiments showed that most thrushes regurgitated the seeds, still viable, within 5–35 minutes. When offered ginseng berries mixed in with corn and other seeds commonly eaten by birds, the thrushes preferentially chose to eat the ginseng fruit first. In the wild, it is likely that thrushes would regurgitate the seeds within 100 meters of the feeding site.

The height of the plant, the number of leaves, and the number of leaflets all increase as the plant ages. The plants are long-lived if left undisturbed, with some reportedly living for seven or more decades, but occasionally remaining dormant for a year or more during their lifetime. The oldest plant on record, collected in southern New York State, was estimated to be 132 years old, based on the number of leaf scars on its rhizome. Plants older than 50 years, though, tend to be in a state of decline.

I can think of only two local native plants that might be confused with American ginseng; one, wild sarsaparilla (*Aralia nudicaulis*), is a relative in the same family (Araliaceae), which also has a whorl of three compound leaves arising from the top of the stem. Closer inspection, however, reveals that rather than having the leaflets all attached at the same point (palmately compound) as in ginseng, four leaflets of wild sarsaparilla are attached opposite each other along the leaf stalk (pinnately compound), with the fifth leaflet at the apex of the leaf; they are also more finely serrate. The inflorescence differs as well, with wild sarsaparilla's 3–4 umbels of white flowers arising from the

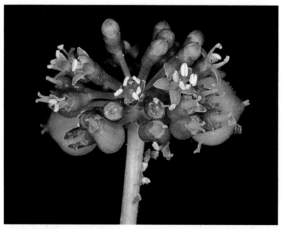

Fig. 47. By July, the inflorescences of ginseng are in various stages of fertility. Flowers usually begin blooming from the base of the inflorescence upward. Here, the lowest flowers are developing into fruits, which are flattened at this stage; the green, five-lobed corollas of the midlevel flowers are fully open, with bright white anthers and stigmas; and the uppermost flowers on the umbel are still in bud. Note that the flowers' two styles are persistent in the developing fruits, and sometimes still visible in the mature fruits (see fig. 53).

American Ginseng

ground on a separate stalk (fig. 48). Wild sarsaparilla grows more often in the acidic soils of hemlock forests. The other possible imposter is Virginia creeper (*Parthenocissus quinquefolia*), a climbing vine in the grape family (Vitaceae), which, before it has begun to climb, might be mistaken for an herb. Its leaves are palmately compound with pronounced serrate margins (but only on their apical halves), and each leaflet is sessile or has an extremely reduced leaflet stalk (fig. 49). The leaves of Virginia creeper are not atop a single upright stem as are those of ginseng, but rather alternate along a creeping stem.

The part of the ginseng plant responsible for its popularity—and its potential extinction—is the tuberous root. Though the roots of 1–2-year-old plants are thinner than the diameter of a pencil, mature ginseng roots are thickened and fleshy. The roots increase in size as the plant ages, branching frequently, which results (with a bit of imagination) in their coming to resemble a human form. For this reason, the Chinese refer to the plant as "man-plant" (*schinseng*), from which our word, ginseng, was derived. Roots of wild ginseng plants do not reach harvestable size until at least their fifth year, with another five years needed to produce a truly prime root. Since ginseng roots do not have many fine root hairs to aid in the absorption of soil nutrients, they benefit from an association with mycorrhizal fungi, which are also connected to the roots of trees in the surrounding forest. The fungal mycorrhizae contribute to the vigor of the ginseng plants by infiltration of the roots of ginseng and neighboring trees, thereby allowing for the transport of minerals from the trees to the ginseng plants.

The manlike resemblance of ginseng root indicated to the ancient Chinese that the plant would be efficacious in treating all conditions afflicting man, both physical and mental. The larger and more humanoid the root, the more powerful its properties were believed to be—and the more valuable it was. Such roots were said to be worth their weight in gold in China, and today a "perfect" root (meaning one that is large and very manlike) can sell for $10,000. The Chinese use of ginseng as a medicinal herb was documented in the medical journal of a Chinese emperor who reigned more than 2000 years ago, but it is presumed to have been in use long before that time.

Fig. 48. (Left) In the same family as ginseng (Araliaceae), wild sarsaparilla (*Aralia nudicaulis*) is a plant of the forest floor that has three leaves arising from the same point on the stem, each with five leaflets, as does ginseng. However, the leaflets of wild sarsaparilla are arranged with two pairs opposite each other on the leaf stalk and a fifth leaflet at the leaf's apex. Additionally, the inflorescence, usually bearing three to four umbels of flowers, does not arise from the same stalk as the leaves, but separately—the basis for its epithet, *nudicaulis*, meaning "naked stem." **Fig. 49.** (Right) These leaves belong to a juvenile plant of Virginia creeper (*Parthenocissus quinquefolia*). Its five palmately arranged leaflets might cause the leaves to be mistaken for those of ginseng, but they are arranged alternately along a creeping stem, and only the apical portions of the leaflets' margins have teeth. Virginia creeper matures to be a woody vine that climbs high into the forest canopy. The small seedlings beneath the Virginia creeper are those of the invasive Japanese stiltgrass (*Microstegium vimineum*).

Ginseng Family

Several Chinese and Korean researchers have published papers showing that the principal compounds in *P. ginseng* responsible for effecting positive health benefits in humans are ginsenosides. Little research has been done on ginseng's medicinal potential in the United States, but in the few double-blind studies carried out, investigators have recorded no beneficial effects from ginseng when compared with a placebo and, in some cases, found that the patients suffered detrimental side-effects. Nevertheless, ginseng is on the U. S. Public Health Service GRAS list (Generally Recognized as Safe) and has been used by some doctors to help cancer patients cope with the stress of chemotherapy and radiation. It has also shown some positive effect in stopping the growth of breast cancer cells in vitro.

Currently, 13 species in the genus *Panax* are recognized, but only two are considered commercially important, the Asian species, *P. ginseng*, and American ginseng, *P. quinquefolius*. It is *P. ginseng* that has a long history of use as a pharmaceutical, which came to a near halt when the species was almost extirpated in China during the early 1700s—due both to overcollection of the roots and to cutting of forests (ginseng requires shade to grow). Fortunately for the Asian market, at about that time a Jesuit cartographer, Father Jartoux, was sent to China to draw the first map of Manchuria. While there, Jartoux became aware of the importance of the ginseng plant to the Chinese, and he sent drawings and descriptions of the plant to the procurator of the missions in China and India. The letter was widely published in Europe, and four years later, a Jesuit missionary, Father Joseph-François Lafitau, happened upon the letter while on a visit to Quebec, in what was then New France. He reasoned that there was a possibility that a similar species might exist in the Montreal area based on the similarity of climate and topography to that in the range of *P. ginseng* in China; it was already well-known that it is not uncommon for morphologically and taxonomically similar plants to be found in both eastern Asia and eastern North America—relics of the once vast temperate forest thought to have covered much of the Northern Hemisphere during the Tertiary Period. Lafitau was a French missionary/anthropologist living across the St. Lawrence River from Montreal and studying the customs of the Iroquois. He soon found a plant that matched Jartoux's description, and which the Iroquois confirmed was used by them for medicinal purposes, in this case as a febrifuge. In 1719, the French established the Company of the Indies to collect and market ginseng to China. Trade began in 1720. This spurred a frenzy of collection, both in New France and in the British Colonies to the south that were to become the United States. Export of the plant from Boston to China began in 1733, launching a multimillion-dollar business between the Colonies and China. Such was the extent of collecting, that by 1750, American ginseng was already becoming scarce in New France. Ginseng (listed as *Panax quinquefolium*) was included in a list of 380 collections made in North America by Peter Kalm, whose mentor, Linnaeus, dispatched him to the New World in 1748 to collect plants and seeds of plants, especially those of potential economic use. Kalm returned to Sweden in 1751, bearing not only his plant collections but also insects, shells, and amphibians. Linnaeus ultimately described as new, 60 species of plants based on Kalm's collections and named the genus of mountain laurel *Kalmia* to honor him.

Among the notable people who participated in ginseng's heyday of the late 1700s was John Jacob Astor, who, in addition to his trade in beaver pelts and other furs, also acquired ginseng roots during his travels in upstate New York. He then sold them by the shipload to China, realizing a profit of $1.14 million (in today's dollars) on one such 1784 cargo. At about the same time, Daniel Boone, an avid ginseng collector, was not as lucky. His year-long ginseng digging efforts—variously claimed to be between 12 barrels and 12 tons of root, with the first figure being more feasible since barrels were once referred to as "tuns"—fell into the water when the boat hired to take his harvest across the Ohio River overturned in a strong current. Boone's ginseng was soaked, ruining most of it before help arrived to salvage the remainder. Boone tried to sun-dry what was left, further damaging it and forcing him to sell it for next to nothing.

Of the thirteen species of *Panax*, only two are native in eastern North America, *P. quinquefolius* and *P. trifolius*, which, as you might suspect from the Latin translation, has three leaves, each of which bears three leaflets attached to the leaf stalk (petiole) in a palmate fashion. As with the leaves of American ginseng, leaflet number may vary. *Panax trifolius* is commonly called dwarf ginseng since it is a much smaller plant than American ginseng, reaching only about 20 cm (7–8 in.) in height compared with 50 cm (19–20 in.) for American ginseng.

American Ginseng

Fig. 50. The only other member of the ginseng genus in the Northeast is dwarf ginseng (*Panax trifolius*), which generally has three leaflets on each of its three compound leaves (there are some exceptions, as seen here). It is a small spring ephemeral that inhabits moist forested areas. Its solitary umbel of white flowers quickly ripens into small yellowish fruits. No medicinal claims are made for its small, tuberous, globular root.

Fig. 51. The fruits of this mature ginseng plant have turned red in early September; soon the leaves will turn yellow. Note that one of the three leaves has been eaten by deer.

Its leaflets are sessile, with the middle one occasionally having a very reduced stalk. They are serrate and have blunt, rounded tips, rather than the sharp, pointed ones seen in American ginseng. The inflorescence of dwarf ginseng is a spherical umbel of small, white flowers borne on a stalk (peduncle) at the apex of the stem, where the three leaves are attached (fig. 50). The flowers' ovaries develop into small yellowish fruits. Dwarf ginseng is an early spring ephemeral favoring wet forests; its leaves wither and disappear before its taller relative, American ginseng, has even produced its flowers. Dwarf ginseng stores carbohydrates for the next year's growth in its small, globular root.

The alarming decline of *Panax quinquefolius* in the wild resulted in the species being included under the CITES (Convention on International Trade in Endangered Species of Wild Fauna and Flora) treaty, as an Appendix II protected species (one that is not currently threatened with extinction but is at risk of becoming so unless trade is carefully monitored). In most states where ginseng grows, it is included on the state's list of protected or vulnerable plants. Many states, however, do allow the seasonal harvesting of ginseng, but under strict regulations usually overseen by the state's department of conservation. In those states where harvesting is permitted, mature plants (those with three "prongs") may be harvested only between September 1 and November 30 (December 31 in some states). The fall dates were chosen to ensure that the ginseng fruit had time to ripen (fig. 51), but before the plant had completely died back to the ground for the winter; the following year's plant will grow from an underground bud at the base of the current year's stalk. The plants are easy to find in autumn because the leaves turn yellow early in the season, making them easy to spot in the dark forest understory, especially since mature plants will have a cluster of contrasting bright red fruits (fig. 52) at that time. It is mandated that harvesters immediately plant the seeds of all fruits (usually two per fruit) (fig. 53) in the locality where they were found. Doing so helps to maintain a sustainable population that can be revisited in the future for subsequent harvesting. Ginseng seeds have immature embryos when the fruit ripens. Germination occurs only after the seeds have undergone a cold period (the first winter), followed by a warm period (the next spring and summer), followed by another cold period (the second autumn/winter). Thus, it takes approximately 18 months for germination to actually take place. Digging ginseng on state or federal lands is prohibited, and landholder permission must be obtained before harvesting on private property. The middlemen who export the roots are required to have a permit from the U.S. Fish and Wildlife Service allowing them to sell

Fig. 52. When the fruits of ginseng are ripe, they turn a brilliant red. Not all the flowers on this inflorescence were successfully pollinated and hence did not develop into fruits.

Ginseng Family

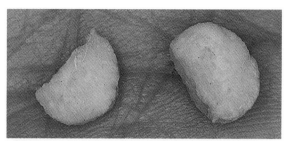

Fig. 53. Ginseng fruits are drupes that comprise two carpels; each will produce one seed if the ovule has been fertilized. The two stones (each containing a seed) are dispersed while the embryos are still immature. They must undergo two cold periods, separated by a warm period, before germinating a year and a half after dispersal.

Fig. 54. This area of forest is the same as that shown in fig. 45. Note that the maidenhair ferns are still present, but all the ginseng plants have been poached from the area. This photo was taken in early September of the same year that fig. 45 was taken in June.

ginseng roots abroad. Permits are granted only if the state where the plants were harvested has a program in place, overseen by a state agency that is compliant with all the above regulations. If all the rules are followed, ginseng and its traditional harvesting could continue well into the future.

In spite of the stringent laws in place to protect ginseng, the lure of quick cash entices some harvesters (variously known as 'sangers, 'sang hunters, or diggers) to disregard the rules and dig plants that do not meet the minimum size standards (having at least three leaves), or dig out of season, or on public lands, or on private lands without permission. Even in national parks it is difficult to patrol such wild areas, and, thus, many poachers are not caught. Illegitimate diggers know where to look for optimal ginseng habitat and often keep their eye on their secret localities for years until the plants are ready to harvest, which, many times, is done in the dark of night. Catching poachers is difficult, and arrests are few, making illegal harvesting worth the risk for many. Indeed, when I returned to the site on private land where I took many of these photos at the end of the season, much of the mature ginseng had been poached (fig. 54). In 2005, the U. S. Fish and Wildlife Service reported that close to 15 million ginseng plants had been harvested from the wild in the prior year—not including the illegally harvested plants that were sold "under the radar." Not only are such large harvests unsustainable, but by leaving very few populations untouched and widely scattered in the wild, inbreeding can result, leading to a decline in species vigor.

While American ginseng is easy to collect, it is slow growing, taking from 5 to 10 years before the plant reaches a size that bears a root worth harvesting. The price of a 6–10-year-old American ginseng root can reach $500–$1000 (the price for dried wild root, which is about one-third the weight of freshly harvested root), making it an attractive means of supplementing income, especially in financially depressed regions like the Appalachians. The knowledgeable buyer can tell wild-harvested from cultivated ginseng with ease. The older, wild American ginseng is darker, and more branched, contorted, and wrinkled than cultivated ginseng. Scars can be counted showing where each year's stalk was once attached, allowing the buyer to estimate how old the plant was when harvested.

This forest-grown botanical requires no maintenance, only patience and secrecy to keep the site from being poached by another digger.

By the early 1800s, as American ginseng became more and more difficult to find in the wild, methods of cultivating the species were developed in an attempt to meet the growing demand for the root. Field cultivation of ginseng began in upstate New York in the 1880s. American ginseng requires canopy shade (at least 70%) in order to grow. George Stanton from the Elmira/Cortlandt area is credited with first realizing that shade was the key to successful ginseng cultivation.

Ginseng farmers now can purchase stratified seed (seed that has already been through the required cold-warm-cold cycle needed to break dormancy), but the cost for good quality seed may be in excess of $100 per pound. The easiest method of cultivating ginseng, termed "simulated wild," mimics the natural growing conditions of wild ginseng, with the exception that the

American Ginseng

seeds have been purchased and manually sown in forests with habitat similar to that of native ginseng. The natural cover of leaves is raked up, the seeds scattered and lightly covered with soil, and the leaves raked back over the planting area. The "crop" requires no further care until harvest time. Seeds must not be planted too closely together in order to prevent the spread of pathogens that may attack the plants. Some seeds fall prey to small mammals, and a root-rot fungus, *Phytophora*, can be a problem for ginseng, especially in wet years. The foliage of ginseng plants is sometimes eaten by slugs, turkeys, and white-tailed deer (fig. 55). In many cases deer eat the fruits as well as the leaves. A study to determine whether deer might disperse the seeds at some distance from where they were eaten found that the seeds in the deer's fecal pellets were no longer intact (and thus not viable). Protective fencing is not used to protect the simulated wild grown areas not only because of its expense but also because of the fear that fencing would alert poachers to the presence of something worthy of protection. Ginseng grown in this way brings a higher price than the farm-grown product. Ginseng is one of the most valuable nontimber forest crops, more so than many trees.

More labor intensive is a method called "woods cultivated." In this case, actual tilled beds are created in naturally shaded areas, but not necessarily typical ginseng forest, and maintained by weeding during the first few years. The roots are generally harvested after 6–9 years. They are considered less valuable than either strictly wild or "cultivated wild" roots.

To grow ginseng in an open field environment requires shading the crop artificially with black shade cloth. Fertilizer is necessary because of the lack of rich humus normally found in typical ginseng forest habitat. Closely grown plants are more prone to fungal disease and other pathogens and, therefore, must be sprayed several times a season. Cultivation of ginseng is expensive and prone to losses due to unusual weather events (late snowfalls, hail, etc.), prompting the farmers to harvest their crops after just three or four years. Plants grow more quickly under these standardized conditions, but the roots are more uniform, tending to look like small, whitish carrots rather than wizened, little old men. Accordingly, field-grown ginseng roots fetch only about $20 per dried pound, a fraction of the price of the wild ginseng. American ginseng farming was negatively impacted when the Chinese began farming cultivated ginseng. Farming became a necessity in China, since many of their ginseng-producing forests had been logged, depriving the ginseng of the shade needed to thrive. Given the lower cost of labor in China, the American market was unable to compete. Yet, some ginseng farms (including one of 1000 acres) still exist in Marathon County, north of Wausau, Wisconsin, where more than 90% of ginseng farmed in the United States is grown. More than 300 tons of ginseng roots were exported to Hong Kong in 2000.

Whole, dried ginseng root can be purchased in the United States today in Chinese markets (fig. 56), but most Americans buy packaged products containing ginseng in liquid or powdered form in health food stores, pharmacies, and supermarkets. It has become one of the most popular herbal supplements in the United States and in Europe. However, one study of these products has shown that 15% of them contained no ginseng at all, and 50% had less than 2% ginsenosides. In addition, many were contaminated by pesticides or heavy metals. *Caveat emptor*.

Fig. 55. Deer browse the leaves, and sometimes the inflorescence or infructescence, of ginseng, causing the plant to lose a year of reproductive potential. Even losing just its leaves (as seen here) is harmful to the plant, since it can no longer photosynthesize and store carbohydrates in its root (see also fig. 51).

Fig. 56. In Chinese markets, such as this one in New York City's Chinatown, ginseng roots can be purchased whole and dried, or in powdered or liquid form in various herbal preparations.

Lotus Family

Fig. 57. A flower of American lotus viewed from below. The two small greenish sepals are evident as are the somewhat larger greenish petals that enclosed the flower bud.

American Lotus

Fig. 58. A close-up view of one flower in a pond nearly filled with lotus plants. The small dots on the central structure of the flower (the receptacle) are the flower's stigmas; they are still yellow and receptive. The darker yellow portions of the stamens surrounding the base of the receptacle are the anthers, each bearing lighter yellow hooked appendages at their tips.

American Lotus
Nelumbo lutea (Willd.) Pers.
Lotus Family (Nelumbonaceae)

American lotus is one of our most strikingly beautiful aquatic plants (fig. 57). Its exquisite yellow flowers and large orbicular leaves would have inspired Monet. The edible rhizomes and seeds are sold in Chinese markets, and its woody fruit receptacles add interest to dried floral arrangements.

Habitat: Still water, commonly less than a meter deep, in ponds, lakes, swamps, and marshes (fig. 58).

Range: Minnesota and Nebraska, south to Texas and east to Ontario and the Atlantic coast, with the exception of Vermont and New Hampshire. American lotus is thought to have originated in the southeastern United States, Mexico, the Caribbean, and Colombia. It was transported north by Native Americans and has more recently been introduced into our Pacific coastal states.

Nelumbonaceae is a family of only two species (some taxonomists consider them both subspecies of *Nelumbo nucifera*): the native American lotus (*Nelumbo lutea*), and the sacred lotus of Asia (*Nelumbo nucifera*). Fossils of *Nelumbo* and several related genera have been found in various parts of the world, all in the Northern Hemisphere, other than one from Patagonian Argentina, indicating that the family was once both more widespread and more diverse. The family is thought to be among the most ancient angiosperms (flowering plants).

The Asian *Nelumbo* is the better known of the two species. It is venerated by people of many cultures and is the national flower of Egypt, India, and Vietnam. *Nelumbo* is derived from *nelum*, a Sinhalese (a language spoken in Sri Lanka) word for this species; *nucifera* refers to the "nutlike" fruits. Sacred lotus is widely cultivated in botanical gardens around the world (fig. 59) for the beauty of its large pink and white flowers (fig. 60), and its unusual fruits. The flowers are frequently depicted in artwork, most notably as a "throne" for the seated Buddha. The lotus signifies

Lotus Family

Fig. 59. (Left) Sacred lotus (*Nelumbo nucifera*) is grown in many botanical gardens including the New York Botanical Garden, as seen in this photo. **Fig. 60.** (Right) The sacred lotus is admired for the beauty of its large pink and white flowers and its even larger peltate leaves.

the growth of the human soul from materialism to enlightenment and symbolizes purity and detachment in the Hindu culture and many other religions of Asia. Its flowers also depict an idealization of feminine qualities: purity, beauty, perfection, grace, and elegance.

Native Americans managed stands of American lotus for its value as a food plant. The starchy rhizomes are nutritious and were harvested each autumn; they could be stored for several months. Harvesting was no easy task as the rhizomes are buried about a foot deep in the mud. When baked, the tubers become sweet and soft, much like sweet potatoes. The marble-sized fruits (often mistaken for seeds) (fig. 61) were also harvested and eaten after the fruit wall was removed. Lotus fruits are high in protein and can be dried and ground into a type of flour used to thicken soups and stews and to make a kind of bread. Immature fruits are eaten either raw or cooked and are said to have a taste and texture much like chestnuts. Young leaves can be boiled and eaten, and in Asia older leaves of the sacred lotus are used as a wrapping for cooking other foods. The plant is still an important part of the diet of Osage Indians (originally from the Osage River area in western Missouri), who reserve it for special occasions. They consider the lotus a sacred food and a symbol of the gift of life. Asian Americans, who enjoy eating the tubers and fruits of the Asian species of lotus, *N. nucifera*, are also fond of eating our native species. In addition to its use as a food, American lotus is popular today in the florist trade. Its hard, brown fruiting structures (the receptacles) are an attractive addition to dried plant arrangements and wreaths in autumn.

American lotus fruits and rhizomes serve as an important food source for marsh animals and birds as well. In fact, near the city of Monroe, Michigan, on the shores of Lake Erie, large numbers of tourists were once attracted to the sizable beds of lotus each year to see or hunt the birds and animals that inhabited or visited the lotus marshes. When game laws were changed to protect native muskrats from hunting and trapping, the muskrats multiplied and their feeding on lotus plants caused a marked decline in the local lotus population—and a corresponding decline in the tourist revenue that the lotus marsh had generated for the surrounding community.

Beyond the uses mentioned above, lotus leaves have served as an inspiration for scientists in the field of nanotechnology. One of the delights of lotus is the way that water beads up on its leaves and rolls

Fig. 61. A mature fruit of an American lotus is brown and about one-half inch in diameter. The fruits are edible when immature and can be ground into flour when mature.

about like silvery balls of mercury. This is true of both American lotus and the sacred lotus of Asia. The superhydrophobic property of the leaves is due to a combination of the very fine ultrastructure of the leaf surface, made up of minute projections, and a waxy coating. As a result, only 2%–3% of a droplet of water is in contact with the leaf surface. The remaining area beneath the water drop is filled with air bubbles, giving the drop its silvery appearance. The leaves repel water even when immersed. They also repel any dirt particles that may land on them by holding the dirt particles above the leaf surface until they are removed by beads of water. Thus, the leaf surface remains clean. This superhydrophobic property is expressed in the leaves of some other species as well (e.g., taro [*Colocasia esculenta*] and lady's mantle [*Alchemilla* spp.]) and is referred to as the "Lotus Effect" by scientists who study the phenomenon. In wet environments where lotus and taro grow, such hydrophobic properties aid in preventing algae and fungi from growing on the leaf. It is this cleanliness that prompted the Buddhists to view the sacred lotus as a symbol of purity. Corollaries in the animal world would be the highly irregular scaled wings of butterflies and the oiled surface of the feathers of ducks, which allow these organisms to repel water.

It was a German botanist, Wilhelm Barthlott, who discovered that the superhydrophobic properties of the leaves were produced by the combination of minute bumps and waxy surfaces. He later patented the idea of applying that property commercially in the manufacture of surfaces that would be self-cleaning. Such a quality can be desirable in a variety of man-made products and has led to efforts to mimic artificially the nanostructure of the leaf. To give an idea of the minute scale at which scientists are working, a nanometer is one-billionth of a meter. Greatly magnified, the leaf surface appears as a multitude of closely packed conical projections, each textured with irregular papillae. Scientists have managed to reproduce the microstructure of the leaf surface with a laser structured silicon surface. This technology has been employed in designing exterior paints (marketed as StoLotusan), coatings for microwave antennas, and self-cleaning fabrics. Among potential uses of such technology are hydrophobic paints for the hulls of ships, which would allow them to glide more efficiently through the water, and airplane coatings that might be able to shed ice. Medical applications include the development of stents that repel various adhesive substances that could potentially lead to clogging of the stent (such as blood platelets). Using natural phenomena as models for man-made products is called biomimicry. Perhaps one of the best-known examples of a product that has resulted from mimicking a biological phenomenon was that invented by a Swiss engineer, who in the late 1940s saw the potential for developing a reversible fastener comprising tiny hooks and loops after pulling the burrs of a burdock plant (*Arctium* sp.) from his clothes and his dog's fur. The multiuse product is, of course, the widely used Velcro.

Water gardeners plant American lotus to provide summer beauty to their ponds and lakes. This practice, however, should be discouraged since lotus reproduces rapidly, primarily by proliferating rhizomes. The plants are capable of spreading at a rate of 45 feet radially in one growing season, quickly causing ponds to become choked with lotus plants (fig. 62).

Fig. 62. The rapid spread of *Nelumbo*, primarily by its underground rhizomes, can quickly choke quiet bodies of water.

Lotus Family

Lotus also reproduces by seeds. Indeed, the seeds are astoundingly long-lived. Until 2005, seeds of sacred lotus found in an ancient Chinese lakebed held the record of 1000+ years old for radiocarbon-documented seed viability. Among factors in the longevity of lotus seeds are seed coats that are impermeable to water and embryos that contain chlorophyll, which thereby enables them to germinate quickly once the proper conditions are present. Amazingly, ancient seeds germinated within four days, equal in time to fresh lotus seeds. However, lotus seeds' longevity record for viability was surpassed in 2005 with the germination of a seed recovered from a cache of seeds stored at Israel's ancient fortress of Masada during the Romans' siege and ultimate conquest of the Jews who were defending the fort in 73 CE (Common Era). Documentation of the age of the seed is based on records of contemporary eyewitness accounts regarding where the food stores were hidden. The recovered seeds were stored for a period of time before attempts at germination were carried out. One seed of three tested yielded a tree—a date palm.

That record, in turn, was surpassed in 2012 when Russian scientists germinated seeds of a Siberian species, *Silene stenophylla*, which had been radiocarbon-dated to 32,000 years BCE (Before the Common Era). The seeds had been discovered buried 124 feet beneath the permafrost. The resultant plants developed through their full life cycle and produced new seeds. This discovery may lead to better ways to preserve seeds for future use and perhaps even to the development of methods of resurrecting extinct species, should that be judged wise.

Nelumbo was formerly included in the water-lily family (Nymphaeaceae), but current classification systems such as that followed by the Angiosperm Phylogeny Group now place it in a distinct family of a single genus, differing from members of the water-lily family in its floral reproductive parts, its fruits, and its leaves. Both the leaves and the flowers of the lotus are raised high above the water rather than floating on the surface as do those of water lilies. The presence of a staminal ring from which the 200+ stamens arise is another feature that differentiates *Nelumbo* from members of the Nymphaeaceae. The flowers' ovaries, and thus the fruits of *Nelumbo*, are distinguished by their sunken arrangement within a large spongy receptacle (fig. 63), which becomes woody upon maturity. The leaf margins

Fig. 63. A lateral section of a still spongy, immature receptacle with its dark, no longer receptive stigmas protruding from the top surface. Note the white latex seeping from the cut section.

of the almost circular leaves of *Nelumbo* are entire, and their petiole (leaf stalk) is attached in the center of the leaf (fig. 64), which differentiates them from the leaves of *Nymphaea*, which have a slit running from the margin into the center of the leaf with the petiole attached at that point. Young leaf margins of *Nelumbo* unroll scroll-like, in two directions from the center (fig. 65). The veins radiate from a central point, forking near the margin of the leaf, with the exception of thwe main, more robust, vein (fig. 66). Unlike most plants, other than aquatics such as water-lily (*Nymphaea*), the stomata are located on the upper surface of the leaves, and gas exchange occurs when the leaf surface is above the water surface as well as when water levels are high enough that the leaves are floating on the surface.

Fig. 64. The orbicular leaves of lotus are peltate (having their stalk attached at the center rather than on the margin of the leaf) with veins radiating from that point of attachment. All veins except one branch as they near the margin of the leaf.

American Lotus

Fig. 65. (Left) The new leaves of American lotus unfurl from the center like a scroll being opened.
Fig. 66. (Right) The underside of a leaf of American lotus showing the venation pattern. All veins radiate from the central stalk of the peltate leaf. One principal vein is unforked; all others fork toward the margin of the leaf.

Both flowers and leaves of *Nelumbo* can rise up to 3 or 4 feet above the water on long peduncles and petioles (see fig. 58). Peduncles, petioles, and rhizomes all contain air channels (fig. 67). When leaf stems are cut underwater, gases can be seen bubbling from the cut ends. These bubbles may even be observed within the large droplets of water that accumulate in the center of the living leaves. Air channels in the tubers (rhizomes thickened into storage organs) of lotus can be seen in the sliced "lotus roots" used in many Chinese and Japanese recipes. The cut surfaces of petioles and of other parts of the plant exude white latex (a milky white sap; see fig. 67), a property found in many other plant families as well. The composition of the latex of lotus is mostly water (91%) with sugars making up a small part of the balance. The latex may contain defensive compounds that protect the plant from herbivores.

Nelumbo lutea flowers have two small greenish sepals and a few slightly larger outer petals that surround the flower bud. These petals become greenish as they open, giving the appearance of additional sepals. There are several to many inner, pale yellow petals (*lutea* is Latin for yellow). American lotus has the largest flower of any wildflower native to North America, measuring 15–25 cm (ca. 6–10 in.) in diameter. The flowers have a three- to six-day life span. When the petals first open, they form a narrowly tubular opening downward to the broad receptacle, the structure on which all the other floral organs are borne. The stigmas protrude only a few millimeters above the circular depressions arranged over the surface of the broad receptacle in the center of the flower and are receptive at this time, secreting a sticky substance that promotes adhesion and germination of pollen deposited on them. At this stage the stamens are pressed tightly against the receptacle by the upright petals so that only the starch-bearing anther appendages at the tips of the innermost stamens are exposed, curling over the

Fig. 67. A cut across the peduncle (flower stem) of a lotus flower showing the air channels and the milky latex.

Lotus Family

margin of the receptacle. Insects (primarily bees, flies, and beetles; fig. 68) are attracted to the flowers by the bright yellow color of the receptacle and by the pleasant, fruity aroma emanating from the receptive stigmas and the hooked appendages of the anthers. The insects are searching for pollen, which is still unavailable when the flower first opens, but some bees linger to forage on the sticky exudates from the stigmas. The insect visitors often arrive covered with pollen from previously visited flowers and can thus effect cross-pollination during this early stage of floral development.

From pre-opening bud to petal drop, lotus flowers are capable of maintaining a steady temperature ranging between 30°C and 35°C (86°F–95°F) despite fluctuating diurnal air temperatures. The increased warmth serves both to increase the volatility of the attractive fragrance and to provide a warm refuge for insects. Although other species have demonstrated the ability to generate heat, only two of those have been shown to maintain such a similarly precise range of temperature throughout the flowering period: a South American species of *Philodendron* and our Northeastern skunk cabbage, *Symplocarpus foetidus* (see the chapter on skunk cabbage in my earlier book, *Spring Wildflowers of the Northeast: A Natural History*).

Flowers begin to close by early afternoon, occasionally resulting in the accidental trapping of visiting insects. Bees trapped overnight are dead by morning, overcome by the strong aromatic compounds. Beetles, however, can survive the night and are released the following morning when the flowers reopen. Diurnal opening and closing of the flowers, heat production, and the presence of starchy food bodies are generally an indication of modification of a flower for beetle pollination. Beetles may have coevolved to be the original pollinators of *Nelumbo*, but this does not appear to be the current situation. Bees are far more frequent visitors and are capable of transporting greater quantities of pollen.

By day two, the stamens are exposed and releasing massive amounts of pollen. Any beetles trapped overnight, as well as second-day visitors, are dusted with pollen. Self-pollination of the still receptive stigmas can occur on the second day, as both the stigmas and the fresh pollen are still producing aroma. By the third day, however, petals and stamens begin to fall, and the browning stigmas are no longer receptive.

Fig. 68. (Left) A *Typocerus velutinus* beetle and a bumblebee (*Bombus* sp.) have been attracted to a second-day lotus flower by the bright yellow color of its central receptacle and the fruity aroma. The stigmas are still receptive, and the anthers are releasing pollen, so both self-pollination and cross-pollination are possible.
Fig. 69. (Right) On the third day after opening, the flowers of American lotus drop their petals and stamens, and the stigmas become brown and are no longer able to be pollinated.

American Lotus

Initially, the floral receptacle is a yellow conical structure with a slightly rounded top. As it matures and increases in size, it becomes green, and the top surface flattens (fig. 69). The exterior surface of young receptacles is covered with groups of calcium oxalate crystals called druses that resemble the spiky end of a medieval weapon known as a mace. These cells serve to protect the receptacles and the enclosed fruits from herbivores. These same crystal-like cells are also found on the leaf surfaces. The fruits are held in circular cavities, numbering up to 40 on the surface of the receptacle. Within a few days after the petals have fallen, the receptacle turns 45° and faces east. The reason for this orientation is not known, but it occurs in response to light striking a small section of the flower stem about an inch below the receptacle. Over a period of several weeks, the receptacle and its fruits mature with the receptacle enlarging until it resembles a green showerhead, and the green fruits turn dark purple. Because of its interesting appearance, many common names have been given to this attractive plant, among them: rattle nut, yockernut, watering can, duck acorns, alligator buttons, and great yellow lily. When fully mature, the receptacle is at a 90° angle from its original orientation (fig. 70). Eventually, as it becomes brown and woody, it resumes its original upright position. The fruits also

Fig. 71. A *Nelumbo lutea* fruit has been cut vertically to show the green embryo embedded in the white endosperm of the single, large seed.

become hard and brown and shrink in size, becoming loose within the enlarged cavities. The receptacles ultimately change position again, this time turning a full 180° until they face the water and detach, floating face down and releasing the fruits into the water as they decay. The fruits sink to the bottom, each bearing a single seed with a fully developed, chlorophyll-containing embryo (fig. 71). The seeds' impermeable seed coats allow them to lie buried in mud until conditions are right for germination, which occurs after some external agent or condition has abraded the seed coat. As previously described, the seeds, even after lying dormant for hundreds of years, can germinate within only a few days. Once the receptacles have fallen, the leaves also turn brown and face downward (fig. 72), gradually sinking into the water and disappearing until new shoots arise the following year.

Fig. 70. As the receptacle matures, it reorients 90° to become lateral-facing, turns from green to brown, and increases in size. The openings that hold the fruits enlarge, and the fruits themselves shrink so they rattle if the receptacle is shaken. When fully mature, the receptacles are completely brown and again face upward.

Fig. 72. By November the leaves are dry and brown and, in most cases, have twisted so that they now face the water.

Aster Family

Fig. 73. New England aster (*Symphyotrichum novae-angliae*) is one of the showiest asters in the Northeast.

Asters

Fig. 74. Another northeastern aster with large (ca. 1 in. diameter) flower heads is New York aster (*Symphyotrichum novi-belgii*). The flower heads look similar to those of New England aster, but the narrow, non-clasping leaves differ.

Asters
Various genera formerly included in *Aster* L.:
Symphyotrichum Nees, *Oclemena* Greene,
Sericocarpus Nees, *Doellingeria* Nees, *Ionactis*
Greene, and *Eurybia* (Cassini) Cassini
Aster Family (Asteraceae)

Habitat: Fields, roadsides, and other sunny open areas, with a few exceptions that grow in shady forests.

Range: Common throughout the range of this book and throughout the remaining United States and Canada, Mexico, and the West Indies; a smaller number of species are found in Central and South America and Eurasia.

As many readers already know—and others may be appalled to learn—there are no longer any asters in the Northeast, or at least not any native members of the genus *Aster*. The loss of our *Asters*, if not our asters, is the result of a splitting-off of New World members of the genus *Aster* into six separate genera (*Symphyotrichum* [figs. 73–74], *Oclemena*, *Sericocarpus*, *Doellingeria*, *Ionactis*, and *Eurybia*), most of which are names formerly applied to these groups of species before they were lumped together in a single, more broadly defined, genus: *Aster*. This recent taxonomic decision, though based on very small botanical differences, is confirmed by molecular evidence. Such differences (e.g., the presence or absence of stalked glands on the ovary or leaves; the length of the pappus bristles [the "fluff" that later serves to disperse the fruits]; the relative measurements and shape of the involucral bracts [phyllaries] that subtend the flowers; etc.) are difficult to discern in the field and are thus of little use as field characters for identifying species to genus, not only for the layman, but even for the botanist who is not a specialist in this family of plants. Yet I feel somewhat less inept at my aster identification skills after reading the letters of Asa Gray, one of America's preeminent nineteenth-century botanists

Aster Family

Fig. 75. Asa Gray in a portrait by J. J. Cade. Gray was one of America's great nineteenth-century botanists. http://ihm.nlm.nih.gov/images/B13543, Popular Science Monthly, vol. 62, Public Domain, https://commons.wikimedia.org/w/index.php?curid=19264287

(fig. 75), written to his contemporaries. In his later years, trying to finish his work on asters and goldenrods for the publication of his *Flora of North America*, he frequently mentions his difficulty in defining the species of *Aster*:

> Dear Redfield—If you hear of my breaking down utterly, and being sent to an asylum, you may lay it to *Aster*, which is a slow and fatal poison.
>
> Apparently, it will take a year or more for me to finish it, with the greater chance that it finishes me before that time.
>
> Letter to John H. Redfield, April 1880

> I am half dead with *Aster*. I got on very fairly until I got to the thick of the genus, around what I call the Dumosi and Salicifolia. Here I work and work, but make no headway at all. I can't tell what are species and how to define any of them, nor what the nomenclature is, i.e., what are the original names.... I was never so boggled.
>
> Letter to George Englemann, April 1880

> My dear old Friend...
> First of all, I am to make complete as I can my manuscript for *Solidago* and *Aster*. *Solidago* I always find rather hopeful. *Aster*, as to the *Asteres genuine*, is my utter despair! Still I can work my way through except for the Rocky Mountain Pacific species.
>
> I will try them once more, though I see not how to limit species, and to describe specimens is endless and hopeless.
>
> Letter to George Englemann, December 1881

> My wife now excuses me to her friends for outbreaks of ill-humor, the excuse being that I am at present "in the valley of the shadow of the Asters."
>
> Letter to Sir Edward Fry, November 1883

The only presence of a member of the genus *Aster* in the Northeast is the introduced *Aster tataricus*, a showy, lavender-flowered ornamental from Siberia that occasionally escapes from gardens (fig. 76). The only *native* species of *Aster* in all of North America is the alpine aster (*Aster alpinus*), a species found in the western mountains of Idaho, Wyoming, and Colorado, ranging northward to Alaska and parts of Canada; *Aster alpinus* is also found in the mountains of Europe (fig. 77). In fact, there are still about 180 species retained in the genus *Aster* in Eurasia. However, the common name "aster" is still used as an umbrella term for the entire group in both North America and Eurasia, and the "aster" is still the birthday flower of September. The asters are among the most numerous species of wildflowers in the Northeast and range throughout North America. They occur in habitats ranging from salt marshes to deserts (fig. 78). Indeed, worldwide, the aster family, Asteraceae, is second in number of species, with an estimated 25,040 species, surpassed only by the Orchidaceae with 26,000 species (numbers from the Angiosperm Phylogeny Group). The ubiquity of Asteraceae may be attributed in part to their method of dispersal. In many species the dispersal unit that carries the seed is equipped with a fluffy "parachute" (the pappus in the case of

Fig. 76. *Aster tataricus* 'Jindai' is the only member of the genus *Aster* that might be encountered in the northeastern United States. This native of Siberia is widely cultivated in gardens and sometimes escapes into the wild.

Asters

Fig. 77. Alpine aster (*Aster alpina*) is a pretty European aster that is also native to North America; it is the only remaining native North American species in the genus *Aster*. Alpine aster grows in our western mountains, Alaska, and parts of Canada. (Photographed on Monte Baldo in Italy)

Fig. 78. Some asters, such as the alpine leafybract aster (*Symphyotrichum foliaceum* var. *apricum*), grow in arid soil, as seen here in the intermountain area of Wyoming.

asters) that can be transported long distances by the slightest breeze.

Aster is derived from the Greek for "star," an apt term for the starlike arrangement of the heads of asters (fig. 79). An ancient Greek myth ascribes the origin of asters to the tears of Asterea, the Greek goddess of the starry sky. Asterea was sad when she looked down upon earth and was unable to see any stars. When a great flood covered the earth, it took the lives of all but two people—the only two who had been loyal to the gods and were thus spared by Jupiter. Asterea saw the two survivors left stranded on a mountaintop and

Fig. 79. Purplestem aster (*Symphyotrichum puniceum*) demonstrates the starlike arrangement of the flower heads of aster. The genus *Aster* (pertaining to "star" in Greek) was named for this resemblance.

surrounded by a morass of mud. She wept for them, her tears falling as stardust that, upon striking the earth, caused aster flowers to grow (or perhaps it was the tears of early botanists frustrated at trying to identify the asters). Yet another Greek myth ascribes the origin of asters to the result of the actions of King Aegeus of Athens. When Aegeus mistakenly believed that his son, Theseus, had been killed during a battle with the Minotaur, he killed himself. The blood from his mortal wound fell to the ground whence purple asters grew.

The floral parts of an aster, or for that matter any member of the aster family, must be examined carefully, and the botanical terminology mastered, when one is attempting to identify a plant to species with the aid of a technical manual. The perceived "flowers" of asters differ from those of most other families in that they are actually heads (capitula; singular capitulum) comprising many small flowers called florets. Because of these heads of many flowers, members of this family are often referred to as composites. The general term for a group of flowers that give the appearance of being one flower is pseudanthium, meaning "false flower." The florets of the aster family are most often of two types: small central florets (disk flowers), which are short, tubular flowers with regular symmetry, and the outer flat, straplike florets, termed ray flowers, which are asymmetrical (fig. 80). Although both types of floret generally are found in the same flower, some members of the Asteraceae have flowers only of one type

Aster Family

Fig. 80. The composite head (capitulum) of stiff aster (*Ionactis linariifolia*) showing the two types of flowers that make up many composite flower heads: short, tubular, central disk flowers (yellow in this case) surrounded by straplike, highly asymmetrical ray flowers (lavender). Three ants and a beetle are on the flower head.

Fig. 81. A close-up view of a tubular, green disk flower of panicled (also called lance-leaved) aster (*Symphyotrichum lanceolatum*). Note that the style has emerged from the staminal tube, but the lobes of the stigma are still closed.

(e.g., chicory and dandelion have only ray flowers; tansy and thistle have only tubular disk flowers). Ray flowers are commonly larger than disk flowers, leading to their being mistaken for the "petals" of the flowerlike head. Like the disk flowers, ray flowers are tubular at the base but are markedly asymmetrical and thus appear straplike. In some Asteraceae each of the "teeth" at the tip of a ray flower represents a petal of the fused corolla; however, in asters, these teeth are indistinct. Ray flower measurements range from as small as 1.0–2.0 mm in some species, such as small-headed aster (*Symphyotrichum parviceps*) or aromatic aster (*Symphyotrichum oblongifolium*) to many times that (up to 2.0–2.5 cm) in the largest-flowered species in our most showy asters, New England aster (*Symphyotrichum novae-angliae*) and low showy aster (*Eurybia spectabilis*). Ray flower color ranges from white through pink and lavender, to deep purple. In the six genera discussed here, the ray flowers are pistillate; that is, they have only female reproductive organs. The central disk flowers are generally yellow to yellowish-green and, in some species, turn reddish-purple as they age. The bases of the five stamens are fused to the corolla and, although their filaments are separate, their anthers are united into a ring called a staminal tube through which the style grows (fig. 81). Within the tube, the anthers open along their inner sides, and as the style pushes past them it picks up some of the pollen and pushes it out of the anther, thus making it more readily accessible to visiting insects. This process is called secondary pollen presentation. In asters, the stigmatic lobes of the style remain closed until all the pollen has been released by the anthers; thus, the flowers do not self-pollinate. Disk flowers of all these former asters are bisexual, having both stamens and pistils. Pollination in Asteraceae flowers results primarily from insect transfer of pollen from one flower to another.

The fact that the number of flowers per head varies among aster species constitutes a useful identification tool; a 10× hand lens is useful for accurate counting of them. Each capitulum includes the tubular flowers (ray, disk, or both), and the capillary bristles (pappus), which are attached to the top of the ovary. An involucre of bracts surrounds the base of the entire inflorescence, creating the appearance of a calyx (fig. 82). The plant heads are grouped into arrangements (capitulescences) that may have an elongate or flattened inflorescence-like appearance—or they may occur in the axils of the leaves along the stem.

The asters' copious seeds are tiny and well adapted for long-distance dispersal by the wind (fig. 83). Perhaps the best-known example of an aster family "seed" is that of the dandelion, one which most children have aided in dispersal by blowing on the puffy "seed heads" and making a wish as the parachute-bearing "seeds" fly off on the wind. The term seed appears here

Asters

Fig. 82. A lateral section of a flower head of New England aster showing outer purple ray flowers and inner yellow disk flowers (some with stigmas exserted and others not yet open) attached to the receptacle, hairy pappus at the base of the flowers, and green bracts composing the surrounding involucre.

Fig. 83. (Right) Two immature fruits of New England aster (*Symphyotrichum novae-angliae*) with the old disk flowers still attached. The white filaments are the pappus that, when dry, will help to disperse the fruits on the wind.

in quotes because the dispersal unit (diaspore) of a member of the aster family is actually a tiny fruit (a ripened ovary)—that is, a dry, indehiscent fruit called a cypsela. Cypselas differ from achenes in general in that their fruit develops from an inferior ovary (where the petals, sepals, and stamens attach to the top of the ovary), rather than a superior ovary (where the other flower parts are attached below the ovary). The parachute that carries the fruit long distances comprises the pappus (see fig. 83)—analogous to the sepals of a more typical flower—that remains attached to the fruit. Some other members of the aster family are dispersed when hooked barbs on their fruits stick to the fur of an animal or to the socks of a passerby.

My favorite aster (as well as that of many of my acquaintances) is the New England aster, with its numerous, many-rayed, flower heads that range in color from pale lavender to deep purple (fig. 84; and

Fig. 84. New England aster (*Symphyotrichum novae-angliae*) is a favorite fall flower because of its numerous showy flower heads with many colorful rays.

Aster Family

Fig. 85. In addition to its colorful flower heads, New England aster (*Symphyotrichum novae-angliae*) is easily recognized by its hairy stems and leaf bases that clasp the stem.

see fig. 73). The color contrasts superbly with the bright yellow flowers in a field of goldenrod (see fig. 367 in the chapter on goldenrods). However, if you crush a flower head of this lovely species in your fingers, you will find that the aroma given off is not at all a pleasant floral scent, but rather a sharp and pungent odor, somewhat reminiscent of turpentine.

The pubescent stem and clasping leaves (fig. 85) of New England aster are helpful aids in identifying this species. Cultivated varieties of New England aster are sold in the horticultural trade as Michaelmas daisies, a name given to them because the flowers bloom around St. Michael's Day, celebrated on September 29th. Inspired by the beauty of the New England aster, George Lansing Taylor, a nineteenth-century clergyman, was moved to compose a poem, which begins:

> Born to the purplest purple, deep, intense,
> Mocking the gentian's fringe with hue more rare,
> New England Aster!–What can be more fair!

Asters are fed upon by the strikingly beautiful larvae of owlet moths. Hoping to see and photograph an adult owlet moth, I brought home a larva of the asteroid paint (aka golden-hooded) owlet (*Cucullia asteroides*) on one of its host plants, New England aster (figs. 86–87), and placed it in a vase on the kitchen counter. These caterpillars are more commonly green with a yellow-green stripe running down the back flanked by several black pinstripes, but mine was

Fig. 86. (Left) The attractive astroid owlet caterpillar (*Cucullia asteroides*) feeds on the leaves and flowers of asters and goldenrods. These caterpillars are more commonly green with yellow stripes. **Fig. 87.** (Right) An astroid owlet caterpillar feeding on the flowers of New England aster. Note the hairy pappus that has been exposed by the caterpillar's eating of the ray flowers. The pappus is the equivalent of the sepals in most other flowers.

a variation having a red base color with black pinstripes flanked by two yellow stripes. Rather than a hood, as suggested by its common name, it appears to me that the caterpillar is wearing a striped turtleneck pulled up to its "ears." Certain that the caterpillar was still in its avid feeding stage and would not wander, I did not enclose the plant immediately and then made the further mistake of getting distracted by my e-mail. When I returned to the kitchen a half hour later, the caterpillar was gone! I searched the kitchen and the basement below, where it might have escaped by crawling down the adjacent stone wall—with no luck. The following summer I kept an eager eye open for a midsized brown and beige moth flying about the house (the asteroid paint overwinters in the ground as a pupa, and we do have a dirt-floored crawl space in the basement), but I never did see the adult. This was my first such rearing failure and a lesson learned not to try to out-think the "intentions" of an insect.

Asters are also a food source for the caterpillars of many other moths, including the familiar black and

Fig. 88. Woolly bear caterpillars (*Pyrrharctia isabella*) feed on asters and other members of the aster family. This one is feeding on pilewort (*Erechtites hieraciifolius*).

rust-colored one that we know as the woolly bear (fig. 88). In addition, many other insects from different orders visit the flowers for pollen and nectar or to prey upon other insect visitors. (figs. 89a–f).

Fig. 89. Visitors to aster flowers: a. A large syrphid fly (*Helophilus fasciatus*) on panicled aster (*Symphyotrichum lanceolatum*) (syrphid flies are common visitors and pollinators of many species of wildflowers). b. A pearl crescent butterfly (*Phycoides tharos*) on an unidentified aster. c. A monarch butterfly (*Danaus plexippus*) nectaring on the flowers of New England aster (*Symphyotrichum novae-angliae*). d. An orange sulphur butterfly on the flowers of an unidentified aster. e. Predators such as this Chinese mantis (*Tenodera aridifolia sinensis*) visit asters in search of prey. f. A crab spider waits poised near an aster flower head to capture the next insect to visit.

Broomrape Family

Fig. 90. Beechdrops (*Epifagus virginiana*) plants are most noticeable after they are dead, when seen against a snowy forest floor.

Beechdrops

Fig. 91. Beechdrops are always found in proximity to their host species, the American beech (*Fagus grandifolia*). Those located at some distance from a tree trunk have an underground attachment to one of the tree's fine roots that may be many meters from the base of the tree.

Beechdrops
Epifagus virginiana (L.) W. P. C. Barton
Broomrape Family (Orobanchaceae)

Beechdrops is an often-overlooked plant in the forests of eastern North America. It is small and inconspicuous, spending most of its aboveground life looking much like a cluster of dry brown sticks and thus, camouflaged in its natural setting, the leafy forest floor.

Habitat: Temperate mesic deciduous forest or mixed northern hardwood-hemlock-white pine forest with American beech trees (*Fagus grandifolia*); overstory tree associates frequently include sugar maple.

Range: Prevalent in New England and New York, south through Pennsylvania, West Virginia, Virginia, and North Carolina, west into Ohio, Michigan, eastern Wisconsin, and southern Indiana, then scattered southward as far as northern Florida, Oklahoma, and eastern Texas, with a strong concentration in west-central Louisiana. In Canada, beechdrops occurs in southern Quebec, southern Ontario, and Nova Scotia. Disjunct populations are found in northeastern Mexico.

Beechdrops (*Epifagus virginiana*) is an obligate parasitic plant, which means that it parasitizes only one host—the American beech (*Fagus grandifolia*). It is a member of the broomrape family, Orobanchaceae, which has a number of other genera that are also parasitic or hemiparasitic. The scientific name of the genus, *Epifagus*, reflects this relationship; *epi*, meaning "on" in Greek, and *fagus*, the Latin name for "beech," thus "on the beech." It is the only member of its genus. The plants are annual, growing each year from seeds that germinate in the vicinity of beech trees and quickly penetrate one of the beech tree's fine rootlets. This process may begin as early as June with development proceeding slowly for the first six weeks or so. There is then a period of rapid growth until maturity is reached in mid- to late August. The plants are usually branched and reach an average of 30 cm (about 12 in.) in height. September and October are the best months to find the plants in flower. By November the plants have begun to die back, but their dried brown remains last throughout the winter, becoming particularly noticeable against the snow (fig. 90).

Not many people notice beechdrops when the plants are in flower. The plants are small, appear

Broomrape Family

leafless, and are more or less colorless. Their fleshy stems are whitish to light brown when fresh and break easily; they wilt quickly if cut. Beechdrops are always found beneath, or in the vicinity of, beech trees (fig. 91). Beech trees are an indicator of high-quality mesic forest, and their presence may be used in assessing the overall state of the forest. Beechdrops' leaves are reduced to tiny triangular, scalelike structures that are closely appressed to the stem (fig. 92). Like normal leaves, the scale-leaves have stomata (though greatly reduced in number) for gas exchange. All plants of beechdrops have small (4–6 mm long), budlike cleistogamous flowers (ones that never open, as discussed in the jewelweed account in this book) situated on the lower portion of the stems (fig. 93). These flowers are fertile and self-pollinate within the closed flower, developing soon into fruits filled with tiny seeds (fig. 94). Some larger beechdrops plants will also develop relatively larger (ca. 1 cm) and "showier," open (chasmogamous) flowers on the upper part of the stems (fig. 95) (a few additional cleistogamous

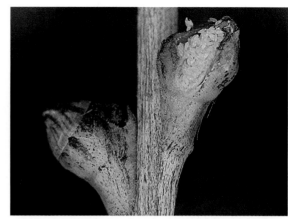

Fig. 94. The right-hand cleistogamous flower (now in fruit) has been artificially opened to show the seeds developing inside; the flower on the left still has its membranous corolla that forms a sheathing cap (the calyptra), which will be pushed off as the ovary expands; the exposed fruits open along a top seam to reveal the seeds naturally.

Fig. 92. (Left) The leaves of beechdrops, not being green and therefore unable to function as photosynthetic organs, have been reduced to small triangular scales, as seen here at the base of the cleistogamous flower on the upper left stem. **Fig. 93.** (Right) Some plants have only cleistogamous flowers (flowers that never open), as seen here. On plants having both closed (cleistogamous) and open (chasmogamous) flowers, the cleistogamous flowers are generally found on the lower portion of the stem.

flowers may be produced above the chasmogamous flowers as well). Both types of flowers have four stamens and a pistil; in the cleistogamous flowers, the style is shorter, and the stigma branched. There are always more cleistogamous flowers than chasmogamous flowers; one study of this species in Indiana found the ratio to be 20:1. To come across a plant with open (chasmogamous) flowers in full bloom is a rewarding surprise. When viewed closely, the tubular chasmogamous flowers are subtly colorful with

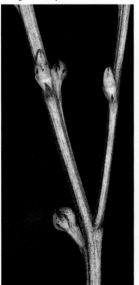 

Fig. 95. (Right) This plant has both cleistogamous flowers (looking like buds) on the lower part of its stem and chasmogamous flowers on the upper part.

Beechdrops

white and magenta stripes on their asymmetrical corollas (fig. 96). There is some disagreement among botanists regarding whether the chasmogamous flowers are fertile or not; many authors state that they are sterile, while others say that it is unknown if the open flowers are capable of producing seed, and still others claim that chasmogamous flowers are *occasionally* fertile and probably infrequently pollinated by insects—bumblebees or ants. The flowers have a semicircular functional nectary at the base of the ovary (fig. 97), and I have watched bumblebees visiting the chasmogamous flowers of beechdrops, probing deeply into them as though searching for nectar, and apparently collecting pollen as well (fig. 98), an activity that could result in cross-pollination. However, such bumblebee visits are infrequent.

Ants, in general, are not considered effective pollinators. They are small with generally smooth, hairless bodies ill-suited for carrying pollen; thus, they have a limited ability to carry pollen from one plant to another. Furthermore, they secrete antibacterial and antifungal compounds that are damaging to pollen. Within 30 minutes of exposure to these ant secretions, many pollen grains are rendered inviable. Despite all these drawbacks, some plants appear to benefit from pollination by ants. Such plants are generally low-growing plants with small flowers that provide a sweet, but small, reward for ants, thus forcing them to travel from one flower to another to garner enough nectar. Beechdrops fit this description.

A study published in 2013 by Abbate and Campbell makes a case for ant pollination of beechdrops. The authors observed and collected insects that visited the flowers of *Epifagus* during September of two consecutive years. The insects they collected included the Eastern bumblebee (*Bombus impatiens*) and at least two species of ants: numerous winter (or beech) ants (*Prenolepis imparis*), and a few ants belonging to the genus *Crematogaster*. Both the Eastern bumblebee and the winter ant had, adhering to their bodies, sufficient pollen to effect pollination, but visits by bumblebees were rare, whereas more than 96% of the visits to the flowers of beechdrops were made by ants (*P. imparis*).

Fig. 96. A close-up of a chasmogamous flower showing the capitate stigma exserted from the tubular corolla, which is magenta-and-white striped.

Fig. 97. Here, a chasmogamous flower has been sectioned to show the reproductive structures: a large, ovoid ovary sits at the base, half-surrounded by an orange semicircular nectary; the style, capped with its stigma, arises from the ovary; and three of the original four stamens can be seen (the fourth was lost as the flower was cut open).

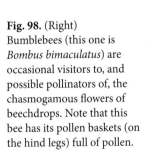

Fig. 98. (Right) Bumblebees (this one is *Bombus bimaculatus*) are occasional visitors to, and possible pollinators of, the chasmogamous flowers of beechdrops. Note that this bee has its pollen baskets (on the hind legs) full of pollen.

Broomrape Family

Based on the extreme dominance of visitations by this species, the authors investigated the possibility that these ants might be effective pollinators of *Epifagus*. They removed anthers from the flowers and dusted the bodies of the ants with fresh pollen, allowing it to remain on the ants for at least 30 minutes. The pollen from the ants did not appear to suffer any loss of viability from the ant secretions; when it was tested for viability, no significant difference was found from viability of pollen taken directly from the flowers. Thus, the pollen carried by ants as they crawl from flower to flower to imbibe nectar could successfully pollinate the flowers of beechdrops. Ants were also observed to move between plants growing in close enough proximity to have touching branches (fig. 99), thereby allowing for potential cross-pollination between plants. Although beechdrops is able to maintain itself by self-pollination, the additional, if only slight, possibility of cross-pollination would benefit the species by increasing genetic diversity.

As far as any value to humans, little is known. It has been reported that Native Americans used the plants to make a bitter tea that was used medicinally.

The cleistogamous flowers of beechdrops have a membranous corolla (see left flower in fig. 94) that, when the fruits are mature, splits in a circumscissile manner such that the apical caplike portion (the calyptra) is forced off by the expanding ovary. The fruits

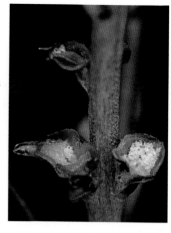

Fig. 100. The cleistogamous fruits of beechdrops open at a suture that extends along the top of the fruit. The seeds remain in the cuplike fruit until they are splashed out by rain drops.

open along a top seam to expose the seeds (fig. 100). Each fruit may contain an average of ca. 800 seeds. The tiny, ridged seeds, measuring only 0.5 mm in length, are somewhat sticky and adhere to each other in small clumps. They are, thus, ill-suited for dispersal by wind and remain in the open fruit until splashed out by the force of raindrops striking them—a method referred to as splash-cup dispersal. As such seeds do not travel far from the mother plant, it is thus ensured that they will remain in the vicinity of the required host species, the American beech. The seeds filter down into the soil where, the following spring, chemicals released by the roots of the beech tree trigger their germination. The seedlings live a brief, subterranean, independent existence, utilizing the minimal food reserves in the seed. They quickly develop a structure called a haustorium (derived from the Latin *haustor*, meaning "to drink," and *ortum*, meaning "a drinking vessel"), which penetrates a fine root of a nearby beech tree, thereby establishing a physiological bridge with the host. All the parasite's water and nutrients will henceforth be derived from the host. Beechdrops do not appear to harm their host trees.

A tuber begins to form around the haustorium and the beech root, and the haustorium shrinks in size so that it is no longer visible within the mature tuber. By excavating a beechdrops plant late in the season, one can find one of these tubers (fig. 101). There is usually no visible connection of the tuber with the host tree, but the initially parasitized tree root may be seen buried within the tuber if it is sectioned and viewed microscopically. Projecting from the tuber are short, stiff, branched rootlike structures referred to as "grapplers"

Fig. 99. When several plants of beechdrops grow in close proximity such that their branches overlap, it would be possible for visiting, nectar-seeking ants to move easily from one plant to another and possibly effect cross-pollination of the flowers.

Beechdrops

(see fig. 101), which appear, when studied anatomically, to be nonfunctional roots of the beechdrops.

Once the fruits are emptied of seeds, they remain open; when viewed closely, the capsules look like wooden mini-flowers (fig. 102). For the most part, the dried, dead remains of beechdrops blend easily into the leafy backdrop of the forest floor and go unnoticed (fig. 103).

Having a parasitic lifestyle has its pros and cons: on the positive side, a parasitic plant does not need to expend energy on the manufacture of its own "food"; in fact, since true parasitic plants contain no chlorophyll, they are unable to use the process of photosynthesis to make carbohydrates. The water and mineral needs of a parasite are also fulfilled via its connection to the host species. There is a negative aspect to this type of relationship, however. The parasite is totally dependent on being in the proximity of a host plant. If the seeds of parasitic plants do not land in an environment where the host species is present, they will not be able to grow beyond the newly germinated seedling stage. Hence, the geographic range of beechdrops is essentially restricted to that of the host tree,

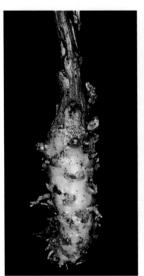 

Fig. 101. (Left) A subterranean tuber of beechdrops measures 1–4 cm long. The small projections on the tuber are nonfunctional roots. Since beechdrops receives all of its water and nutrients from the host tree (a root of which is now embedded in the tuber), it has no need for its own functional roots. **Fig. 102.** (Right) In winter, the remains of the now empty fruits of beechdrops' cleistogamous fruits appear like little wooden flowers.

Fig. 103. Once beechdrops plants have died, they turn dark brown and look like a cluster of sticks nearly hidden in the leafy substrate.

American beech (*Fagus grandifolia* var. *grandifolia*), in the United States and southern Canada.

A disjunct population of beechdrops occurs in relic deciduous forest in eastern Mexico, where a different variety of American beech (*Fagus grandifolia* var. *mexicana*) occurs and serves as the host for plants in the Mexican population. As we look toward the future in this time of climate change and attempt to forecast possible migration patterns and range changes for different species, information from previous plant migrations could inform our predictions. In studies that looked at the impact of northward migratory expansion of *Fagus grandifolia* and *Epifagus virginiana* during the last postglacial warming period, examination of fossil pollen showed that the parasite lagged behind its host in expanding its range northward. Examining this lag resulted in the conclusion that it was not solely the *presence* of the host that was necessary for the colonization of the same territory by its parasite; rather, it was the *density* of the host tree species that was the critical factor. American beech migrated at a faster rate than beechdrops, but it did so, perhaps, in a spotty way that didn't provide an area with sufficient density for newly germinated beechdrops seedlings to encounter nearby host roots to establish itself. It wasn't until beech trees had begun to fill broad areas of the migrating deciduous forest that beechdrops could be assured of enough host-root interactions. This same incongruence in migration rate might be expected to have occurred in other host-parasite pairs as they began to migrate in response to warming temperatures, a phenomenon that might also apply to such pairs in the future, thus altering their current ranges as global warming continues.

Iris Family

Fig. 104. The shiny black seeds that persist long after the capsules of blackberry-lily (*Iris domestica*) have split open are responsible for the first part of its common name.

Blackberry-lily

Fig. 105. The lily-like flowers of blackberry-lily have three petals and three sepals that are similar in appearance (and therefore called tepals), with the outer three just a bit larger than the inner three; this differs from the typical iris flower of three (usually) upright petals (standards) alternating with three sepals (falls) that are spreading or reflexed.

Blackberry-lily

Iris domestica (L.) Goldblatt & Mabb.
Iris Family (Iridaceae)

Blackberry-lily is a good example of why the common names of plants can be misleading. This plant is related to neither blackberries (in the rose family) nor lilies (in the lily family), but rather it is a member of the iris family. However, because it bears a cluster of shiny black seeds that together resemble a blackberry (fig. 104) and flowers that resemble those of lilies (fig. 105), it was given the descriptive moniker, blackberry-lily. Before molecular studies were done within the iris family, blackberry-lily was considered to belong to a monotypic genus (a genus having only a single species) within the Iridaceae—*Belamcanda chinensis* (L.) Redouté. As its former specific epithet, *chinensis*, implies, it is native to China—as well as to Japan, Taiwan, Korea, and a portion of the Russian Far East. It has long been cultivated in gardens and has escaped and naturalized in many regions of the world.

Habitat: In the northeastern United States, blackberry-lily is a perennial generally found naturalized in moist, well-drained soils having a sandy component. It grows in full sun to partial shade.

Range: This East Asian native, while not considered invasive, has colonized suitable habitat in a scattershot manner from southern Vermont to northwestern Florida, west to southern Minnesota and southeastern South Dakota, and south to eastern Texas.

The iris family is found nearly worldwide but is rare in lowland tropical regions. Iris means rainbow in Greek and is derived from the myth that the goddess Iris carried messages from Mount Olympus to earth along a rainbow. Because of their colorful flowers, many members of the iris family are popular as garden plants (e.g., *Iris*, *Crocus*, and *Crocosmia*) and in the cut flower trade (e.g., *Gladiolus* and *Freesia*). The genus *Iris*, itself, is confined to the Northern Hemisphere. In northeastern North America, the

Iris Family

family is represented by members of only two native genera: *Iris* and *Sisyrinchium* (blue-eyed grass).

Our native irises belong to a subset of the genus known as the beardless irises (subgenus *Limniris*). In the Northeast are the well-known blue-flags, *Iris versicolor* and *I. virginica*, as well as such delicate species as crested iris (*Iris cristata*) (fig. 106), dwarf lake iris (*I. lacustris*), and dwarf violet iris (*I. verna*). The latter three of these species are low-growing plants bearing flowers that display crests or patterns on their sepals, rather than the hairy beards typical of the showy introduced irises of our gardens.

The red-spotted orange flowers of blackberry-lily attract butterflies, making it a welcome addition to the perennial garden. Plants can be grown easily from seed or from sections of the rhizome planted just under the soil surface, as one would with an iris rhizome. The bright orange of the late summer-blooming flowers blends nicely with late-flowering yellow daylilies, purple-flowered asters, and other late bloomers. When naturalized in an open area, blackberry-lily is often accompanied by Queen Anne's lace (*Daucus carota*), viper's bugloss (*Echium vulgare*), common mullein (*Verbascum thapsus*), and other introduced "weedy" species (fig. 107).

The beauty of blackberry-lily was not lost on our third president, Thomas Jefferson, an early American polymath who included among his many interests that of gardening. While still in office, Jefferson thought often of his Virginia gardens and spent time devising elaborate plans for their renovation—from adding favorite trees to the woodlands to increasing the area devoted to his already large, terraced kitchen garden. By the end of his second term as president, Jefferson was so eager to return to his home—and especially to his gardens—that he rode from Washington on horseback for more than five days (including through a heavy March snowstorm in the Blue Ridge Mountains), in order to reach Monticello as quickly as possible. Jefferson had a great appreciation for plants native to the region and left much of Monticello's woodland in its natural state. But he was also eager to add species from other parts of the country, including some brought back by Lewis and Clark from their exploration of the Louisiana Territory. In addition, through commerce or trade with other gardeners, Jefferson obtained seeds and plants from other areas of the world. The seeds of blackberry-lily were one such acquisition. Known then as Chinese ixia (*Ixia chinensis*), the seeds were provided to Jefferson by nurseryman Bernard McMahon in 1807. McMahon was an Irish-American horticulturist best remembered for the first comprehensive book on gardening in America—his *American Gardener's Calendar* (to which he appended a sizable seed list). Jefferson's respect for McMahon's abilities as a horticulturist led to his being chosen as one of only two nurserymen to receive (and subsequently grow) the living botanical collections brought back by Lewis and Clark.

Fig. 106. The irises native to North America belong to a group known as the beardless irises. This small crested iris (*Iris cristata*) is typical of those native to the Northeast. In place of beards our native irises have ridges or markings on their sepals (contrast with *Iris reichenbachi* in fig. 114).

Fig. 107. The flowers of blackberry-lily, seen here naturalized at a New York farmstead, resemble those of a lily, giving rise to the second part of its common name. They often grow with other introduced species such as Queen Anne's lace and viper's bugloss.

Blackberry-lily

According to Jefferson's personal, detailed "Garden Book," the Chinese ixia seeds were sown in the East Front Oval garden at Monticello. To this day, descendants of those original plantings of blackberry-lily are naturalized throughout the Monticello property, especially in the West Lawn.

Despite the beauty of blackberry-lily's flowers and its widespread use in gardens, I could find little published about the pollination of the species—nothing about pollination within its native range, and little else from areas of the world where it has been introduced. One paper described a study conducted on an introduced population in Brazil. The researchers documented that the flowers have osmophores (areas of scent production) on their tepals along with UV patterns along the tepal margins and on the anthers. Both the scent glands and the UV patterns function as attractants for pollinators. They also found that blackberry-lily was self-compatible (its ovules could be fertilized by pollen from the same flower), but that cross-pollination was more frequent. Many species of small bees were observed visiting the flowers of blackberry-lily at the Brazilian research site, with three species accounting for more than 600 of the 1100 total visits. Few butterflies were observed. The bees visited primarily in the morning before the stamens had spread away from the style, and they continued to visit until the early afternoon. They appeared to be foraging for pollen rather than nectar. When the flower first opens, the anthers are close to the stigmatic surface, and the movement of the bees could readily result in pollination. However, since the flowers secrete nectar from grooves on the upper side of the base of the outer whorl of the perianth, it is likely that in its native range, the flowers of blackberry-lily are pollinated by insects seeking *that* sweet reward. The Brazilian researchers discovered that the nectar is less sweet than might be predicted, in that it is very dilute and contains only 3% sugar. Although butterflies need nectar low in viscosity in order to draw it up efficiently through their long proboscises, their preferred sugar concentration is more commonly in the 20% range.

My own observations of blackberry-lily plants that have naturalized on an old farmstead near my home in New York State indicate that, at that site, butterflies are the most frequent visitors. I've seen several species—from monarchs (fig. 108) to swallowtails (fig. 109)—alighting on the flowers and inserting their proboscises into the central part of the flower. Small flies (fig. 110) and bees (fig. 111) land on the anthers and on the tips of the style branches, perhaps gleaning pollen grains deposited there.

Not only is the *common* name of blackberry-lily confusing, but the scientific name of the species is in

Fig. 108. (Left) A monarch butterfly has its proboscis inserted deeply into the corolla of this blackberry-lily flower. **Fig. 109.** (Right) This spicebush swallowtail butterfly is also seeking nectar from the nectaries at the base of the outer tepals of blackberry-lily.

Iris Family

Fig. 110. A small syrphid fly has landed on an anther of blackberry-lily, probably to collect pollen.

Fig. 111. This small bee is probably gleaning pollen dusted onto the stigmatic surfaces of the style branches of this flower. Note the older, twisted flower below.

a state of controversy as well. The nomenclature of blackberry-lily provides a useful case study in the difficulty of plant classification. The taxonomic placement of blackberry-lily has long been debated among botanists. The species was originally described in 1712 as *Epidendrum domesticum*, a member of the orchid family (this unfortunate error was based on an illustration of a mixed collection of two species, one an orchid; the other actually the blackberry-lily). Blackberry-lily was subsequently moved in and out of the genera *Vanilla* (another orchid), *Gemmingia* (a synonym of *Belamcanda* published without knowledge of its previous publication as *Belamcanda*), and *Pardanthus* (an invalid name published without a description), then later as *Ixia*, and finally as *Belamcanda* (the latter two are accepted genera in the iris family). Linnaeus, in his 1753 *Species Plantarum*, used the name *Epidendrum domesticum* for this species.

Fig. 112. This lovely illustration of blackberry-lily was painted by Pierre Joseph Redouté to accompany his 1807 description of the plant when he renamed it *Belamcanda chinensis*. Attribution: Pierre-Joseph Redouté. https://commons.wikimedia.org/wiki/File:Belamcanda_chinensis_in_Les_liliacees.jpg.

More than 50 years later, in 1807, Pierre Joseph Redouté, a Belgian born botanist, perhaps better known for his exquisite botanical paintings of roses and other flowers (fig. 112), saw reason to transfer blackberry-lily from the genus *Ixia* (a large genus of bulb plants native to South Africa) to its own monotypic genus, *Belamcanda*. His decision was based on a number of features that distinguished *Belamcanda* from *Ixia* (*Belcamanda* has a wheel-like corolla that coils after flowering for only a day and is frequently marcescent [remaining attached] to the developing ovary; its tepals are fused at their base rather than free as in *Ixia*, and its round seeds persist on a central column after the capsule has split open).

In 2002, when Missouri Botanical Garden botanist Peter Goldblatt treated the iris family for *Flora of North America*, he retained *Belamcanda* as a distinct, monotypic genus; however, three years later, Goldblatt and David Mabberley reviewed new molecular information for the Iridaceae. The molecular studies of two regions of chloroplast DNA provided evidence showing that *Belamcanda* was nested within the genus *Iris*. Thus, Goldblatt and Mabberley concluded that *Belamcanda chinensis* should actually be included as a somewhat unusual member of the genus *Iris*.

Goldblatt and Mabberley made further detailed observations of the morphology of *Belamcanda chinensis* as compared with that of various species of *Iris* and were able to see morphological similarities in the two genera as well. The overall appearance (the gestalt) of the vegetative parts of the plants can be recognized immediately as similar. Both have sword-shaped basal

Blackberry-lily

Fig. 113. Like those of *Iris*, the sword-shaped leaves of blackberry-lily are arrayed in one plane, in a fan-shaped arrangement; unlike most irises that have only basal leaves, the leaves of blackberry-lily continue upward along the stem.

leaves (with some arising from along the stem in blackberry-lily) that are spread into a fanlike arrangement (fig. 113). However, unlike most (but not all) species of *Iris*, the leaves of blackberry-lily are deciduous; there is no sign of them in winter. In contrast, the leaves of the bearded iris grown in our gardens often remain throughout the winter. Both blackberry-lily and *Iris* have the same type of inflorescence, referred to as a rhipidium, in which the branching flower stalks are arranged in a fan-shaped plane comprising one to six flowers (see figs. 104–105). As with *Iris*, the underground portion of the plant is a rhizome (an underground stem), which is found at or just below the surface of the soil.

It is the flowers of blackberry-lily that differ markedly from those of other species of iris. Unlike more typical members of the genus *Iris*, the six perianth parts of *Belamcanda chinensis* are very similar in appearance to one another as they would be in a lily (see fig. 105), rather than arranged in the typical iris flower form of two distinctly different whorls, an outer whorl of three spreading sepals (falls) and an inner whorl of three upright petals (standards) (fig. 114). The flower is fleeting, opening for only a day and then tightly wrapping its tepals into a coil (fig. 115; note left flower), which eventually turns black and remains clinging to the apex of the fruit as it develops (fig. 116)—sometimes even until the capsule splits open to reveal the shiny black seeds (which give the cluster of seeds the appearance of a blackberry). The blackberry-lily flower also has a united style that splits into three parts only near the tip (fig. 117). The spotted flowers give rise to two other common names: leopard-lily and leopard flower.

Despite these marked differences, careful examination of the separate tepals of blackberry-lily reveals that, as in *Iris*, they *do* taper into a more narrowed "claw" at their bases (note narrowed upper left tepal base in fig. 117), just not as conspicuously as in *Iris*.

Fig. 114. (Left) A typical European iris flower (*Iris reichenbachi*) in which three upright petals alternate with three sepals that are reflexed (and bearded). The reflexed sepals are referred to as "falls." Fig. 115. (Right) After flowering for a single day, the tepals of blackberry-lily coil tightly closed, as seen to the left of the open flower (the white flowers in the background are those of Queen Anne's lace).

Iris Family

Fig. 116. (Left) The coiled flower of blackberry-lily turns black as it dries, and often the withered perianth remains attached to the apex of the expanding fruit. **Fig. 117.** (Right) Blackberry-lily's united style divides into three parts near its apex. Each of the three branches bears a tiny stigmatic area; two upper flaps surrounding it resemble, in miniature, the stigmatic crests at the tips of the petaloid styles of *Iris*. The three stamens are free and diverge widely from the style tips soon after the flower has opened.

The floral parts of *Iris* arise from the summit of the ovary as a fused tube, while those of blackberry-lily form merely a vestigial tube at their base. The three stamens of *Iris* are free and are appressed to the undersides of the style branches, whereas those of blackberry-lily have freely spreading filaments that arise from the base of the three outer tepals (= sepals) and diverge widely from the style after the flower opens. A major difference in the flowers is the lack of the rather elaborate petal-like style branches so characteristic of typical *Iris* (see fig. 114). Yet Goldblatt and Mabberley found that even this feature could be observed in miniature with the aid of a 10× hand lens. At the tip of the united style, where it splits into three minute branches, is found a small stigmatic lobe just beneath the apex of each lobe with two small flaps flanking it above (see fig. 117). Although they are only about 2 mm in size, these two flaps are thought by Goldblatt and Mabberley to be homologous to the crestlike appendages on the styles of *Iris*. It appeared that perhaps *Belcamanda* has evolved its morphologically different flower structure to accommodate a different group of pollinators—butterflies (rather than bees as in *Iris*).

The fruits of both blackberry-lily and iris are loculicidal capsules, meaning that they split open midway between each of the three locules (compartments) of the fruit rather than along the seams (septa) of the three segments (fig. 118) The angularly flattened seeds of *Iris* are generally brown and are easily shaken from the open capsules. In contrast, the seeds of blackberry-lily are round, hard, shiny, and black (ripening from orange) (see fig. 118) and persist on the central axis (placenta) long after the capsule has split open (see fig. 104). Because of their persistence, the attractive seed-bearing stalks are a nice addition to dried bouquets in autumn. Goldblatt and Mabberley hypothesize that the appearance of blackberry-lily's compact cluster of seeds (which resembles a fleshy blackberry) may be an adaptation that has evolved to render them attractive to birds, thus facilitating seed dispersal.

Deciding on the proper name for this new species of *Iris* required some deep delving into early botanical literature. Another species of *Iris* already bore the epithet *chinensis*; thus, a simple change of the scientific name from *Belamcanda chinensis* to *Iris chinensis* would result in two species with the same name. The oldest name applied to blackberry-lily was the

Blackberry-lily

1753 Linnaean misnomer, *Epidendrum domesticum*, (from the mixed collection cited above); thus that earliest epithet, *domesticum* (changed to *domestica* to agree in gender with the new genus placement), was required to be used in the naming of the new combination, in accordance with the rules of botanical nomenclature—resulting in the new scientific name, *Iris domestica* (L.) Goldblatt & Mabb.

But that is not the end of the story. In 2014 an international group of botanists published a paper titled, "At Least 23 Genera Instead of One: The Case of *Iris* L. s.l. (Iridaceae)," in which they reexamined both molecular data and various taxonomic arrangements of the family and determined that not only should 18 species of *Iris* formerly classified as distinct genera be re-segregated from *Iris*, but an additional five *Iris* species should be removed from the genus as well. Since I have chosen to follow the taxonomy of *Flora of North America* for the Iridaceae—and the subsequent update of the name for blackberry-lily by the author of that treatment—I have used as the scientific name for blackberry-lily, *Iris domestica* in this essay.

As evidenced by the relatively simplified explanation presented here, taxonomists do not make name-changing decisions lightly. A name change requires thorough research and careful consideration of the history of the taxon's nomenclature in conjunction with new information. The next time you are tempted to roll your eyes about having to learn yet another new name for a well-known plant, take a moment to reflect on the exhaustive deliberation that went into making that change.

In China the plant has been the focus of studies of its medicinal properties. While listed as having toxic properties, the rhizomes in particular have been used to treat many conditions in Chinese traditional medicine. Known as *she gan*, the drug is most commonly used in the treatment of respiratory disease.

"*Belamcanda*" has long been known to contain phytoestrogens—13 of which have been found in its rhizomes. Such estrogen-mimicking compounds have been shown to have a positive therapeutic effect on cell lines of hormone-dependent prostate cancer. This has led medical researchers to investigate the potential of blackberry-lily compounds for use in treating this cancer. In vitro studies have demonstrated that an increase in the dosage of this rhizome extract is associated with a decrease in the expression of the androgen receptor, a concomitant decrease in the secretion of PSA (an androgen-regulated protein), and a decline in tumor cell proliferation. These results were obtained in laboratory studies in 2007 but had not yet been tested in living subjects. Further research of the efficacy of extracts from the rhizomes of blackberry-lily, both in China and elsewhere, continues to show the compound's ability to inhibit tumor growth. In a Chinese study in 2012, 10 compounds never before isolated from the rhizomes of blackberry-lily were discovered. One, ursolic acid, was found to have a potent effect against the growth of human carcinoma cell lines. Those studies are ongoing, with reviews of the literature on this topic published in 2015 and 2016. Although there is much yet to be learned, the rhizome of *Iris domestica* (syn. *Belamcanda chinensis*) is now included in the *European Pharmacopeia*, a reference of plants with known pharmacological activity. Perhaps blackberry-lily will one day be better known for its medicinal use than for its horticultural beauty.

Fig. 118. This capsule has begun to split along the central lines between the carpels. The round orange seeds will become black when fully mature.

Orchid Family

Fig. 119. Two stunning flowers of dragon's mouth (*Arethusa bulbosa*) growing in a sphagnum bog.

Bog Orchids

Fig. 120. Rose pink (*Pogonia ophioglossum*), pictured here, is more often a paler shade of pink than dragon's mouth or grass pink (*Calopogon tuberosus*), and it has a bright pink ruffled lip.

Dragon's Mouth
Arethusa bulbosa (L.) Ker Gawl.
Grass Pink
Calopogon tuberosus (L.) Britton, Sterns & Poggenb.
Rose Pogonia
Pogonia ophioglossoides (L.) Britton, Sterns & Poggenb.
Orchid Family (Orchidaceae)

These three lovely pink orchids contribute to the lure of the bog habitat for naturalists willing to risk wet feet (and sometimes more). Each has more than one common name, and when a common name is shared with another species, it tends to promote confusion. All are early summer bloomers that may inhabit the same boggy wetlands. However, once one learns the characteristics of each, they are easy to distinguish.

Habitat: All three of these species are associated with sphagnum bogs, although *Arethusa* and *Calopogon* will also grow in swampy, peaty tamarack wetlands. *Arethusa* may even be found in limy, marly wetlands, while *Calopogon* also inhabits wet sandy areas. Aside from sphagnum bogs, *Pogonia* is found growing in low, wet sandy habitats such as dune swales of the Atlantic and Gulf Coastal Plains and shores of the Great Lakes, as well as sedge mats and peaty marsh edges.

Range: Dragon's mouth (fig. 119) has the broadest range of the three species discussed here, occurring from eastern Canada west to Ontario, Manitoba, and Saskatchewan, and south to northeastern Minnesota, northern Illinois, northern Ohio and Delaware, and extending in the mountains to northwestern South Carolina. Within all portions of its range, dragon's mouth is considered vulnerable, imperiled, or possibly extirpated with the exception of Ontario, Minnesota, Newfoundland, and Nova Scotia. The range of rose pogonia (fig. 120) is nearly identical to that of grass pink (fig. 121); however, within each state or province where rose pogonia is found, it tends to be restricted to smaller, more localized regions than is *Calopogon*. The range of rose pogonia and grass pink extends from parts of almost all provinces in eastern Canada (excluding Labrador) west to Manitoba, and south in the United States to Minnesota, eastern Iowa, Arkansas, Oklahoma, Texas, and Florida.

Orchid Family

Five chapters of this book are devoted to orchids. Orchid flowers are not only beautiful and unusual in form, but many species have interesting habitat requirements and have developed intricate relationships with their pollinators. The orchid family is one of the most diverse plant families on earth, numbering an estimated 26,000 species, or about one-tenth of the approximately 250,000–300,000 known plant species. Orchid species found in North America number 250 (with 60 species native to my home state of New York alone), and surprisingly only a single adventive orchid, the weedy broad-leaved helleborine orchid (*Epipactis helleborine*) (see helleborine account in this book) from Europe. The only other plant family that comes close to this number of species is the aster family. With so many species yet to be discovered and described, there is some controversy regarding whether Asteraceae or Orchidaceae is the larger family. Orchids inhabit a wide range of habitats—from tundra north of the Arctic Circle to rainforest treetops in the tropics, and south to the tip of Tierra del Fuego; indeed, orchids are found everywhere but in deserts and Antarctica. Orchid plants may be terrestrial, epiphytic, and lithophytic (growing on rocks). One species is even subterranean—*Rhizanthella gardneri*, an odd orchid discovered in western Australia in 1928. Researchers found that the flowers of this tiny underground orchid were pollinated by gnats and termites! Aside from typical photosynthetic plants, some orchids lack chlorophyll and are parasitic, and others are mycoheterotrophic (dependent on a fungal relationship for a part of their nutrition). Most orchid flowers are diurnal and are pollinated by a range of insects from mosquitoes and ants to bees and butterflies; in some cases, birds provide the pollination service. Other species, being nocturnal, are pollinated by moths and even a cricket. It is common for orchid flowers to be adapted for pollination by a single species of insect or by only one group of like pollinators.

I have grouped these three orchid species into the same essay because of their similarity of flower form and color, their partially overlapping flowering times, the presence of ultraviolet-reflecting hairs on the enlarged lip (the labellum, a modified petal), and their similar modes of bee pollination. The three bog orchids occupy similar habitats and have similar distributions as well. Do not be surprised if you encounter more than one of these species in the same bog or wetland. All three are perennial, though individual plants may not appear every year.

Because of these similarities and the fact that bog orchids have multiple common names, these three species may be confused with one another. Dragon's mouth (*Arethusa bulbosa*) is also known as swamp pink or grass pink, while grass pink (*Calopogon tuberosus*) is sometimes called tuberous grass pink, and rose pogonia (*Pogonia ophioglossoides*) is known as snake mouth in some parts of its range). All three have pink flowers, with those of *Pogonia* always being at the paler end of the pink range. The convergence of characters exhibited by the three species may aid in attracting pollinators since visitation has been demonstrated to increase when several plants of similar size and color, whether of the same species or not, grow in close proximity. Bees are the primary pollinators of all these orchids, but differences in flower morphology have resulted in the flowers being adapted for pollination by different species of bees. Despite all their similarities, the three species of pink bog orchids do not hybridize in nature. This is due to the varied placement of pollen on parts of the "wrong" bees such that if that bee does visit another of the other two orchid species, the pollinium does not come into contact with the stigma. Discussion of each species separately should clarify these variations.

Dragon's mouth (*Arethusa bulbosa*)

Arethusa is a monotypic genus, meaning that *Arethusa bulbosa* is the sole species in its genus. The species is considered by some to be our most beautiful wildflower. The solitary flower (occasionally two) of dragon's mouth is generally a rich, magenta-pink (see fig. 119), but occasionally flowers with a lighter, more delicate pink coloration may be found (fig. 122); rarely are the flowers white. As with other orchids, there are three sepals and three similarly colored petals, one of which is modified into an expanded lip (labellum) and is wider when compared with the two lateral petals. In *Arethusa* the two lateral petals converge to form a hood over the column (where both male and female reproductive structures are located).

Widely known as dragon's mouth, the common name for *Arethusa bulbosa* is derived from its elongated profile, which, with a bit of imagination, resembles the gaping mouth of a dragon. The dragon's long,

Bog Orchids

Fig. 121. (Left) The flowers of grass pink (*Calopogon tuberosus*) are generally a striking magenta pink. Note that the lip of the flower is in the uppermost position, unlike in most other orchids (it is, thus, nonresupinate). **Fig. 122.** (Right)Dragon's mouth plants with paler pink flowers. This cluster of *Arethusa* orchids in a New Jersey fen is growing with Atlantic white cedar, pitcher plants, sundews, huckleberry, cranberry, bog asphodel (in bud), and sphagnum moss.

downward curving tongue (the orchid's highly embellished lip or labellum) has a rippled pink border and is patterned with deep pink spots and lines on a white background. It is also adorned with three longitudinal rows of yellow to white tufts of hairs, which absorb UV light, further increasing their visibility to insects, and functioning to orient the insect into the proper position to probe the floral tube for nectar (fig. 123). Dragon's mouth has a relatively large opening into the throat of the floral tube of the flower. Effective pollination will result only if an insect of large enough size (usually a queen bumblebee or, less commonly, a carpenter bee) comes into contact with the reproductive parts as it backs out of the flower. If the bee is already carrying pollen from a visit to another flower, the pollen sacs (pollinia) are caught by the sticky stigma. As the bee continues to back out, the back of the bee's thorax then presses against the flower's twin pollinia, which are cemented to the bee by the sticky viscidium of the pollinia, and subsequently may be transported to another flower. The flower has a sweet aroma, but the bare trace of nectar contained in the deep recesses of the floral tube offers scant reward to those visitors in exchange for their pollination services. After a few visits, even naive queen bumblebees learn not to repeat these fruitless efforts. Thus, in spite of their multiple attractants, *Arethusa* flowers receive few visitors, and only about 5% produce capsules with viable seeds (fig. 124, previous year's remnant fruit). As a result, dragon's mouth reproduces primarily by vegetative shoots that arise from its bulbous corm.

Fig. 123. The "hood" formed by the two lateral petals of dragon's mouth covers the reproductive structures. Three rows of hairs ornament the lip and help to attract insect visitors.

Orchid Family

Fig. 124. An old capsule from the previous year's flower of dragon's mouth still remains in late June when this flower was photographed.

Linnaeus named the genus for the Greek mythological river nymph, Arethusa (meaning "the waterer"), a reference to the watery habitat of the plant. The legend is a bizarre tale of pursuit, escape, and ultimate transformation of the young river nymph, Arethusa, into an underground stream that emerges eventually as a freshwater fount or spring on a small island off Sicily (near today's city of Syracuse). The species epithet of the plant, *bulbosa*, denotes that the plant arises from a small bulblike corm. At the time of flowering, dragon's mouth has a short, inconspicuous, grasslike leaf that arises from the corm and is hidden by a sheath until flowering has finished. The leaf may then grow to more than 7 inches in length. Once the flower has withered, the plant is almost impossible to find, as its solitary grasslike leaf is intermixed with other vegetation. Even when in flower, the small plant can be difficult to spot (fig. 125).

The plant typically grows in sphagnum mats with other wetland plants, including: pitcher plant (*Sarracenia purpurea*); sundews (*Drosera* spp.); Atlantic white cedar (*Chamaecyparis thyoides*), cranberry (*Vaccinium macrocarpon*); buckbean (*Menyanthes trifoliata*); bog asphodel (*Narthecium americanum*); and others, depending on the bog's locality (fig. 126 and see fig. 122).

Grass Pink (*Calopogon tuberosus*)

The genus *Calopogon* is found only in North America, with just one of its five species (*C. tuberosus*) venturing beyond the southeast United States into our northeastern region and into the Caribbean (considered part of North America according to WorldAtlas.com). The scientific name of the genus is derived from the Greek *kalos*, meaning "beautiful" + *pogon*—a reference to the hairs ("beard") on the labellum. The species epithet, *tuberosus*, is from the Latin name for the plant's tuberous corm.

Calopogon is a showier plant than its fellow pink bog dwellers. It is taller and often has a number of flowers in bloom simultaneously (fig. 127). The flowers attract purely with their beauty; they have no aroma and offer no nectar. Like *Arethusa*, the flowers

Fig. 125. A single flower of dragon's mouth is almost lost among other vegetation at the edge of a beaver pond (note the beaver lodge in the background).

Fig. 126. In addition to dragon's mouth orchids, this Adirondack bog is habitat for pitcher plants, cranberry, buckbean, marsh fern, and sphagnum moss.

Bog Orchids

Fig. 127. (Left) Several flowers may bloom simultaneously on a plant of grass pink.

Fig. 128. (Right) As with all the orchids discussed in this chapter, a range of colors can be seen, such as this pale pink flower of grass pink.

of *Calopogon* are generally magenta pink, but they occur as well in a paler pink (fig. 128) and rarely in white (fig. 129). Because of its brilliant display grass pink is more successful than the other two pink bog dwellers in attracting pollinators—midsized bees— and about 15% of plants set seed. In *Calopogon* it is the weight of the bee that is important. The flowers are nonresupinate, meaning that the lip remains as it was in bud—in the topmost position (see fig. 121), rather than twisting 180° to the bottom position (resupinate) as commonly encountered in most other orchids (e.g., see figs. 119–120). The lip is ornamented with white or yellow-tipped hairs, giving the impression of clusters of stamen filaments. If a bee lands on this attractive part of the flower and is heavy enough, it is in for a shock. The lip is hinged at its base, which when triggered by the weight of a bee, swings forward over the rest of the flower (see lower flower in fig. 121) and flings the bee backward onto the column (where the *true* stamens and pistil are housed). Small bees will not cause this action to occur (fig. 130). A bee that has enjoyed such a thrilling ride on a previous flower likely will have a pair of pollinia (fig. 131) from that flower adhering to the upper part of its abdomen.

Fig. 130. (Right) The small bee investigating the stamen-like hairs on the lip of grass pink is not heavy enough to trigger the hinge that would cause the lip to close downward onto the reproductive parts.

Fig. 131. (Below) These pollinia (pollen packets) of grass pink would become stuck to the upper side of a larger bee's abdomen when the bee is thrown backward onto the column of the flower.

Fig. 129. White-flowered flowers of grass pink (known as forma *albiflorus*), or of either of the other two species included here, are unusual.

Orchid Family

Fig. 132. The lip of this grass pink flower has remained closed against the column after being triggered by a bee that most likely escaped before the lip was fully closed.

park ride does not result in a nectar reward does it learn not to waste further time visiting other flowers of this species. Usually the lip will spring back up into its upper position after the bee escapes, but in some cases, it remains in the closed position (fig. 132), indicating perhaps that a large bee has avoided entrapment or that the flower is old and the lip has just fallen into the closed position.

Rose Pogonia (*Pogonia ophioglossoides*)

Pogonia is a genus of about 10 species distributed primarily throughout East Asia, with only one, *Pogonia ophioglossoides*, being found in North America. As you might surmise from the etymology of *Calopogon*, the generic name, *Pogonia*, makes reference to the bearded appearance of the lip of this species. The species name, *ophioglossoides*, was applied to denote the similarity between the leaf of this species and that of the single-leaved fern genus, *Ophioglossum*.

Rose pogonia is the most delicate of the three bog orchids discussed here. Not only is it small in stature, but it more often has a paler pink color rather than the flashy magenta commonly seen in the two previously discussed orchid species. Yet, in contrast to the other two species, rose pogonias are often found with hundreds of plants in flower in a sedge-dominated area of a bog or peaty sedge meadow (fig. 133). Typically, the plants grow among grasses and sedges in wet meadows, as described by Robert Frost in the following stanzas from his poem, "Rose Pogonias":

Those pollinia will be deposited on the next-visited flower's stigma as the bee slips down the column and escapes from the second flower, picking up yet another pair of pollinia in the process. Since weight correlates with size, the bees must be fairly large, but not so large as to be able to push the lip back up and fly off before completing the task of pollination. In this case, queen bumblebees would be too large and strong to be effective pollinators of *Calopogon*. Only when a bee "wises up" to the fact that this floral amusement

Fig. 133. Large clonal colonies of rose pogonia develop when their stolons spread through the sphagnum substrate. The opposite leaves of marsh St. Johnswort and the orange-tinged leaves of huckleberry are also present.

Bog Orchids

A saturated meadow,
Sun-shaped and jewel-small,
A circle scarcely wider
Than the trees around were tall;
Where winds were quite excluded,
And the air was stifling sweet
With the breath of many flowers,–
A temple of the heat.

There we bowed us in the burning,
As the sun's right worship is,
To pick where none could miss them
A thousand orchises;
For though the grass was scattered,
yet every second spear
Seemed tipped with wings of color
That tinged the atmosphere.

We raised a simple prayer
Before we left the spot,
That in the general mowing
That place might be forgot;
Or if not all so favored,
Obtain such grace of hours,
that none should mow the grass there
While so confused with flowers.

Pogonias also grow at the edges of acidic bogs, which in early summer often have water levels that can be quite deep—requiring me to wear chest-high waders to photograph them (fig. 134)! They sometimes inhabit wet, sphagnum-covered swales in sandy soil, or the wet areas interspersed in flat, rocky riversides known as ice meadows along the upper reaches of New York's Hudson River (fig. 135).

Rather than growing from a tuberous corm, rose pogonia arises from a cluster of fibrous roots, and produces stolons that spread laterally through the sphagnum mat to produce additional flowering shoots. As a result, pogonia can form large colonies in sunny sphagnum bogs. As with the other pink bog orchids, there is just one leaf (up to 1 inch wide), but it is located on the floral stem, subtending the flower, rather than at the base of the plant (fig. 136). The position of the leaf is one means to differentiate rose pogonia from the other two species included here even when not in flower.

Plants typically have one pink flower, rarely two or three (fig. 137). The surface of the lip, as well as its margin, is embellished with several rows of fleshy hairs; those in the center are typically pink and yellow and those on the margin a darker purple pink. Thus, as in the previous two species, the same false stamen ruse is employed to attract visitors. The flowers are successful in attracting bees, both worker bumblebees and other midsized bees of different genera (fig. 138), as well as the occasional other insect (fig. 139). Sepals and petals arise from the top of a swollen, inferior ovary (see lower flower in fig. 137) that, if the ovules are fertilized, will develop into a typical orchid capsule (fig. 140) containing thousands of dustlike seeds. Capsules are abundant in this species and are often seen in early fall.

Fig. 134. The author photographing rose pogonias in a New York bog. (Photo by Taro Ietaka)

Orchid Family

Fig. 135. Rose pogonia is found growing in wet areas among the bare rocks lining the edges of the upper Hudson River. These river margins are scoured clean of vegetation each winter by the movement of a slushlike mixture called frazil ice.

Fig. 136. Leaves of rose pogonia are broader than those of dragon's mouth and grass pink and arise partway up the floral stem rather than from the base of the plant.

Fig. 137. A two-flowered plant of rose pogonia. The green structure beneath the lower flower's pink floral parts is the flower's ovary.

Bog Orchids

Fig. 138. Only bees of medium size, such as these two (multitasking) *Megachile* (leaf-cutter) bees, can access the scant amount of nectar in the base of rose pogonia flowers.

It has been observed in experimental trials that ultraviolet patterns are used by bees to guide them to the locality of the nectar. Thus, the prominence of UV patterns on the flowers of all three of these species amounts to false advertising. *Calopogon* has no nectar, while dragon's mouth and rose pogonia have just trace amounts.

Because of the similarity of color and morphology (especially the striking brushlike hairs on the lips), one might think that these three species would compete for the same pollinators and eventually form hybrids; however, certain characteristics almost always allow them to remain reproductively distinct. A gap in flowering time at either end of the bloom period at a particular site affords each species a period during which the other species is not in flower and, therefore, not competing for pollinators during that period.

The distinct positions of the pollen masses also result in the attachment of pollinia on different parts of a bee's body. Thus, an insect with pollen on its head would be ineffective for pollinating a flower with reproductive structures designed to receive pollen from a bee's abdomen, etc. Therefore, in spite of the flowers' similarities, different insects are involved in effective pollination. In many orchids, *artificial* (man-made) crosses between genera will sometimes yield capsules with seeds, and interestingly, a rare *natural* cross between grass pink and dragon's mouth has been found.

It is certainly worth getting your feet wet to seek out these beauties in their various wetland haunts. But DO tread lightly around the borders or on a boardwalk so as not to harm the delicate vegetation, and do NOT, as Robert Frost wrote, "pick where none could miss them a thousand orchises."

Fig. 139. A little glassywing skipper butterfly repeatedly attempted to probe this rose pogonia flower for nectar.

Fig. 140. A mature capsule of rose pogonia has split open, allowing the minute seeds to be shaken out and transported by air currents.

Orchid Family

Fig. 141. A flowering plant of broad-leaved helleborine. Note how flowering progresses upward from the base of the inflorescence.

Broad-leaved Helleborine

Fig. 142. The leaves of the plant spiral around the stem, continuing even into the inflorescence, where they become smaller and bractlike.

Broad-leaved Helleborine

Epipactis helleborine (L.) Cranz
Orchid Family (Orchidaceae)

Broad-leaved helleborine is a European orchid introduced into northern New York in the late nineteenth century; it has become naturalized throughout much of northeastern North America. Because its flowers are relatively small and somewhat dull in color, they do not inspire the same degree of admiration as many of our native orchids.

Habitat: In its native range, broad-leaved helleborine is primarily a denizen of woodlands, but as an introduced plant in North America, it is found in a range of habitats—from woods to disturbed lands, and in lawns and gardens.

Range: *Epipactis helleborine* is native throughout much of Europe and Asia and occurs in Israel and Syria as well. It has naturalized in most states of the East and Midwest—with the exception of the Deep South—and occurs sporadically along the West Coast from British Columbia to California, in southwestern Montana, and in localized areas of north central Colorado and New Mexico. The eastern Canadian provinces have also been colonized by broad-leaved helleborine.

Broad-leaved helleborine was first introduced in the Syracuse area of upstate New York in 1879 and quickly naturalized there. It was brought from Europe because of its use as both a medicinal herb and a garden plant. In natural areas of North America, it is frequently found growing in shaded deciduous, coniferous, or mixed forest; in disturbed areas of cedar swamps; and along woodland streams. Because of its broad tolerance for a variety of habitats and soil types, there is some concern that helleborine might outcompete some of our native forest species, including orchids. Thus, attempts are underway to eliminate it from parts of the Northeast, where it is the most common woodland orchid. Probably as a result of its original introduction in the Syracuse region, the population of helleborine is especially robust in New York State, but it is also common in New England and the Great Lakes region. It is New York State's only alien orchid, although a few other introduced orchids are occasionally found elsewhere in the Northeast. Of the

Orchid Family

other 24 species of *Epipactis* worldwide, two others have been reported in North America. One, a Eurasian species of *Epipactis* (*E. atrorubens*), was found naturalized in northern Vermont in the 1990s, but it apparently has not persisted there, and the other, *Epipactis gigantea*, is native to the northwestern United States. In Wisconsin, where broad-leaved helleborine has proliferated—particularly in the botanically rich Door Peninsula, *E. helleborine* has been declared invasive.

Plants of broad-leaved helleborine may grow to be 1–3 feet tall and have numerous clasping leaves that spiral upward on the stem, becoming smaller and smaller near the apex of the plant (fig. 141); they continue into the inflorescence as narrow bractlike structures (fig. 142). The leafy plants bear a resemblance to those of yellow lady-slipper (*Cypripedium parviflorum* var. *pubescens*). The stems and ovaries of the plants are described by different sources as either hairy or smooth—I've seen both. The flowers are ¾–1 inch wide and arranged on a raceme, with larger plants having up to 50 flowers. While the inflorescences are nodding in bud (fig. 143), they become upright when flowering, with the flowers opening sequentially from the bottom up (fig. 144). As with the majority of orchids, the flowers are resupinate, meaning that the flower twists 180° from its position in bud, resulting in the lip of the orchid (the upper petal) being lowermost when the flower is open. The broad, spreading sepals are most commonly greenish with a central line of purple. The two uppermost petals are smaller than the sepals and generally more colorful—varying from yellow, to dull pink, to reddish-purple; the lip is highly modified into a deep glistening purple bowl (known as the hypochile) that narrows toward the front and then flares out into a recurved flange (the epichile) at the front edge (fig. 145). This flange serves as a landing platform for visiting insects. Albino plants are occasionally encountered. The plants bloom in midsummer.

The floral morphology and dull color of *Epipactis* flowers renders them somewhat unattractive to most insect pollinators. Rather, it is thought that scent produced by chemical compounds in the nectar serves as the lure for insects. Studies of the nectar components of various orchid species have identified more than 1500 chemicals, mostly alkaloids. By use of gas chromatography and mass spectrometry, the nectar of *Epipactis helleborine* has been analyzed and shown to contain (in addition to the main constituent—sucrose) an array of other chemical compounds. Among these are methyl eugenol (having the scent of

Fig. 143. Inflorescences are nodding when in bud.

Fig. 144. As the flowers open, the inflorescence becomes erect.

Broad-leaved Helleborine

Fig. 145. Although the flowers of broad-leaved helleborine are generally small and dull in color, some plants have more colorful flowers. In this rather pretty example, the three sepals (the upper one is partially hidden in this view) are greenish with a purplish midline, the two upper petals are reddish-purple, and the bowl-like lip has a deep purple interior that glistens with nectar. The pink tip of the lip is extended, providing a landing platform for insects. Arching over the lip is the column bearing the reproductive structures of the flower.

cloves) and vanillin (with the aroma of vanilla), both known to be highly attractive to insects. Other compounds known as "green volatiles," which mimic the scent of wasp prey (insects or spiders), are wafted into the air, adding to the allure of the flowers. Also present are ethanol (an alcohol, which is probably produced by fermentation of the nectar by microorganisms), oxycodone, and other opioids (morphine derivatives) capable of causing narcotic effects. It is likely that the presence of these compounds explains the erratic flight and/or sluggishness of insect visitors frequently reported by observers.

Both in its native range and where it has been introduced in North America, broad-leaved helleborine has been documented as having social wasps as its principal pollinators (figs. 146–147). Although reportedly self-compatible, *Epipactis helleborine* appears to be primarily an outcrossing species, reliant on pollinators to bring pollen from one plant to another. When a wasp visits an *Epipactis* flower

Fig. 146. Social wasps, such as this species (*Dolichovespula arenaria*), are the pollinators of this orchid. The wasps visit the flowers to sip nectar held in the lip. They often brush against the column as they fly off, resulting in the pollen-bearing sacs becoming stuck to their heads. In this case, the wasp arrived at the flower with pollinia (white structure) already glued to its face. Pollination will result if the pollinia brush off on the stigmatic structure of a subsequent flower as the wasp flies from the flower

Fig. 147. *Vespula consobrina*, another species of social wasp, is drinking nectar from the lip. Note how its position will likely result in its head contacting the tip of the column as it leaves the flower, thus the pollinia becoming attached to the wasp's head.

Orchid Family

Fig. 148. This *Dolichovespula arenaria* wasp, with pollinia glued to its face from a recent helleborine visit, wandered about the leaf litter for a few minutes before flying off—perhaps intoxicated by the alcohol or opioids found in helleborine nectar.

to drink the nectar from its bowl-like lip, its trajectory as it flies upward from the flared section of the lip causes its head to contact a viscid substance on the tip of the overarching column and the adjacent anther, resulting in the pollinia becoming stuck to the wasp's head (note the positions of the wasps and the orchids' columns in figs. 146 and 147). Wasps are generally meticulous groomers, but perhaps because of the intoxicating properties of the nectar, they become less so after imbibing helleborine nectar and thus fail to remove the pollinia stuck to their heads. Indeed, I watched one wasp depart from a flower with pollinia attached to its face and crawl aimlessly in the leaf litter for several minutes before taking flight (fig. 148). The sticky pollinia are carried by the wasp until it visits another helleborine flower where the pollinia come into contact with the stigma of that flower, resulting in pollination. As far back as 1869, Charles Darwin wrote of his own observations of wasps visiting *Epipactis helleborine* (= *E. latifolia*) in his English garden. He noted that despite the presence of numerous bees, it was only wasps (*Vespula sylvestris*) that visited the flowers and carried away the pollinia. Darwin posited that should wasps become extinct, then surely the orchid would as well.

Other types of insects also visit the flowers of *Epipactis helleborine* to drink the alluring nectar (particularly flies, as seen in figs. 149 and 150), but they rarely contact the pollinia and therefore do not serve as effective pollinators. Unlike most other species of *Epipactis*, the red-flowered Eurasian *E. atrorubens* (fig. 151) is pollinated by bees, and the western *E. gigantea* by syrphid flies.

I first noticed helleborine in the area where I live about 25 years ago. Although I recognized it as an orchid, it was not one I was familiar with. Luckily for me, friends visiting from Europe immediately

Fig. 149. Aside from wasps, flies, like this syrphid (hover) fly, visit broad-leaved helleborine flowers, but their actions do not result in the removal of pollinia from this species. However, syrphid flies *are* the pollinators of our native *Epipactis gigantea*, found in the western part of the United States.

Fig. 150. The dull purplish or greenish colors of most *Epipactis helleborine* orchids are attractive to flies. These small flies are probably scavenging pollen grains or taking a sip of nectar. They are not effective pollinators.

Broad-leaved Helleborine

Fig. 151. (Left) A Eurasian species of the same genus, *Epipactis atrorubens*, was found growing in a serpentine-rich area of northern Vermont in the 1990s, but it has not persisted there. In its native range (Europe, as in this photograph), it is pollinated by bees. **Fig. 152.** (Right) Two plants of broad-leaved helleborine emerging among the low-growing junipers in our garden. Note how the leaves of the *Epipactis* resemble those of yellow lady-slipper orchids.

identified it as one of "their" orchids: *Epipactis helleborine*. Once I had the helleborine "search image," I began finding the plant in many places I botanized—from Maine to Virginia to Wisconsin. Not only was helleborine in woodlands, its natural habitat in Europe, but it seemed to grow almost anywhere; it has since taken up residence in my gardens (fig. 152), my lawn, and even my gravel driveway (fig. 153)! Because of this great adaptability, helleborine has become invasive in some regions. Although many of our native orchids are browsed by deer, I see evidence of this herbivory on *Epipactis* only occasionally (fig. 154).

Initially, I was torn about pulling broad-leaved helleborine from my garden—it is, after all, an orchid—with interesting, and sometimes even pretty, flowers. However, as it popped up in more and more inappropriate places, I decided that it must go. It may be an orchid, but it's also a weed! Although broad-leaved helleborine is the common name usually applied to this species, it is also widely known as "the weed orchid." Helleborine orchid is perennial, arising from a compact rhizome having a cluster of fleshy roots (fig. 155).

Fig. 153. *Epipactis helleborine* even manages to colonize gravel driveways and small fissures in cement walkways!

Orchid Family

Fig. 154. Although deer seem to have a fondness for orchids, they don't often chew on the plants of helleborine in my garden; this is an exception.

Fig. 155. The pale stems of broad-leaved helleborine arise from short rhizomes, which also produce clusters of fleshy roots.

One must dig deeply enough to remove the rhizome completely; otherwise a flowering stem, or even several, may reappear from the buried rhizome remnant in a subsequent year. Helleborine orchid doesn't spread by vegetative means, only by seeds.

The seeds of orchids are minute and are often referred to as "dust seeds" (fig. 156); as such, they can be transported easily by wind. Orchid fruits may contain thousands of such seeds (fig. 157)—an advantage in terms of the number of plants that could result if large numbers of these seeds were to germinate successfully. But because such tiny seeds contain no endosperm (a food source for the embryo plant found in most other seeds), and the embryo itself has minimal food reserves (fig. 158), the seeds must rely on external factors in order to germinate and develop into seedlings—thus small size also has its disadvantages.

The nutrients that orchid seeds require to develop into seedlings must be supplied via their relationship with mycorrhizal fungi in the soil. Orchid seeds, once infiltrated by these threadlike fungal hyphae, break them down to obtain the carbon, amino acids, and other nutrients they contain. These fungi are also involved in a partnership with neighboring photosynthetic plants (usually trees and shrubs) from which they obtain carbon and minerals. Thus, the fungus serves as a conduit of nutrients from the trees or shrubs to the tiny orchid seedlings, allowing them to prosper. Obviously, many seeds land in places that lack the fungi necessary to supply their nutritional needs—thus the species benefits from a surplus of seeds. The seeds of some orchid species require a specific fungus or a group of related fungi to fulfill this role, but other species (*Epipactis helleborine* being one of them) are

Fig. 156. I shook the split capsules of *Epipactis helleborine* over a white paper to show the thousands of tiny "dust seeds" they contain. In the wild, these seeds are easily carried long distances by the wind. Helleborine is a perennial plant and will resprout from the same rhizome when the conditions are favorable; new plants result only from reproduction by seed.

Fig. 157. This immature seed capsule has been cut open to show the mass of developing seeds (white) contained inside. Seeds become tan when mature.

Broad-leaved Helleborine

Fig. 158. A photo of seeds of *Epipactis helleborine* taken through a dissecting microscope; each increment of the scale bar on the left is 1.6 mm. The dark area at the center of each winged seed is the embryo. As there is no endosperm, the seeds must quickly be infected by a fungus from which they can obtain the nutrients necessary to grow.

not as selective about their fungal partners, and therefore can flourish in many environments—even when introduced into other continents.

In some species of orchids, the fungal partners are members of the basidiomycetes (those fungi that produce fruiting bodies that we recognize as mushrooms). But others have as their fungal associates, members of the ascomycetes (fungi that contain their spores in sacs [asci] and include yeasts and molds, as well as edible morels and truffles). *Epipactis* is commonly associated with an ascomycete in the genus *Tuber*, the same genus that includes truffles.

While all orchids are dependent upon an association with compatible fungal partners in order to germinate, it was long believed that this relationship was important only up to the stage when most orchid species develop green leaves. Once the plant could photosynthesize and produce its own carbohydrates, it was thought that the fungal relationship was no longer necessary. There are a relatively small number of nonphotosynthetic orchids, which, because they contain no chlorophyll (and are therefore unable to produce the carbohydrates needed to support their growth), *must* continue this relationship throughout their life span; such plants are termed mycoheterotrophs, meaning they obtain their nutrients via the fungal association. Before the mycorrhizal relationship between plants and fungi was known, such plants were believed to derive all their nutrients from decaying leaf litter on the forest floor and were termed saprophytes.

The great majority of orchids *do* develop green leaves at maturity, and, in theory, those species are capable of fulfilling all their own nutritional needs through the process of photosynthesis. However, in the first decade of this century, groundbreaking research carried out by Duncan Cameron and collaborators on another green-leaved orchid species, *Goodyera repens* (rattlesnake-plantain), showed that a fungal relationship remained important throughout the life span of the plant, even when in full leaf. Many orchids have fleshy roots that are poorly suited to absorbing nutrients from the soil as easily as those with fine, threadlike roots; thus, the mycorrhizal relationship is essential to them as an aid in nutrient acquisition. This relationship is particularly important in times of stress such as drought or other adverse environmental conditions. Not only do such orchids continue to receive water and nutrients, including carbon, via the fungal connection, but surprisingly, carbon produced by the orchid has been documented as passing from the plant to the fungus as well. Thus, rather than the orchid being parasitic upon the fungus, the association could now be considered a mutualistic one in which both partners benefit. Green-leaved orchids that maintain this fungal association to augment their carbon needs are in a category now classified as mixotrophic orchids, meaning that they utilize a mixed system of obtaining nutrients throughout their lifetimes, partly through their fungal association and partly through their own photosynthesis (e.g., *Goodyera repens*). It has been shown that in soils with a greater amount of fungus (as measured by looking at the fungal DNA in the soil), more orchid plants are likely to emerge in any given year rather than having to spend a year or more in dormancy in order to build up enough reserves to produce a flowering plant. Thus, paradoxically, the soils in old growth forests, with closed canopies, generally have more fungi. Since most orchids require a good amount of light to thrive, a balance of these factors is necessary for optimal growth.

Moreover, additional studies have shown that the carbon transferred from fungus to orchid is the result of ectomycorrhizal relationships of fungi with the surrounding trees. Therefore, efforts to protect various species of orchids must include the protection of the fungal network in the soil as well as the neighboring trees, which are the indirect suppliers of the orchids' carbon needs. While this might not be a consideration in the case of *Epipactis helleborine*, it is an important lesson to apply to orchid conservation in general.

Buckbean Family

Fig. 159. A flowering shoot and the three-parted leaf of *Menyanthes trifoliata* growing from a submerged rhizome in a sphagnum bog in the Adirondacks.

Buckbean

Fig. 160. Several long-styled (pin) buckbean flowers near the apex of an inflorescence.

Buckbean (Bogbean)
Menyanthes trifoliata L.
Buckbean Family (Menyanthaceae)

Buckbean, the only member of the genus *Menyanthes*, is one of only two members of the aquatic/wetland family Menyanthaceae native to the northeastern United States (fig. 159). Flowers of buckbean are small (ca. 1.2–2.5 cm in diameter), but because of their elaborately fringed petals (fig. 160), they are among our prettiest wildflowers.

Habitat: Still waters of bogs, fens, lake margins, and slow-moving rivers.

Range: Throughout most of Canada and the more northern continental United States, extending southward along the mountains to Virginia and North Carolina in the East and in the mountains southward to northern New Mexico, northern Arizona, and central California in the West. Buckbean's circumboreal range encompasses Greenland, Asia, and Europe, extending north of the Arctic Circle and south to the mountains of Morocco in Africa.

The Menyanthaceae is a small family of aquatic and wetland plants with only five genera (which some taxonomists believe should be segregated into additional genera based on recent morphological and molecular studies). Though few in species, the family's range includes all continents with the exception of Antarctica. At one time *Menyanthes* was considered a member of the gentian family (Gentianaceae), with which it shares certain similarities: a tubular corolla, a bilobed stigma (see fig. 160), and capsular fruits. However, based on differences in leaf arrangement, chemical makeup, and other more technical characteristics, Menyanthaceae is clearly a unique family, separate from Gentianaceae.

The scientific name, *Menyanthes*, is derived from the Greek *menyein*, meaning "disclosing" (a reference to the sequential opening of the flowers from the bottom of the raceme to the top), and *anthos*, meaning "flower." The epithet, *trifoliata*, from Latin, alludes to the three leaflets (see fig. 159). In addition to buckbean, the plant has many other common names: bogbean, marsh trefoil, and water shamrock among them. The first name is indicative of both the habitat and

Buckbean Family

the shape of the small seeds, and the latter two names reflect the similarity of the three leaflets to those of the legumes, trefoil and shamrock (the common name shamrock may also be applied to members of the genus *Oxalis*).

Buckbean requires a sunny habitat but has a wide tolerance of wetland conditions, from acidic bogs to alkaline fens (fig. 161). While the plants are generally found in shallow water, they can grow in water more than 2 feet in depth and are capable of surviving periods of drought without being submerged in water, providing the soil remains moist.

Menyanthes trifoliata is the only member of its family to have compound leaves (fig. 162). The leaves have special pores (hydathodes) that secrete water from the apices of the veins that end along the margins of the leaflets. This is a feature commonly found in aquatic plants, but it is sometimes seen in other plants as well (e.g., strawberry). The droplets of water can be seen especially in morning at the tip of each serration of the leaflets). The leaves are bitter and are

Fig. 162. One inflorescence among many other nonflowering, three-leaved buckbean plants. Note that the lower flowers have already turned brown and that there are still some unopened buds at the apex of the inflorescence.

occasionally used to flavor the strongly alcoholic beverage known as *snaps* in Scandinavia (equivalent to *schnapps* in Germany), which is traditionally drunk at festive meals. In addition, buckbean leaves sometimes are substituted for hops in the flavoring of beer. Bitter herbs, such as buckbean, have also found use in medicine, with a preparation from buckbean leaves used as a cathartic, a digestive, and a febrifuge; as well as for treating muscular weakness, rheumatism, jaundice, and other conditions. However, as with any herbal medicinal product, warnings are issued regarding possible negative side effects.

Fig. 161. A large fen filled with buckbean, among other plants.

Buckbean

The rhizomes, too, have a long history of medicinal use and were even recommended by Linnaeus in 1749 for the treatment of nephritis and rheumatism. A study of the chemistry of the rhizomes carried out in China in the 1990s showed that three of the compounds isolated from the rhizomes demonstrated a far greater potency for treating inflammation than did aspirin. There is, thus, some credence for its longtime use in treating rheumatism and in quelling fever. Another of the rhizome compounds showed inhibitory effects against HIV replication in lymphocyte cells, but it's unclear whether this was further investigated for pharmaceutical potential.

The aspect of the buckbean plant that drew me to it was the unique flower. I marveled first at the intricate beauty of the elaborately fringed petals (fig. 163)—and then I wondered, "What is the purpose of the fringe?" I could find only one plausible hypothesis in the literature, which agreed with my conjecture: that the intertwined, twisted fringes serve to prevent small insects (which play no role in terms of pollination services) from reaching the nectar. This resource, therefore, is protected against pilferage, and conserved for the larger bees and flies that can negotiate the fringes without becoming entangled in their twisted mass and are of a size to be effective pollinators. Bumblebees are the most frequent visitors, followed by flies, a few butterflies, and, rarely, a honeybee. With their larger, hairier bodies, bumblebees and flies are the most effective transporters of pollen.

The buckbean we see in North America was formerly segregated into a variety (*Menyanthes trifoliata* L. var. *minor* Fernald). It differs from the Old World *Menyanthes trifoliata* var. *trifoliata* in the smaller size of the plant and its flowers, when compared with those in other parts of its geographic range. The flowers of our buckbean have fringe only on the lower part of the petal lobes (see fig. 163), whereas, in the flowers of plants in other parts of the world, the entire petal lobe is fringed.

With respect to reproduction, buckbean is a classic example of a self-incompatibility mechanism known as heterostyly. The flowers of buckbean, with few exceptions, are of two types (distylous)—those having long styles and short stamens (called pin flowers), and those with short styles and long stamens (called thrum flowers). On my most recent visit to a stand of buckbean, I could find only pin flowers—a possible indication that the stand is, in fact, a single clone. If a visiting insect picks up pollen from a pin flower (short stamens/long style), the pollen is usually positioned on the insect in such a way that when it visits a thrum

Fig. 163. A close-up of a buckbean flower. Note the bilobed stigma, on which there are pollen grains, and the petal lobes, which are not fringed all the way to their tips.

Buckbean Family

flower (long stamens/short style) the pollen is more likely to be deposited on the stigma than if the insect were to visit another pin flower. However, even if the pollen were to brush off onto the stigma of a long style in another flower of the same type (i.e., short stamens and a long style), the pollen will not germinate and grow to fertilize the egg. That is why it is called an "incompatible" reproduction system: fertilization can occur only if the second plant is of the opposite type (pin to thrum or thrum to pin—i.e., short stamens to short style and long stamens to long styles). Buckbean is self-sterile and cannot be fertilized by pollen from the same plant or from any other plant within its clone of genetically identical plants, so reproduction often is vegetative through the spreading of the rhizomes.

Other floral features that correlate with pin or thrum characteristics are stigma type and pollen size. The stigmas of thrum flowers are larger than those of pin flowers and therefore provide a greater surface area for deposition of large pollen grains (see the small stigmatic surface—bearing a few pollen grains—on the pin flower shown in figure 163). In addition, thrum flowers produce pollen that is smaller than pin flower pollen. Although pin flowers have smaller stigmas, the stigma surface is held above the rest of the floral parts and is more likely to be contacted by a nectar-seeking insect. The pin stigma has longer papillae and is thus better suited to capture more of the smaller pollen grains brought from thrum flowers. Occasionally flowers with stamens and styles of equal length (homostylous) are encountered, but they are still self-sterile (self-incompatible).

Successful fertilization results in the production of seeds, but only about one-third of the ovules are fertilized. (This low success rate most likely relates to the low transfer of the right kind of pollen.) The capsular fruits (fig. 164) do not dehisce until autumn. Seeds are small (ca. 2.5 mm in diameter) and smooth. They can remain buoyant for up to a year due to air-filled tissue (aerenchyma) that is supported and protected by tough, hard, sclerenchymal tissue in the seed coats. Seeds are transported by birds (primarily geese and ducks), which carry the seeds to more distant places, where they can begin to colonize other wetlands. Since buckbean is an obligate outcrosser, it is difficult for populations to become established through such long-distance dispersal. The timing

Fig. 164. The immature green fruits on this plant will remain on the plant until autumn when they will turn brown, dry out, and split into two parts to release their small seeds. Note the long, persistent styles that remain on the fruits, indicating it was a pin flower.

must be synchronous so that seeds of both types of plants arrive at the distant location at about the same time, and in close enough proximity to allow plants that bloom simultaneously to cross-pollinate. Only if all these criteria are met can a sexually reproducing population be established.

In addition to reproduction by seeds, *Menyanthes* increases through the proliferation of its rhizomes. Such increase is clonal, and all the flowering stems are genetically alike. The flowers will be fertilized and set seed only if pollen is brought from another nearby clone.

The other native member of the Menyanthaceae that occurs in the Northeast is the white-flowered little floating heart, *Nymphoides cordata* (fig. 165), named

Buckbean

Fig. 165. (Left) A flower of little floating heart (*Nymphoides cordata*) has a crest running vertically down the center of each petal. **Fig. 166.** (Right) The flowers of big floating heart (*Nymphoides aquatica*) have a yellow center and petals with a margin that is both winged and fringed at the base.

for its more-or-less heart-shaped floating leaves. The petals of this species are ornamented with a comblike crest that runs vertically up the middle of each petal. In southeastern Virginia and southward along the Coastal Plain, one may encounter the big floating heart, *Nymphoides aquatica*, a more southern-based species, the flowers of which have wings and basally fringed petals (fig. 166). All these species require good light to thrive.

Unfortunately, the attractive European yellow floating heart, *Nymphoides peltata*, which was introduced to this country to beautify private and public lakes and ponds, has now escaped; it can multiply rapidly and is considered an invasive species. Yellow floating heart has pretty flowers with petals that are fringe-tipped (fig. 167).

Fig. 167. Yellow floating heart was introduced into the United States from Europe as an aquatic ornamental. It has escaped and is sometimes found naturalized in the Northeast.

Dogwood Family

Fig. 168. Three mature shoots of a bunchberry clone with the apparent whorl of six leaves. Each plant has an inflorescence of four white bracts surrounding a compact cluster of small flowers

Bunchberry

Fig. 169. In late summer, clusters of red fruits (from which the common name, bunchberry, is derived) are attractive to birds and mammals.

Bunchberry
Cornus canadensis L.
Dogwood Family (Cornaceae)

Bunchberry is one of the few herbaceous members of a family otherwise comprising small trees and shrubs. The dwarf plants carpet the understory in northern forests, brightening the dark forest floor with their white "flowers" in June (fig. 168). In late summer, globose clusters of fruits turn bright red, giving rise to the plant's common name, bunchberry (fig. 169).

Habitat: Most commonly in acid soils of cool, moist, evergreen or mixed forests in boreal regions.

Range: Circumboreal; in North America from Newfoundland and Nova Scotia to Alaska, extending southward to the Appalachians, the Rockies, and northern Idaho and Montana; also, in the northern latitudes of Europe and Asia.

You may wonder why the word "flowers" is in quotation marks in the above introductory statement. Like the lovely "flowers" of our spring-blooming tree, flowering dogwood (*Cornus florida*) (fig. 170), the apparently single white flowers are actually inflorescences comprising many small flowers surrounded by four showy, white (or sometimes pink in flowering dogwood) bracts. The attention-getting bracts make the flowers more visible to insects. Such an inflorescence is termed a pseudanthium, meaning false flower. The many small flowers (average = 22), clustered in the center of bunchberry have a densely hairy floral cup (hypanthium), four tiny creamy white petals that are attached to the summit of the floral cup and are strongly reflexed downward, four 1 mm-long stamens and a 1 mm-long style, the base of which is surrounded by a dark purple nectar-secreting disk (fig. 171).

As an herbaceous perennial plant, bunchberry grows from a woody underground rhizome (fig. 172) and reaches 20–25 cm (8–10 in.) in height aboveground. Such plants are often referred to as suffrutescent (half-shrubby—herbaceous above and woody

Dogwood Family

Fig. 170. (Left) A branch of flowering dogwood (*Cornus florida*), a large-bracted, red-fruited tree in a group of *Cornus* that is most closely related to the herbaceous bunchberry (*C. canadensis*). **Fig. 171.** (Right) Four showy, white bracts advertise the presence of the small flowers to potential pollinators. A dark nectar-secreting ring surrounds the base of the ovary.

at the base, with the upper part dying back in winter). Vegetative reproduction via spreading rhizomes results in the formation of large clones of closely packed, upright shoots (fig. 173) (up to 300 shoots per square meter), each of which appears to be a separate plant. In some cases, underground rhizomes may grow for more than 60 cm (2 ft.) before sending up another shoot. Since clonal plants are all genetically the same, pollen must be transported to flowers in another clone for successful fertilization to occur.

The leaves of bunchberry vary in shape, having either blunt or pointed tips. Looking at the undersurface of a leaf with a 10× hand lens will allow you to see the bunchberry's characteristic T-shaped hairs. *Cornus*, with few exceptions, has leaves that arise from nodes opposite each other on the stem. In *C. canadensis*, the distance between the upper nodes has been reduced to the point that the short-petioled leaves

Fig. 172. Two shoots of an excavated portion of a clone in the Adirondacks showing characteristics intermediate between *Cornus canadensis* and *C. suecica*, but closer to *C. canadensis* in overall appearance (the lower green leaves are typical of *C. suecica*). This specimen most likely represents a backcrossed hybrid between the two species.

Bunchberry

Fig. 173. A dense clone of bunchberry in flower.

appear in a whorl, usually consisting of six leaves in a reproductive plant, but only four if the plant is not yet fertile. In autumn the leaves turn reddish-purple and are deciduous. The veins, as in other dogwoods, are strong; if the leaf is carefully torn perpendicular to the midrib, and gently pulled apart, the vascular structure of the veins remains and stretches to hold the two halves of the leaf together (fig. 174). The lower part of the stem has a few small, colorless scalelike leaves. The four white, petal-like bracts have either blunt or pointed tips.

The flowers attract many types of insects—from ants and beetles to flies and bees. Flies are particularly important pollinators at higher elevations and colder temperatures (fig. 175). Larger, hairier insects are more effective pollinators for a few reasons: they move rapidly from inflorescence to inflorescence (unlike smaller insects that tend to linger on one inflorescence); their hairier bodies can hold and transport more pollen; and their greater weight is important for triggering the small flowers to open. *Cornus*

Fig. 174. A leaf of *Cornus kousa* gently torn so as not to break the strong vascular strands of some of the main veins.

Dogwood Family

Fig. 175. A tachinid fly visiting the flowers of bunchberry. The flower between the fly's center and rear legs has sprung open.

canadensis, along with other dwarf dogwoods, is notable for the fastest movement recorded in the plant world. In bud, the stamens are held under tension in a bent position inside the tightly apposed flower petals, which are joined at their tips. The anthers have already dehisced (split open) while inside the bud. As the flower bud matures, the bent section of each stamen protrudes from between the sides of the petals, spring loaded and ready to be triggered by the slightest touch. A long awn on only one of the petals serves as the trigger mechanism. When the petal tips are displaced enough to break their attachment (such as by a visiting insect) the stamens spring upward, their anthers explosively shooting the pollen onto the insect, or (if the petals have opened without the aid of an insect) into the air. Each hinged anther rotates freely at the tip of the staminal filament, releasing its pollen only when it has reached the optimal point for directing the pollen straight upward. Thus, each stamen works much like a miniature trebuchet (fig. 176), a weapon used in medieval times for launching stones or other missiles from—or into—castle walls. In the case of the bunchberry flower, this movement optimizes pollen broadcast direction to ensure that the pollen is more thoroughly distributed over the insect. It is thus less likely to be completely groomed off and eaten and more likely to be transported by the insect to another flower. Recall that fertilization can occur only if that other flower is a member of a different clone. Lacking insect visitation, the continued growth of the petals may trigger this mechanism and the anthers can shoot the pollen 10 times above the height of the plant, where air currents can carry it to nearby flowers. This movement has been captured on high-speed video by the authors of a fascinating study on plant movement conducted at Williams College by Joan Edwards and colleagues. By viewing the video at a slower speed, the explosive mechanism can be seen and measured. When triggered, the flowers were seen to open in less than a third of a second, with the anthers expelling their pollen at an astounding 4

Fig. 176. A replica of a trebuchet, a medieval weapon used for hurling stones and other objects at the enemy. (Photographed at Château de Castelnaud-la-Chapelle, France)

Bunchberry

Fig. 177. A typical northern mixed forest floor scene. Bunchberry often grows on or near decaying wood. Note how the two flowering shoots have six leaves, while the immature (nonflowering) shoots have only four.

meters per second—more than 2000 times the acceleration of gravity. The authors also documented that pollen was carried more than 22 cm (8.7 in.) even in basically still conditions—thus, although the self-sterile bunchberry was long believed to be reliant solely on insect pollination for sexual reproduction, we can now add airborne distribution as an alternative method of pollen transport, especially in somewhat open forest habitats on days when a sufficient breeze sweeps across the plants at ground level to carry the pollen even further.

Bunchberry is typically an understory plant of boreal forests (fig. 177), but it is also found in acidic coniferous swamps and in the alpine vegetation zones such as that of the Presidential Range of the White Mountains in New Hampshire. In particular, it grows at the edge of krummholz vegetation and in snowbank communities (see alpine wildflowers essay in this book for definitions of these communities). Bunchberry can grow in dense shade, but the plants flower and set seed best with ample light in more open forest or edge habitat (fig. 178).

The bright red fruits may look like berries, but they are technically drupes, fruits in the same category as plums and cherries. Drupes have a thin skin surrounding a fleshy interior that, in turn, surrounds a hard stone or pit. Inside the stone is the seed.

Fig. 178. Bunchberry can be found flowering profusely when growing in an open, sunny area on the edge of forest.

In *Cornus*, there are one or two stones per drupe. The flesh of bunchberry fruits is edible, but almost

Dogwood Family

tasteless, and the smooth, ovoid stones (with their seeds) are too hard to chew. Birds (especially during fall migration) consume the fruits, which provide an important source of energy as they prepare to embark on their long migratory flights. Finches and other birds with large, strong beaks may crush the seeds and are considered seed predators rather than seed dispersers. Ground birds, including different species of grouse, consume many of the fruits of bunchberry in autumn, and the seeds can be found in their droppings. These seeds, undamaged by passage through the bird's gut, may germinate the following spring. I have seen starlings strip the fruits from my flowering dogwood (fig. 179) in less than an hour. Mammals, too, eat the fruits as well as the vegetative parts of the plants. Depending on where in North America the plants are growing, ungulates such as moose, deer, elk, and caribou browse on bunchberry. Bears relish the fruits, as do small mammals such as chipmunks, and rabbits. Seeds that have had the surrounding flesh removed by having passed through the digestive system of an animal have a better chance of germinating. The seeds need a period of cold in order to germinate (cold-stratification) and, thus, have greater success if they remain at the surface of the forest duff that covers the soil. Seeds can remain viable in the soil bank for at least three years.

Similar in appearance to *Cornus canadensis* is another dwarf dogwood, *C. suecica* (Lapland [or Swedish] cornel), a species of far-northern coastal areas and of wet inland areas in eastern and western North America. As its common names suggest, it is also an inhabitant of the colder regions of Europe (e.g., Sweden) as well as Asia. It can easily be mistaken for bunchberry, but by examining the plant carefully, the differences between the two species are easily seen. Most noticeable is that the petals of the small flowers of Lapland cornel are a deep purple color (fig. 180), and the hypanthium (floral cup) is glabrous (without hairs) whereas the hypanthium of *C. canadensis* is covered with hairs. The leaves differ as well: the secondary veins of *C. canadensis* arch upward from the lower one-third of the midrib (see fig. 168), but those of *C. suecica* begin at or near the base of the leaf, and the midrib is not prominent. In addition, leaves of Lapland cornel do not have petioles; they are in opposite pairs rather than appearing whorled; and there are sometimes small leafy lateral branches arising from the leaf axils. Rather than having colorless scale leaves on the lower portion of the stem, the lower leaves of Lapland cornel are larger and green—true leaves. Where the ranges of these two species overlap, they may hybridize (see fig. 172).

Cornus suecica has a large air pocket in its seeds, which permits the propagule to float and be dispersed in that manner. The fruits are also eaten by seabirds, which may then transport the seeds over long distances. Thus, although bunchberry is often an inland species and Lapland cornel is primarily coastal in habitat, the latter is found growing along the shores of

Fig. 179. A cluster of red fruit on a flowering dogwood branch. Next year's onion-shaped flower bud is already formed at the tip of the branch.

Bunchberry

Fig. 180. The flowers of *Cornus suecica* (Lapland cornel) are purple and the leaves are paired, with an indistinct midvein and secondary veins that arise from the base of the leaf. Note the small, leafy branch growing from the leaf axils of the leftmost shoot.

Hudson Bay and other wet inland continental areas. Lapland cornel could have come into contact with *C. canadensis* during the breakup of the ice sheets, when bunchberry migrated further north during a warmer period. Although there is still overlap of the ranges of *C. canadensis* and *C. suecica* in northeastern Canada and Greenland, only in Alaska, western Canada, and the United States, did a fertile hybrid bunchberry develop.

With features intermediate between bunchberry and Lapland cornel, the Alaska bunchberry (*Cornus unalaschkensis*) is found only in the Pacific Northwest. Its flower petals are white at the base and purple at the tips, and the hypanthium is entirely covered with hairs. The upper leaves appear to be whorled, have short petioles, and have a pronounced midrib; the lower leaves may be either scalelike or green. Despite conflicting evidence regarding the possible polyploid status (having more than the usual number of chromosomes) of Alaska bunchberry, morphological characters alone suggest that Alaska bunchberry is probably of hybrid origin, dating back to a time when *C. canadensis* and *C. suecica* were in contact in western North America. Like many polyploids, Alaska bunchberry is a larger plant overall; it is fertile and reproduces by both seeds and rhizomes.

Compounding this confusing situation is the existence of two other dwarf dogwood entities, both of which appear to be hybrids between bunchberry and Lapland cornel that have rehybridized (backcrossed) with one or the other original parent. In the case of one population found in the Northeast, a greater influence of *C. canadensis* is seen, the plants appearing more like those of bunchberry, but having purple-tipped petals and/or leafy lower leaves. The second purported hybrid population has more features in common with *C. suecica*, but with secondary veins arching from the midvein, hairs on the lower half of the hypanthium, and petals that are cream-colored at the base but purple toward the apex (or distally). It has a primarily coastal distribution like that of *C. suecica*, but in western North America extends inland. The plant shown in fig. 172, photographed in the Adirondacks, appears to be of hybrid origin, backcrossed with *C. canadensis*, with upper leaves like those of *C. canadensis*, but green leaves at the lower nodes like *C. suecica*. I did not observe the flower color. However, the lower leaves are further aberrant with more than two arising from each node.

All the dwarf dogwoods and other dogwoods with showy bracts subtending the inflorescence (e.g., flowering dogwood) fall into the red-fruited division of *Cornus* (see fig. 179). The red-fruited species have comparatively smooth stones, with those of the dwarf species being smaller than those of the red-fruited tree species. Shrubby members of the genus have blue, or occasionally white fruits with ribbed, or otherwise sculptured, stones. Surprisingly, there are a two species of *Cornus* in Central and South America, and one in tropical Africa, all at higher elevations.

The classification of the genus *Cornus* has been questioned by some taxonomists. When *Cornus* is treated in an extremely narrow sense, the dwarf dogwood species would be put into the spelling-test genus of *Chamaepericlymenum*, while our familiar flowering dogwood would be placed in the genus *Benthamidia*, and the shrubby dogwoods (e.g., red osier and gray-stemmed dogwood) would be classified as *Swida*. Fortunately, the family treatment has now been published in *Flora of North America*, and the genus *Cornus* remains unified, but diverse, being divided into several subgenera.

Bellflower Family

Fig. 181. The depth of the brilliant red color of cardinal flower is unrivaled in the flora of the Northeast United States.

Cardinal Flower

Fig. 182. Cardinal flower is most often encountered growing in wet soils along streams and on the edges of lakes and ponds.

Cardinal Flower
Lobelia cardinalis L.
Bellflower Family (Campanulaceae)

Unrivaled for the intense color of its oddly shaped red flowers (fig. 181), cardinal flower is a plant eagerly sought by wildflower enthusiasts from midsummer into early autumn. Roger Tory Peterson chose to single this wildflower out as "America's favorite" in his popular field guide. Most modern systems of classification, including the Angiosperm Phylogeny Group, place the genus *Lobelia* in a subfamily of the bellflower family, the Lobelioideae; others give it family status as the Lobeliaceae.

Habitat: Woodland streamsides and other wet areas (fig. 182), and also (in the Midwest) along roadsides and in prairies and meadows (fig. 183).

Range: Southern New Brunswick to Ontario and southeastern Minnesota, south to Florida and west to California in an area south from southern Nebraska and Nevada.

Cardinal flower is named for the scarlet color of the hats and cassocks worn by leaders of the Roman Catholic Church for more than 500 years, yet the red of those garments cannot outshine these flowers for richness of color. Linnaeus named the genus *Lobelia* for a French-born physician and botanist, Mathias de L'Obel, who, like Linnaeus, had "Latinized" his surname to Lobelius. Like its popular name, the species' scientific epithet (*cardinalis*) also reflects the similarity in color to the vestments of Catholic cardinals. The plant is sometimes called scarlet (or red) lobelia. Of the 15 species of *Lobelia* native to the United States and Canada, *Lobelia cardinalis* is the most widespread member with red flowers. The only other is *L. laxiflora*, a species native to Arizona, Mesoamerica, and northern South America (fig. 184). In addition to the cardinal flower, one of our favorite North American birds, the bright red male cardinal, derives both its scientific name (*Cardinalis cardinalis*) and its standardized common name (northern cardinal) from the color of the robes of Catholic cardinals, though the bird's color can't compete with that of the flower.

Bellflower Family

Fig. 183. (Left) In the Midwest, cardinal flower can also grow in meadow and prairie habitats, such as this created meadow landscape in a botanical garden. **Fig. 184.** (Right) Loose-flowered lobelia (*Lobelia laxiflora*) is native from Arizona through Mesoamerica and parts of northern South America. The evolution of this species has proven ideally suited for hummingbird pollination.

The beauty of cardinal flower has inspired poets to immortalize it in verse. The flower's brilliant scarlet is especially striking when glimpsed glowing along a shady woodland stream (fig. 185). John Burroughs, famous nineteenth-century naturalist and writer, expressed this effect of cardinal flower in his short book of poetry, *Bird and Bough*:

> Like peal of a bugle
> Upon the still night,
> So flames her deep scarlet
> In dim forest light.
>
> A heart-throb of color
> Lit up the dim nook,
> A dash of deep scarlet
> The dark shadows shook.
>
> Thou darling of August,
> Thou flame of her flame,
> 'Tis only bold Autumn
> Thy ardor can tame.

Burroughs also described cardinal flower thus: "There is a glow about this flower as if color emanated from it as from a live coal." Emily Dickinson, writing in the late nineteenth century, briefly referred to cardinal flower in her short poem about Indian pipe (see the account of Indian pipe in this book). And Nathaniel Hawthorne,

Fig. 185. A solitary cardinal flower seems to glow in the dark of a wooded wetland.

Cardinal Flower

author of *The Scarlet Letter*, admired the cardinal flower as well, writing of it on an August day in 1842:

> For the last two or three days, I have found scattered stalks of the cardinal-flower, the gorgeous scarlet of which it is a joy even to remember. The world is made brighter and sunnier by flowers of such a hue. Even perfume, which otherwise is the soul and spirit of a flower, may be spared when it arrays itself in this scarlet glory. It is a flower of thought and feeling, too; it seems to have its roots deep down in the hearts of those who gaze at it. Other bright flowers sometimes impress me as wanting sentiment; but it is not so with this.

Hawthorne's observations of the scarlet color and the lack of fragrance in cardinal flower are relevant to the flower's mode of pollination. Nature does not bestow such beauty upon a flower without purpose. Birds, in general, have a poor sense of smell, but keen eyesight. Red is known to be a color attractive to birds, with hummingbirds in particular finding this color irresistible. Indeed, the sound of the whirring wings of hummingbirds near my head has startled me more than once when I happened to wear something red while working in tropical rainforests. Hummingbirds will investigate anything red as a possible source of nectar. With their long bills and longer tongues, hummingbirds are ideal pollinators of the flowers of *Lobelia cardinalis*. The combination of red color and tubular flowers are thought to represent the embodiment of the "hummingbird syndrome." These characteristics are found in flowers that have evolved so as to encourage a constancy of this specific type of visitor. Yet this preference is not an *innate* characteristic of birds, but rather a *learned* response. Red flowers are seldom visited by pollinators other than hummingbirds and butterflies. The common belief that bees are unable to see red is incorrect; bees can detect red, but they cannot easily distinguish it from the surrounding green foliage. Thus hummingbirds, which can spot red at a great distance, have learned that red flowers will usually provide a reliable source of untouched nectar. It is the nectar, accessible to the long tongues of hummingbirds, that is the key attraction; the red coloration is merely the "hook" that advertises its presence. Despite this "preference" for red flowers, hummingbirds visit flowers of many other colors: the lavender flowers of creeping phlox (*Phlox stolonifera*), purple coneflower (*Echinacea purpurea*), great blue lobelia (*Lobelia siphilitica*), and many others in other parts of the Americas.

I have watched and photographed hummingbirds visiting the red flowers of cardinal flower. Their visits are quick, and it is difficult to see just how the pollen is being transferred from flower to bird. Looking at my images enlarged on the computer monitor, I was shocked to see that of those hummingbirds I had photographed,

Fig. 186. This ruby-throated hummingbird is taking nectar through an opening in the corolla at the base of a cardinal flower rather than by inserting its bill into the long tubular portion of the flower. It, thus, fails to touch the reproductive structures above with its head, and will not effect pollination.

Fig. 187. A ruby-throated hummingbird visiting the flowers of great blue lobelia for nectar. Because it is able to reach the nectar without contacting the reproductive parts with its head, it is not a successful pollinator.

Bellflower Family

all had "stolen" the nectar! They had not entered the flower by "legitimate," that is, expected means, which would have dusted their heads with pollen that then could be transported to the stigmas of subsequently visited flowers. Rather they probed their bills and tongues into slits at the base of the corolla to reach the nectar directly (fig. 186). Although this was true in every one of my photographs, I have certainly seen enough photos by others to know that hummingbirds generally *do* follow "proper" procedure. Perhaps I had happened upon a gang of ruby-throated thieves. I have also seen hummingbirds visiting the blue flowers of *Lobelia siphilitica*, which has nectar accessible to the birds, but because the flower tube is shorter than that of the red-flowered lobelia, the birds do not contact the part of the flower that would result in pollination (fig. 187).

If you are an observer of local wildflowers in the Northeast, you will likely have noticed the paucity of truly red flowers. Among native red-flowered species are Indian-pink (*Spigelia marilandica*), in the logania family; beebalm (*Monarda didyma*, in the mint family (fig. 188); fire-pink (*Silene virginica*), in the pink family; and columbine, (*Aquilegia canadensis*), in the buttercup family, all of which can join cardinal flower in this very small "red hat club." A visit to the western United States or south of the border would yield many more candidates. Related to this discrepancy is the presence in eastern North America of only one species of hummingbird, the ruby-throated hummingbird, whereas the rest of the continental United States

Fig. 189. Eaton's firecracker (*Penstemon eatonii*) is one of several members of this genus in the western United States with tubular red flowers.

has ca. 14 additional breeding species of hummers. Thus, in western states many species of wildflowers (across a range of families) have evolved to be attractive to these specialized pollinators, and red flowers are far more numerous, both in species and in number of plants e.g., Eaton's firecracker (*Penstemon eatonii*), in the plantain family (fig. 189); claret-cup (*Echinocereus triglochidiatus*), in the cactus family (fig. 190); scarlet gilia (*Ipomopsis aggregata*), in the phlox family (fig. 191); scarlet sage (*Salvia coccinea*), in the mint family (fig. 192); and many others.

Fig. 188. Looking down on a head of red beebalm flowers (*Monarda didyma*), a summer-flowering northeastern species attractive to hummingbirds.

Fig. 190. Native to deserts of the Southwest, claret-cup (*Echinocereus triglochidiatus*) is a species of cactus with spectacular red flowers.

Cardinal Flower

Fig. 191. (Left) The flowers of scarlet gilia (*Ipomopsis aggregata*) look like exploding skyrockets.
Fig. 192. (Right) Scarlet sage (*Salvia coccinea*) ranges widely from southeastern United States and the Caribbean to northern South America. Its red flowers attract both hummingbirds and butterflies.

In order for hummingbirds to sip nectar from flowers, it must be of a viscosity dilute enough to be sipped with their specialized tongues, but at the same time high enough in sugar content to justify the energy expended in first finding the flower, and then hovering by it while feeding. The main components of floral nectars are water and sugar, with sugar concentrations ranging widely between 10% and 75%. Chemically, nectar sugars are of three different types: sucrose, hexose, and fructose. Most often, the nectar favored by hummingbirds is the highest in sucrose content, with a concentration of 20%–26% sugar (or sometimes higher). Hummingbirds must visit many hundreds of flowers each day to maintain their high energy requirements; they must consume one and a half to three times their body weight in nectar and insects.

J. H. Brown and A. Kodric-Brown conducted a study in an area of Arizona where several species of plants with similar tubular, red flowers occur. Interspersed with these species, they were surprised to encounter a population of *Lobelia cardinalis* that produced no nectar but had the classic cardinal flower color, thus blending in with the surrounding red flowers. Hummingbirds visited the red flowers indiscriminately, expecting all of them to provide a nectar reward and thus found themselves victims of the nectarless *Lobelia*'s deception. In attempting to understand this phenomenon, the researchers hypothesized that because *L. cardinalis* flowered toward the end of summer in Arizona when many young hummingbirds were migrating south, the naive juvenile birds were not experienced enough to recognize that the individuals in this population were deceivers. By failing to produce nectar and thus reducing its energy expenditure, this "clever" cardinal flower accomplished two tasks: it cut the "cost" of attracting pollinators, and it still managed to be pollinated.

Earlier I mentioned that butterflies also frequent red flowers (as well as many non-red flowers). In fact, I probably have seen as many butterflies as hummingbirds visiting cardinal flowers. In my tristate area of New York, New Jersey, and Connecticut, the most prevalent of these is the spicebush swallowtail, *Papilio troilus* (fig. 193). However, the butterflies do not deplete the nectar supply in any meaningful way and, therefore, do not affect hummingbird visits. Hummingbirds, however, are territorial and will often chase the butterflies from "their" flowers.

I have both cardinal flower and great blue lobelia (*Lobelia siphilitica*) growing in my garden, their tall racemes of red or blue flowers providing a contrasting note of color to the yellows of black-eyed Susan (*Rudbeckia* spp.) and goldenrod (*Solidago* spp.) and the white flowers of *Boltonia asteroides*. In the wild, the two *Lobelia* species occasionally hybridize, producing

Bellflower Family

Fig. 193. A spicebush swallowtail butterfly (*Papilio troilus*) nectaring at the flowers of cardinal flower. It occasionally comes into contact with the reproductive parts of the flower and may play a small role in pollination.

plants that have deep pink or magenta flowers. As with many other wildflowers, both species of red and blue lobelia occasionally have white flowers (figs. 194–195), and rarely, cardinal flower may have pale pink flowers and great blue lobelia pale blue ones.

While similar in overall growth form and basic flower shape, there are differences between these two showy *Lobelia* species that result in different pollination systems. The most noticeable of these is flower color, red vs. blue. Rather than attracting hummingbirds and butterflies to perform the job of pollination, blue lobelia relies on its showy blue flowers to attract bees. Bumblebees are the most effective pollinators, being both strong enough to push their way into the flowers to reach the nectar, and large enough to inadvertently come into contact with the reproductive parts while doing so (fig. 196). Bumblebees, not to be outdone by the hummingbirds, are sometimes guilty of taking advantage of the corolla's basal slits to obtain nectar the easy way (fig. 197).

If you compare the flowers of great blue lobelia with those of cardinal flower, you will notice that the former are not as elongate and have a shorter tubular portion than the latter (fig. 198). Both species have openings at the base of their corolla tubes (the slits referred to above), and each has two "lips," the apparent lower one with three lobes, and the upper with two. The word "apparent" is used here because (similar to many members of the unrelated orchid family) as the flower opens, it rotates 180° from its position when in bud. The resulting mature flower is thus inverted from its original position in bud. Even when past flower, the two species may be distinguished by looking carefully at the dry, brown fruits. Those of cardinal flower have a much longer remnant of the staminal tube and style, and their individual flower stalks (pedicels) have small bracts attached at the base, whereas the bracts of *L. siphilitica* are attached midway up the pedicel.

Fig. 194. (Left) Like many other species, cardinal flower will sometimes produce white flowers. **Fig. 195.** (Right) A white-flowered plant of great blue lobelia.

Cardinal Flower

Fig. 196. A large bumblebee (*Bombus* sp.) is pushing into the flower of great blue lobelia to obtain nectar. As evidenced by the masses of yellow-orange pollen on its hind legs, it has also been collecting pollen at other species of flowers.

Fig. 197. Some bumblebees "steal" nectar from the flowers of great blue lobelia by probing between the slits in the base of the corolla to obtain nectar directly without "paying their dues" by entering in such a way as to cause pollination.

Fig. 198. A lateral view of the flowers of great blue lobelia showing the two lips of the flower. The corolla tube is shorter than that of cardinal flower. The staminal tube extends just slightly out of the opening of the flower and between the two lobes of the upper lip.

Bellflower Family

The tall inflorescence stalks of lobelia begin flowering from the bottom and proceed upward, with the bisexual flowers first opening in the male phase and then entering the female phase, with no overlap between the two phases in each flower. Therefore, the flowers on the bottom half of the inflorescence are already in the female phase while the newer flowers on the upper half are in the male phase. Flowers with this pattern of flowering are said to be protandrous (meaning male first). The stamens are fused together into a tube with the anthers opening toward the interior of the upper portion of the tube. When a hummingbird contacts the staminal tube, a valve at the end of the tube is triggered, and the anthers release pollen into the interior of the tube. The loose pollen gets dusted onto the bird's head as it probes for nectar. More than one visit is necessary to release all the pollen contained in the anthers. Hummingbirds usually visit several flowers on an inflorescence, working their way upward toward the male phase flowers, which produce significantly more nectar than the female phase flowers. The birds carry pollen from the last flowers they have visited (those nearest the top) (fig. 199) on one inflorescence and transport it to the first-visited flowers of the next inflorescence (the lower, female phase flowers). During the female phase, the style (the extended part of the female pistil) pushes upward out of the tube and spreads its two red stigma lobes (fig. 200), which at that time are receptive to pollination. Pollination cannot occur prior to this stage, and the plant is almost always outcrossed with pollen from another plant. Experimental crosses have shown *Lobelia cardinalis* to be self-compatible, but self-fertilization rarely occurs in nature. The plant's own pollen can collect on the fringe of hairs surrounding the tip of the stigma when it passes through the staminal tube (fig. 201). It may then be

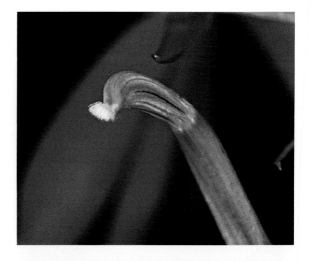

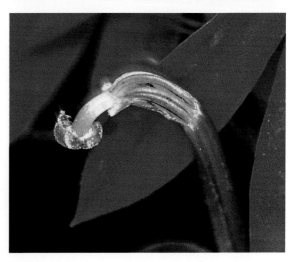

Fig. 199. (Top) The gray tip of a staminal tube of cardinal flower comprises the fused anthers; the anthers open toward the inside of the tube, releasing their pollen when triggered by touch. There is a small white "mustache" of white hairs at the tip of the staminal tube. **Fig. 200.** (Middle) The freshly spread lobes of the stigma (looking like collagen-enhanced lips) have just emerged from the staminal tube. **Fig. 201.** (Bottom) In this photo, the female style has further elongated, and the stigmatic lobes are fully reflexed. The fringe of hairs at the base of the stigma is evident, as are the pollen grains that have accumulated on the stigma.

Cardinal Flower

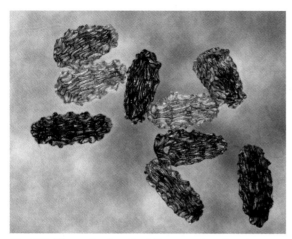

Fig. 202. The seeds of cardinal flower are less than 1 mm in length. This photo was taken through a dissecting microscope, affording a view of the highly sculptured seed coats.

contacted by visitors and rubbed onto the stigma of the same flower or carried to other flowers on the same inflorescence.

Successful pollination and fertilization of the ovules results in hundreds of tiny seeds per fruit (fig. 202)—up to 11,000 per plant! The seeds are light enough to be carried short distances by wind and water. In addition to sexual reproduction, cardinal flower can reproduce by offshoots from its basal rosette (in cultivation, these may be separated and planted individually), and by the process of layering (if part of the stem is bent down and buried, rooting can occur in that section, giving rise to new plants). Cardinal flower is a short-lived perennial (each offspring rosette can survive from as few as three, to as many as a dozen, years). Therefore, reseeding is important, especially so because this beautiful species is in a state of decline as a result of habitat destruction when wetlands are drained and filled for development.

This beautiful, conspicuous plant caught the attention of early European explorers, and by the 1600s, cardinal flower was being cultivated in Europe. Of course, such attention-grabbing plants have had legends developed around them: the roots of both cardinal flower and great blue lobelia are said to serve as love potions, capable of restoring love between feuding couples and even of bringing love to the lives of elderly women. Native Americans utilized various parts of the plants to treat bronchial ailments, digestive problems, and, as reflected in the epithet of great blue lobelia (*L. siphilitica*), in the treatment of syphilis.

Lobelia cardinalis is rarely browsed because of the toxicity of its milky white sap. (Unfortunately, the deer that frequent my backyard did not get that message, and I had to move our plants into a fenced part of the garden.) The principal toxic alkaloid in cardinal flower is lobinaline 19, which is related to a form of lobeline 1, the active component found in the small, blue-flowered *Lobelia inflata* (Indian tobacco [fig. 203]). As its common name suggests, Indian tobacco is a plant long used by Native Americans for its stimulant effect when smoked, and for its medicinal properties. It is currently under study for medicinal uses in the treatment of various drug dependencies and of central nervous system disorders.

Fig. 203. A relative of cardinal flower and great blue lobelia, Indian tobacco (*Lobelia inflata*) has much smaller white-to-blue flowers. It is being investigated for potential medicinal properties.

Aster Family

Fig. 204. The flowers of chicory are a deep blue when they begin to open in early morning. The ray flowers begin to lengthen during the nighttime hours but do not open until the sun rises.

Chicory

Fig. 205. Chicory is a common roadside "weed" often growing in association with Queen Anne's lace. Both are Eurasian species that have become naturalized throughout much of North America and other parts of the world.

Chicory
Cichorium intybus L.
Aster Family (Asteraceae)

Of all the plant families in the world, Asteraceae vies with the Orchidaceae regarding which family has the greatest number of species, with pundits weighing in on both sides. The Angiosperm Phylogeny Group website currently lists Orchidaceae as the leader with an estimated 26,000 species, followed closely by the Asteraceae with 25,040 species. Even these numbers are approximations because new species are yet to be discovered and described in both families. Aster family members are found almost everywhere on earth, from the Arctic to the tropical rainforests of both the Old and New Worlds, and as far south as Tierra del Fuego; they range in habit and habitat, from small alpine plants (e.g., edelweiss [*Leontopodium* spp.] in the Alps), to woody trees (e.g., the "daisy" trees [*Scalesia* spp.] in the Galapagos), to shrubs (e.g., brittlebush [*Encelia farinosa*] in the deserts of the Southwest). The aster family is also known as the composite family, and before family names were standardized by the addition of "-aceae" to the root of the generic name (e.g., *Aster*) to form the family name (resulting in Asteraceae), the family was called the Compositae, a descriptor of its inflorescence type.

One reason for the family's cosmopolitan distribution is that many of its species produce seeds (actually a type of one-seeded fruit called an achene) modified for wind dispersal. In Asteraceae, the achene is specialized by having the calyx fused to the fruit and often modified into a "parachute" that serves as an aid in long-range dispersal, as in a dandelion. The seeds of chicory, however, utilize other means of dispersal, which will be discussed below. Chicory's trans-Atlantic dispersal from its Eurasian homeland to our part of the world as well as to other temperate regions is a result of human movement, either deliberate (for culinary purposes) or accidental.

Habitat: In North America, chicory is found in almost every kind of soil along sunny roadsides, railroad track margins, and other disturbed areas. In its native range, it grows in sunny meadows and along open roadsides.

Range: Chicory occurs in all the lower 48 states in the United States and all provinces of Canada, but not in the colder, northern Canadian territories.

Aster Family

If you have read the chapter on Virginia bluebells in my *Spring Wildflowers of the Northeast* book or the gentian chapter in this volume, you are aware that I am enamored of blue flowers. Thus, the sight of chicory brings joy to my heart—but joy of a fleeting nature since each flower has such a brief time to flaunt its deep blue color (other names for chicory include blue sailors and ragged sailors, referring to the color of the flowers and their ragged tips). The best time to enjoy a chicory flower's intense blue is soon after the sun's rays strike the plant in the morning and the flowers begin to open (fig. 204). As the sun rises higher and the temperature increases, the color begins to fade, until by noon or soon after, it is a pale semblance of its former glorious self—becoming almost white—before closing forever. Chicory flowers might not open at all on a dark, overcast day or, conversely, remain open throughout the day if skies are cloudy-bright. Although the beauty of each individual chicory flower head (capitulum) is short-lived, the plant continues to produce a few flowers per stem on a daily basis throughout the summer season and into autumn.

Chicory plants are particularly attractive when growing intermixed with Queen Anne's lace, which like chicory is a common Eurasian emigrant that sometimes lines road edges for hundreds of yards (fig. 205; and see fig. 601 in the Queen Anne's lace account). Such sights beautify the roadsides in the morning, but when traveling home on the same road in the afternoon, one would never know that chicory's blue flowers had ever existed; their apparently flowerless, wiry stems blend into the grasses and other vegetation and become "invisible." The German name for chicory: *der wegewart*, meaning "wayward" (as in "difficult to control"), is especially apt. One reason chicory does well along highways in the Northeast is that our acidic rainfall leaches calcium from concrete and limestone used in road building and causes the nearby soil to have a higher pH level. Alkaline soils favor the establishment of certain plant species, such as chicory.

Rather than being wind-dispersed, chicory seeds (fruits) are small, wedge-shaped achenes, topped with short bristles. The seeds are relatively heavy, and even if they were equipped with "parachutes" like those of dandelion (*Taraxicum officinale*), they would not be light enough to be transported by the wind. Instead, the seeds of chicory remain within the bracts of the closed floral heads along the stiff, ribbed stems (fig. 206). Goldfinches and other small birds are reported to visit the plants to feed on the fruits. Chicory stems remain upright well into winter, and any seeds not consumed by birds tend to be shaken out by the wind or other movement resulting in colonies of chicory plants developing near the mother plant. If you view chicory as a weed needing to be eliminated, just cutting the plants down to the soil level will not kill them; the stout underground taproot remains viable and will continue to send up new shoots. Such cutting will result only in plants that are capable of flowering when barely a few inches tall. To eliminate chicory that is well established requires deep digging in order to remove every bit of the root; even a small remnant is capable of growing into a new plant.

On the other hand, in large numbers chicory plants can add a welcome touch of blue to a flower garden (occasionally a chicory plant will bear white flowers [fig. 207], and rarely, pink). However, if planted in a garden, flower heads must be snipped off frequently to prevent their going to seed, which can occur in as little as two weeks. The plant easily seeds itself and may take over more than its allotted share of the garden if the seed heads are not removed.

Fig. 206. (Left) Occasionally the flowers of chicory are white rather than blue, and even more rarely, pink.
Fig. 207. (Right) The remnants of several old brownish chicory flowers can be seen on this plant. The green bracts beneath the flower heads will hold the developing fruits until they are either eaten by small birds or scattered by the wind.

Chicory

There are gardeners who cultivate wild chicory (*Cichorium intybus* var. *sativum*), for its long, thick, brown root, which can be dried and ground to be used as an admixture with coffee or even as a coffee substitute (another common name for this plant is "coffeeweed"). The slight bitterness is an acquired taste and is particularly popular in France, where the practice of mixing the two ingredients to prepare a hot beverage began in the nineteenth century and continues today. In New Orleans, chicory coffee is a favorite beverage, especially as served at the popular Café du Monde (figs. 208–209), and one may buy the mixture by the pound for home consumption. When New Orleans was under French rule in the early eighteenth century, coffee from Caribbean plantations was imported to North America through the port of New Orleans, and coffee houses flourished in the old city. During the Civil War, when Union naval blockades prevented ships from entering the mouth of the Mississippi and coffee was difficult to obtain, the mixture of chicory with coffee became an important means of extending the limited amount of coffee available. Ground, roasted chicory root is said to taste somewhat like coffee but, being caffeine-free, it does not provide the energy boost of true coffee. Not being a coffee drinker, I can't speak from experience, but some profess to prefer their "coffee" to be straight chicory or a mixture of chicory and coffee, rather than pure coffee itself. Because chicory is less expensive than coffee, its use again gained in popularity during both the Great Depression and World War II.

Prior to its use as a coffee additive, chicory found use as both a food and a medicine as far back as Egyptian times when it was known as *chicouryeh*. Several iterations of that name, from the ancient Greek *kikhórion*, to the Middle French *chicorée*, have resulted in our English name for the plant: chicory.

Though bitter, the young leaves of chicory are eaten in salads. If the plants are covered with mulch when they first appear in spring, the leaves remain white (they do not develop chlorophyll without the presence of light) and will be less bitter; cooking can remove some of the bitterness in older leaves. A common practice in Europe is to bring the root indoors in winter, pot it up, and place it in a warm, dark spot. The resulting white leaves, called *witlof* (white leaf) in Dutch, are considered a delicacy in spring salads. Chicory is also among the wild herbs collectively called *horta*, gathered in Greece for culinary purposes. Cultivated forms of chicory have been developed, and their leaves, known as Belgian endive and radicchio, are eaten as salad greens. Their wild chicory origin accounts for the bitterness of these greens.

Unverified medicinal claims for the efficacy of the chicory root in treating such maladies as gout, rheumatism, and cancer led to its traditional use as a folk medicine. In veterinary medicine, a preparation made from the root is used in treating livestock suffering from intestinal parasites.

French colonists brought chicory with them to Canada in the early eighteenth century, and by midcentury it was well established in the environs of

Fig. 208. The famous Café du Monde in New Orleans is known for its chicory coffee—and its beignets.

Fig. 209. Coffee-chicory mixture is for sale at Café du Monde and other markets in New Orleans.

Aster Family

Montreal. Although it was deliberately imported into Massachusetts by the then governor, James Bowdoin, in the 1780s, perhaps its most famous advocates were two of our founding fathers, George Washington and Thomas Jefferson. Both Washington and Jefferson were avid gardeners and eager to try new crops when offered seeds from fellow horticulturalists. They found chicory to be an excellent fodder crop, growing quickly and easily without need for cultivation once established. Despite their hearty endorsement, chicory never gained enthusiasm among farmers, and alfalfa became the green forage crop of choice.

Chicory is generally a short-lived perennial plant, beginning as a rosette of (usually) pinnately toothed leaves (fig. 210) that measure 10–20 cm (4–8 in.) long and resemble those of a dandelion (but are hairy beneath); they grow from an underground taproot. From this rosette arise one to several tough, wiry, upright stems, 0.5–1 m (2–4 ft.) tall, bearing small, clasping leaves with clusters of three to five sessile, or nearly sessile, oblong buds in the axils of the upper leaves. A well-established plant may produce up to 150 flowers over the course of a summer. Chicory flower heads are composed solely of ray flowers attached to a common receptacle (fig. 211). These ray flowers have been modified from the more common tubular flower type that has its petals fused into a tubular corolla. Early in the ray flower's maturation, a split develops, extending nearly to the base of the floral tube, thus allowing the tube to spread into a straplike ray that, at its apex, bears five points, each representing what would have been a lobe of the original tubular corolla. While the ray flowers of the head of a sunflower (fig. 212) are sterile (without reproductive parts) and serve primarily as an attractant to insects, those of chicory have stamens attached within the base of the ray, where it remains fused for a very short distance; the anthers of the five stamens are fused to each other, forming an elongate, dark blue tube. These ray flowers are perfect, meaning they are bisexual, with the female part (the pistil), in addition to the aforementioned male stamens.

The reproductive organs of chicory are interesting (and typical of Asteraceae flowers): the anthers are closed when the flower first opens (fig. 213), but they soon dehisce (open), shedding their pollen to the inside of the anther tube; the elongating style grows up through the anther tube, acting as a plunger, pushing the pollen up and out of the anther as the style grows. Hairs on the style aid in this action. In the process, the style becomes coated with pollen. Only then does the style begin to split at its tip to expose the receptive stigmatic surfaces within (fig. 214), thereby presenting a "virgin" surface on which insects can deposit pollen from other chicory flowers, resulting in cross-pollination. As the flower ages, the styles continue to elongate (fig. 215) until the curved stigmas curl even further back such that they may come into contact with pollen remaining on the style. When pollen is available for pollination through a secondary means (other than directly

Fig. 210. The basal leaves of chicory resemble those of dandelion (another member of the Asteraceae, subfamily Cichorioideae), but they do not appear until spring; those of dandelion may remain throughout the winter as a low rosette. The upper leaves of chicory (not visible in this image) subtend the flower heads and are much smaller.

Chicory

Fig. 211. (Left) Chicory flower heads (capitula) grow along the stem in clusters of usually 3–5 heads. What appear to be petals are actually individual bisexual ray flowers, with each strap-shaped corolla united at the very base. Above the flowers in this image, an oblong bud can be seen along the upper part of the left side of the stem (the bud is the one closest to the stem). To the left of the bud is the involucre of a head that has finished flowering and shed its flowers. The fruits will develop within this protective structure.

Fig. 212. (Right) These sunflower heads are another type of composite flower (characteristic of a different subfamily) having numerous small, tubular bisexual disk flowers in the center and many petal-like sterile (lacking reproductive organs) ray flowers surrounding them. (See also fig. 80 in the species account on asters).

from the anther itself) the method is termed secondary pollen presentation. Secondary pollen presentation may serve as a backup mechanism that allows for pollination by self-pollen if a flower hasn't already been cross-pollinated. Although chicory is generally considered self-incompatible, self-pollination can—and

Fig. 213. When chicory flower heads first open, the five stamens of each ray flower are attached at the base of the corolla with their five anthers tightly fused into a tube. The anthers soon dehisce (open) inward, shedding their pollen toward the interior of the anther tube.

Fig. 214. Shortly thereafter, a bifid female style begins to emerge from the tip of each anther tube, carrying with it white pollen from the inside of the anther tube. The two stigmatic surfaces soon become receptive to pollen from another flower.

Aster Family

Fig. 215. By 10:30 a.m., the styles have fully emerged, and the stigma lobes are widely spread.

does—occur in this manner. This was demonstrated in a study carried out in South Carolina by Michael Cichan. About one-third of flowers that were experimentally hand-pollinated with self-pollen yielded viable fruit. Cichan showed that fruit set was the result of sexual reproduction rather than agamospermy (the formation of seeds without fertilization—as in the case of dandelions) by comparing his self-pollination results with those of flowers in which their reproductive parts had been manually removed to prevent pollination; none of those flowers produced fruits. These results held true even for flowers on the same plant that were treated using the two different methods. The mechanism for this variance was not determined.

The numerous ray flowers of chicory form a head that is about 1.5 inches in diameter. Since each of the 15–20 ray flowers has the potential to produce a seed, a plant bearing 150 flower heads is capable of yielding up to 3000 seeds in a season; however, many of the normal-appearing achenes do not actually contain embryos and thus cannot germinate and form new plants. Just beneath the flower head are two rows of bracts, the outer ones shorter and more spreading than the inner (fig. 216). These make up the involucre, a feature often useful in identification of members of the Asteraceae. By looking carefully, you can see that the involucral bracts are covered with hairs, each tipped with a sticky drop (see fig. 216). These drops may serve to protect the flowers' nectar from pilfering ants.

The aster family is so large that it is divided into subfamilies, and further subdivided into tribes. Flower heads with perfect ray flowers, like those of chicory, are called ligulate heads and occur in the subfamily Cichorioideae of the Asteraceae. Other familiar members of this subfamily having ligulate heads include hawkweed (*Hieracium* [fig. 217]), dandelion (*Taraxicum*), and lettuce (*Lactuca*). This group shares in common the presence of white sap, which in chicory may be seen by cutting the hollow stem or the main veins of the leaves.

In other tribes (groups of genera having certain physical features in common) of this large family, flower heads (technically called capitula) may lack ray flowers entirely and consist of only disk (tubular) flowers; such capitula are called discoid heads and include ironweed (*Vernonia* spp.) and blazing star (*Liatris* spp. [fig. 218]). The flower heads of some similar groups have modifications of this arrangement and are termed disciform. The other common type of flower head that occurs in the aster family is termed radiate; it has both pistillate (female) ray flowers and perfect disk flowers as may be seen in ox-eye daisies (*Leucanthemum vulgare* [fig. 219]), with white ray flowers surrounding yellow disk flowers.

Chicory flowers are visited by flies (especially hover flies, many species of which are mimics of yellow and black wasps and bees [fig. 220], and, although hover flies are said to be most strongly attracted to

Fig. 216. Beneath the flower head are two rows of bracts, called phyllaries, which together make up the involucre. The outer whorl of bracts is shorter and more widely spread than the inner. The bracts are covered with glandular hairs, best seen with a 10× hand lens.

Chicory

Fig. 217. (Left) Hawkweeds such as orange hawkweed (*Hieracium auranticum*), are in the same tribe of the aster family as chicory and, accordingly, have flower heads comprising bisexual ligulate (ray) flowers.

Fig. 218. (Right) The flowers of blazing star (*Liatris* spp.) are all tubular and perfect (bisexual). Their heads are termed discoid since they have no ray flowers. The Peck's skipper on the inflorescence is the victim of a crab spider.

yellow flowers, they are frequent visitors to the blue-flowered chicory), bees (which *are* documented as being attracted to flowers with blue corollas), and small butterflies. The flowers offer both pollen and nectar as rewards.

My own observations, made over several hours on a sunny July morning, found hover flies and small, dark, solitary bees visiting the deep blue flowers by 8:15 a.m., presumably for nectar, before the styles had emerged from the anther tube (fig. 221). A metallic green bee was also an early visitor, and one bumblebee was seen flying around the flowers. In Europe, bumblebees have been reported to wait, and even hover, near chicory flowers prior to their opening. At 8:30,

Fig. 219. The flower heads of ox-eye daisy are typical of the tribe of Asteraceae having radiate flower heads. The many tubular yellow flowers in the central disk are bisexual, and the white ray flowers surrounding them are sterile. Thus, only the disk flowers form fruits. The disk flowers open initially from the outer edge of the disk, continuing inward toward the center.

Fig. 220. Hover flies (also called flower flies or syrphid flies) are common visitors to chicory flowers. They are usually observed feeding on pollen. Like many other syrphids, this hover fly (*Eupeodes* sp.) mimics a wasp, which serves to protect it from being attacked by birds or larger insects that fear it will sting. Hover flies are beneficial not only as pollinators in their adult phase, but as larvae they also prey upon plant pests such as aphids.

Aster Family

Fig. 221. This small bee was visiting chicory flowers early in the morning, before the styles had emerged from the anthers; no pollen was available, but the bee was probing for nectar with its proboscis.

Fig. 222. Another species of hover fly, *Toxomerus geminatus*, is inserting its labella into an anther tube in an attempt to scrape out pollen from the interior of the tube.

some of the styles were visible emerging from the anther tubes (see fig. 214). Some hover flies were using their modified mouth parts (labella), which are lined with tiny "teeth," to scrape the pollen from within the anther tube (fig. 222). They also regularly cleaned their legs of pollen, which they transferred to their mouths. Some hover flies feed only on pollen, others on both pollen and nectar.

Fig. 223. A bumblebee is flying from flower to flower carrying loose pollen scattered on its hairy body and can, therefore, act as an effective pollinator. The pollen packed into its pollen baskets (corbiculae) is not available for pollination.

The nectar of *Cichorium intybus* contains almost equal amounts of the sugars fructose and glucose (hexose sugars), but no sucrose. The tongues of these small hover flies are short, only 2–4 mm, but perhaps just long enough for them to reach the nectar held in the very short tubular base of the ray flowers; however, I did not observe them imbibing nectar. By 10:40 a.m., the styles were well emerged and carried on their surface the white pollen they had pushed up out of the anther sacs; the tips of the styles had split in two, thus exposing the inner stigmatic surfaces (see fig. 215). At that time, the flowers were being visited by many bumblebees (fig. 223), some small bees, many small hover flies, one large hover fly, and a single skipper butterfly. The corbiculae of the bumblebees were packed with white pollen (fig. 224). The flowers were still a sky blue, but not as dark as they had been earlier (see fig. 213). When I returned to the chicory roadside patch at 1:00, almost all the flowers were closed and dirty white (fig. 225). I cut a stem to take home, and new blue flowers opened the following morning, only to close that afternoon. This was repeated on the following day.

The tendency of certain flowers to open and/or close at specific times of day (or night) prompted Carolus Linnaeus to hypothesize a plan for a floral clock (his *Horologium Florae*). He envisioned a circular design divided into triangular sections planted with species of flowers that would open or close at particular times.

Chicory

Fig. 224. This bumblebee, its pollen baskets already full of white chicory pollen, is now probing the base of the flowers for nectar. As it does so, some of the loose pollen on its head and body will brush off on the flowers' stigmas, thus effecting pollination.

Linnaeus' clock began with the opening of goat's-beard (*Tragopogon pratensis*) at 3:00 a.m. and ended with the closure of daylilies (*Hemerocallis lilioasphodelus*) between 7:00 and 8:00 p.m. *Cichorium intybus* was the fourth species on his clock, opening at 4:00–5:00 a.m. One must realize that Linnaeus was using the floral opening and closing times as they occurred in his northerly latitude of Uppsala, Sweden (60°N), where the summer sun rose early and set late. Although Linnaeus is not known to have actually planted such a garden, floral clock gardens were later designed and planted at botanical gardens in other parts of Europe. The species used in Linnaeus' plan, of course, required modification according to the hours of daylight at the latitude of each botanical garden where such plantings were attempted. Soil types and altitude necessitated substitutions as well. As might be expected, the results were mixed; not only are flower opening and closing times responsive to day length but to climate and daily weather as well. As we have seen with chicory, the flowers may open late on cloudy days or in areas that are shaded in the morning—or not at all on very overcast and rainy days. Their closing times vary as well. Such was the case in the floral clock gardens—some flowers "behaved" as expected; others did not.

As with many other plants considered "weeds," chicory, when looked at closely, proves to be interesting, even fascinating, from both a biological and a cultural perspective.

Fig. 225. By noon to 1:00 on a sunny day, most flowers of chicory have faded from blue to white and started to close, their mission accomplished. In this image, the top bud on the upper part of the stem will probably open the following day (this is the same plant seen earlier in the day in fig. 211).

Dogbane Family

Fig. 226. A flowering plant of common milkweed with inflorescences at different stages of development. The lower inflorescences flower first; the uppermost inflorescences are still in bud.

Common Milkweed

Fig. 227. A sunny field in the Adirondacks that has been colonized by common milkweed.

Common Milkweed

Asclepias syriaca L.
Dogbane Family (Apocynaceae)

Common milkweed is appropriately named—it is one of our most abundant native wildflowers, has a milky white sap, and is considered a weed by many people. In the last few decades, however, people have begun to realize that common milkweed serves as the linchpin of an ecological guild of insects whose lives are intrinsically tied to it. In addition, milkweed's dusky rose flowers (fig. 226) emit a heavy, sweet scent that pervades the midsummer air with a rich floral perfume.

Habitat: Common milkweed grows in open, sunny meadows (fig. 227), along roadsides, and in waste places, as well as invading crop fields and their margins.

Range: *Asclepias syriaca* is found nearly throughout the eastern half of North America. In the United States it grows along the East Coast (with the exception of Florida) and in eastern Canada; it is found in Newfoundland, Labrador, and Nunavut. From there, its range extends westward to Saskatchewan and Montana in the north and to Texas in the south, with only a limited distribution in Arkansas, Louisiana, and Oklahoma. Its range has been extending southward over the past decades. Surprisingly, common milkweed is also found growing (apparently natively) in just one county (Marion) in Oregon.

The milkweeds and their close relatives were once classified in a separate family, the Asclepiadaceae. Today, with the advent of modern methods of taxonomic classification—those based on genetic analysis—the milkweed family has been demoted to a subfamily, the Asclepiadoideae, nestled within the dogbane family (the Apocynaceae). The two families were always closely allied and, at least in our area, share some obvious characteristics including (usually) opposite, simple leaves, milky sap, and two separate ovaries that are united at the apex and have swollen style heads.

The focus of this essay is the most common of our milkweeds, *Asclepias syriaca*, and the guild of insects dependent upon it. *Asclepias syriaca* is generally known as common milkweed, or simply "milkweed." Some additional milkweed species native to the Northeast are described in the sidebar that

Dogbane Family

Fig. 228. This view down onto a milkweed plant shows how the pairs of opposite leaves are offset from those above by just enough to allow all leaves to receive some sunlight. This leaf arrangement is described as decussate.

Fig. 229. The flower buds of milkweed are tightly compressed into ball-like inflorescences, which expand into umbels as the flowers open.

accompanies this chapter (images of some of them have been used to illustrate certain features of milkweeds throughout this essay). The genus *Asclepias* includes approximately 130 species, all native to the Americas, with more than 100 of them found in North America. Although readily establishing by their great volume of wind-dispersed seeds, patches of common milkweed are often strongly clonal, spreading by thick underground rhizomes. Their strong, fibrous stems may grow to be 3–6 feet tall. Broadly oval leaves (sometimes having a pointed tip) grow opposite each other along the stem, each pair almost, but not quite, perpendicular to the pairs above and below. This slight offset maximizes the photosynthetic potential of the lower leaves since they are not shaded by those above (fig. 228). The inflorescences arise in the axils of the upper leaves, beginning as tightly packed clumps of buds (fig. 229). As the buds mature, each develops into a flower at the end of a lax stalk (pedicel), resulting in a dense umbel of muted pink flowers (see fig. 226).

The flowers of common milkweed are unusual in their appearance as well as in their method of pollination. The five corolla lobes of the flowers are united at their bases and strongly reflexed downward, thus covering the calyx. At the center of the flower is a structure, the gynostegium, comprising fused male and female reproductive parts. The flat-topped style (female) is surrounded by the tightly appressed filaments of the stamens (male). From near the top of the filament column arises a crownlike structure called the corona, consisting of five nectar-containing "hoods," which alternate with the corolla lobes and exceed the height of the gynostegium. These hoods are further ornamented by short "horns" that originate at the base of each hood (fig. 230). Despite the success of milkweed's fragrant, nectar-producing flowers in attracting insect visitors, it is unusual to see more than one or two fruits develop from each inflorescence. Each milkweed flower has two separate ovaries and, thus, could produce two fruits, but generally only one develops to maturity. Milkweed plants within a clone are incapable of self-fertilization, and bees often remain within a single clone for an extended period, resulting in little opportunity for cross-fertilization and, thus, limited fruit production. The fruits, commonly called pods,

Fig. 230. This close-up of common milkweed flowers shows the magenta petals folded downward and the upright pale pink hoods (collectively called the corona). The sepals can be seen on the closed bud in the background.

Common Milkweed

Fig. 231. (Above) One hood of this common milkweed flower has been removed to better show the central gynostegium that consists of the fused male and female reproductive structures. The dark corpusculum joining the pollinia is visible in the slit on the left; the right pollinia have already been removed. The narrow, curved pink structures arching out of the hoods are called horns.

Fig. 232. (Right) A milkweed fruit is called a follicle. It opens along only one seam, which distinguishes it from a pod, a fruit that splits along two seams (as in a pea pod). The overlapping seeds are thus exposed, arranged in a fish-scalelike pattern. As they dry out, wind will blow them from the fruit, their white "parachutes," called the coma, carrying them away.

are more correctly termed follicles, a descriptive term used for a podlike fruit that splits open along only one side (fig. 231)—unlike a pea pod, which opens via two seams. As anyone who has happened by a milkweed field in autumn knows, when the seeds and their tuft of hairs (collectively called the coma), are released from the open fruits, they float on the wind like tiny gossamer ballerinas (fig. 232).

One might wonder why the epithet *syriaca* (having to do with Syria) was given to this plant of North American origin. The original error was made in the 1630s when, upon receiving a packet of plant collections from the Americas, botanists at the Paris Botanical Garden incorrectly identified specimens of common milkweed as being the same species as one collected earlier from the Mideast and named in 1601 as *Apocynum syriacum*. More than a century later when Carolus Linnaeus published *Species Plantarum* (1753), he created a new genus for this plant, and because of its reputed medicinal properties, named it *Asclepias* for the Greek god of medicine (the god Asclepius is depicted as carrying a single snake-entwined staff, the Rod of Asclepius, an icon that remains a symbol of medicine today). Linnaeus lists *Apocynum syriacum* as a synonym of this new name. Because he retained the epithet *syriaca*, the scientific name has caused confusion regarding the origin of the plant to this day.

Milkweeds and monarch butterflies (*Danaus plexippus*) have become inextricably linked in the minds of both schoolchildren and the general public. To raise a monarch caterpillar on milkweed and watch its transformation into a jewel-like chrysalis and then to the familiar orange and black butterfly is to witness a miracle of nature. Monarchs are probably the most widely recognized butterfly in North America, and the story of their reliance on milkweed as a larval food plant is widely taught as a classic example of coevolution in science classes at every level. Americans have become enamored of this butterfly and are distressed by its declining numbers—to the extent that in a survey of United States citizens, many indicated that they would be willing to donate substantial sums toward conservation efforts for this charismatic butterfly.

Dogbane Family

Fig. 233. A freshly laid monarch egg is a thing of beauty when examined through a magnifying lens. In general, a monarch will lay only one egg per plant since the first caterpillar  to emerge would eat any later-hatching caterpillars. Note the hairiness of the underside of the leaf that the tiny caterpillar must contend with before being able to feed.

Although the larvae (caterpillars) of monarch butterflies are dependent solely upon milkweed as their food source, this is *not* a mutual relationship. Indeed, milkweeds have no need for monarchs. On the contrary, milkweeds have evolved several defenses that deter caterpillars from feeding on their leaves. The leaves of common milkweed are hairy, especially on the underside where monarch eggs are laid (fig. 233). These inedible hairs (trichomes) make it difficult for the tiny, early stage caterpillars to access the edible leaf tissue. They must first "shave" the hairs from an area before they can chew into the leaf itself. Most monarch caterpillars don't survive their first day. The time it takes to "mow the lawn" of trichomes on the leaf in order to reach the edible leaf tissue puts the caterpillars at risk of death by dehydration. Of those that do succeed in reaching the leaf tissue, 60% die after their first bite, since that bite is rewarded with a burst of gummy white latex. If not quickly wiped from their mouthparts, the latex can prevent any further feeding by sealing the caterpillars' mouths permanently closed; or the sap may simply entrap them, such that they become stuck to the leaf surface and die. Latex is an important plant defense and has evolved in close to 10% of even distantly related plant families. In milkweeds the milky sap contains a quickly coagulating latex (fig. 234), as well as a selection of toxic cardenolides (cardiac glycosides), saponins (soaplike chemicals), and an enzyme that eats away at the interior of the caterpillar's gut. It's a wonder that any animal has evolved to feed on *Asclepias*!

But this is a coevolutionary arms race—an alternation of reciprocal adaptations to new defenses on the part of the plant with new strategies to overcome those defenses by the caterpillars. Monarch caterpillars have evolved a clever method of first chewing through small sap-carrying veins, forming a trench around an island of leaf tissue, thus cutting off the flow of sap that can enter that isolated section of leaf and preventing the sap from sealing the caterpillar's mouthparts. As the caterpillar matures, this technique involves larger and larger veins, and, ultimately, the main midrib of the leaf (fig. 235). The plant's additional defenses—the chemical weapons—present a further obstacle to the hungry caterpillars. Cardenolides are capable of sickening, or even killing, a much larger animal (including livestock and humans). Some of these chemicals remain in the leaf tissue that the caterpillars feed on. Monarch caterpillars would not normally be immune to their toxic effects (ingesting too much of this

Fig. 234. (Left) Milkweed's white sap (latex) flows freely from a freshly broken-off leaf, but it soon begins to coagulate. It presents another barrier that monarch caterpillars must overcome in order to feed.

Fig. 235. (Right) To feed on milkweed leaves, monarch caterpillars stem the flow of toxic, gummy latex to the leaf tissue by biting though larger and larger veins before feeding on the leaf blade supplied by those veins; in this case the leaf's main midrib is being severed.

Common Milkweed

compound can kill a monarch caterpillar), but because of a mutation in the gene that regulates certain aspects of the monarch's metabolism, the binding of cardenolides is greatly reduced. Monarch larvae are thus able to tolerate up to 300 times the amount of these compounds compared with caterpillars of butterflies that do not feed on milkweeds. Moreover, monarch caterpillars graze selectively, avoiding plants with a higher concentration of cardenolides—perhaps sensing the concentration in a given plant with the sensory organs on their legs. In spite of all this reciprocal back-and-forth, the proportion of monarch eggs that make it all the way through the five stages (instars) of caterpillar growth is a mere 10%.

The end result of this chemical warfare turns out to be positive for the caterpillar: the cardenolide-tolerant monarch caterpillars sequester some of the toxic compounds in their bodies, which thus afford them protection from avian predators throughout their life cycle. Most birds, after attempting to feed on a monarch caterpillar or butterfly, will exhibit an immediate, strong vomiting response. The birds quickly learn to associate insects with brightly contrasting aposematic (warning) coloration with a distasteful experience and seldom repeat their mistake. The toxicity of milkweed plants, even within the same species, varies, and therefore so does the toxicity of the monarch larvae that feed upon them. In a population of monarchs, approximately two-thirds of them are toxic enough to cause birds to become ill when they attempt to eat them. The less toxic one-third, because they are identical in appearance, are protected through a type of "self-mimicry" that deters birds from eating them because more than half the time that a bird has attempted to eat a monarch, it has become ill. Unfortunately for the monarchs, their warning colors do not seem to protect them from attack by invertebrates, and they often fall prey to other insects and spiders.

An unrelated butterfly, the viceroy (*Limenitis archippus*), is also a mimic of the monarch with a pattern and colors so similar to those of the monarch that birds are fooled by the deception and leave them unmolested as well. When examined closely, it is easy to tell the two butterfly species apart: a viceroy is smaller (and thus given the name "viceroy" to denote that its status is beneath that of a monarch), and viceroys have a distinctive black cross band across their hind wings that is lacking on monarchs (fig. 236). This type of mimicry—when a nontoxic species derives benefit from mimicking a toxic one—is known as Batesian mimicry, after the great nineteenth-century Amazonian explorer Henry Walter Bates (1825–1844), who developed this theory of mimicry. Batesian mimicry in monarch and viceroy butterflies has been known since 1902.

Fig. 236. (Left) A viceroy butterfly is smaller than a monarch and has a distinct black band across the center of its hind wings (compare with the monarchs in figs. 237, 253, and 254). Credit: U.S. Fish and Wildlife Service, William Powell. **Fig. 237.** (Right) An adult monarch butterfly (*Danaus plexippus*) visits the flowers of butterflyweed (*Asclepias tuberosa*) to gather nectar, but monarch butterflies are not important pollinators of milkweeds.

Dogbane Family

Fig. 238. The pollen of milkweed flowers is held in two "saddlebag" pollinia attached by wiry arms (called translators) to a small, dark body called a corpusculum, thus forming a wishbone-shaped structure. Here, I partially pulled out one unit of two pollinia (on the left of the flower) to show where it had been enclosed. The spiny legs of insects often catch on this structure if they slip down into the slits where the pollinia are held. The pollinarium (the entire unit of two attached pollinia), containing hundreds of pollen grains, is then carried off by the insect to another flower.

Fig. 239. A bee or other insect may accumulate many of these pollen sacs before depositing some of them at other milkweed flowers. This bumblebee, seen here on swamp milkweed, has amassed quite a collection of yellow pollinia on the legs visible in this image

The sight of monarch butterflies nectaring on milkweed flowers during the summer months (fig. 237, on butterfly-weed) might give the impression that milkweeds do eventually benefit from monarchs by having their flowers pollinated by the adult butterflies; however, this is seldom the case since monarchs are ill-suited for picking up and carrying milkweed pollen. The flowers of milkweeds, as seen in this image of swamp milkweed (fig. 238), are incredibly complex, and the most important pollinators turn out to be bees, particularly bumblebees. Bumblebee visits outnumber those of monarchs by 10–20 times, and their strong legs are ideal for pulling out and transporting the pollen-bearing sacs (pollinia)—each of which contains 100–200 pollen grains—from one flower to another. The two sacs resembling saddle bags are connected by threadlike arms (translators) to a dark-colored sticky gland called the corpusculum just above a groove (a stigmatic slit). If the visiting insect is strong enough (e.g., a bumblebee), the whole two-sac unit (pollinarium) is pulled out by hooks on the insect's legs and carried away to another flower. After a few minutes in transit, the drying translator arms rotate the pollinia, causing them to move into an optimal position for insertion into a stigmatic slit of a subsequently visited milkweed flower. Milkweed is a (near-) obligate outcrossing species, so pollinia deposited within the same clone generally will not produce viable fruit. The few minutes it takes for the translators to change position allow enough time for the insect carrying the pollinarium to travel to another clone. Despite this process, insects may accumulate several pollinia on their legs after visiting many flowers (fig. 239). The pollen can remain viable for up to three days. Weaker insects are sometimes unable to extricate themselves from the tightly appressed slits and dangle from a trapped leg until they die or are eaten by a predator (fig. 240).

Not only monarchs utilize common milkweed as a food source, but 14 other insect species feed upon milkweed as well, some of them exclusively. Because milkweed serves as an important ecological link in an intricate food web, and because some of the insects, even those in unrelated families, have evolved some of the same methods of dealing with the consumption of such a toxic plant, it seems warranted to discuss this

Common Milkweed

Fig. 240. If an insect is not strong enough to wrest its leg (or proboscis) free from the flower's tight stigmatic slit, it will become entrapped there, as has happened with this delicate moth.

group of insects in some detail. The insects that feed on milkweed are said to make up a guild, that is, a group of species that exploit the same resource. In this case the insects belong to widely differing insect orders that include five lepidopterans (the two most widely known being the monarch butterfly and the milkweed tussock moth; the other three are less common feeders), two species of true bugs, three species of beetles (including a weevil), three species of aphids, and a fly. All feed on milkweed, some exclusively, but they do not compete with each other because, depending on the insect species and its mode of eating, the members of the guild feed on different parts of the plant, or at different times of the year, or both, thus avoiding competition with each other. Each must have the capacity to cope with the various herbivore defenses that have evolved in the milkweed plant, especially the toxic cardenolides. The result of the feeding of any one of these species may sometimes evoke a response in the milkweed plant (e.g., an increase in toxicity) that can affect all species that will ultimately feed on the plant. Fourteen organisms sharing the same resource may seem like a large number, but that number pales when compared with the 100 species known to feed on tall goldenrod (*Solidago altissima*)—with 42 of those feeding solely on goldenrod and its close relatives!

The other primary lepidopteran that eats milkweed leaves is the larva of the milkweed tussock moth (*Euchaetes egle*)—actually a tiger moth. Like the monarch caterpillar, the tussock moth caterpillar sequesters cardenolides (employing the same trenching method to avoid becoming a victim of the sap). Their toxicity, along with their extreme hairiness and contrasting colors of orange, black, and white, help to protect the larvae from predation by birds (fig. 241). I have watched larvae of milkweed tussock moths feeding on the related dogbane (*Apocynum cannabinum*) as well, another species with copious latex.

When mature, the adult nocturnal-flying milkweed tussock moth has dark gray wings and retains its unpalatability. Warning coloration is not the deterrent in this case. The bats that prey upon these nocturnal moths are deterred only by very high doses of cardiac glycosides, so the tussock moths use an additional means of defense as well, that of emitting clicks from the tymbal organs on their thoraxes, which serve to confuse the click-emitting bats that might prey upon them.

Warning coloration affords protection to the four-eyed milkweed beetle (*Tetraopes tetrophthalmus*),

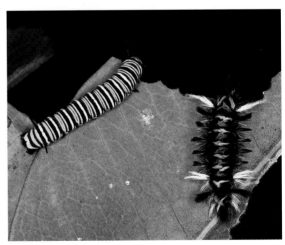

Fig. 241. The main other lepidopteran to feed on milkweed is the caterpillar of the milkweed tussock moth. Here, caterpillars of a monarch butterfly (on the left) and a tussock moth (on the right) are engaged in a speed-eating contest to see which can consume the greatest amount of leaf. Both species sport warning coloration.

Dogbane Family

Fig. 242. Milkweed beetles (*Tetraopes tetrophthalmus*), also bearing warning colors of red and black, feed on the leaves of milkweed, preferring the young tender leaves and sometimes the flowers. Their larvae feed on the roots of milkweed plants.

Fig. 243. Long-horned beetles (referring to their long antennae) have a notch in the perimeter of their eyes to allow for placement of the antennae. In the milkweed beetle this has been taken to the extreme such that the "notch" bisects the entire eye causing the beetle to appear to be four-eyed (the genus name, *Tetraopes*, means four-eyes).

which advertises its unpalatability with black-spotted red wing-covers (elytra) (fig. 242). The beetle is one of the long-horned beetles, meaning that it has long antennae. Long-horned beetles have the peculiar trait of their antennae growing from a notch in the circumference of their eyes. In the case of the four-eyed milkweed beetle, this is taken to the extreme, such that each antenna actually bisects an eye, giving the beetle the appearance of having four eyes (fig. 243). Milkweed beetles are most commonly seen in mating pairs (fig. 244). The adults of this species feed almost exclusively on the leaves and flowers of common milkweed, but the dozen other members of this beetle genus in the United States feed on different species of milkweed. The beetles' larvae burrow into the soil and feed on milkweed roots, which do not have a milky latex, but *do* contain cardenolides. When milkweed's roots are attacked by beetle larvae—or other creatures—they respond by releasing volatile compounds that attract soil-dwelling nematodes, which then prey upon the root-eating larvae.

The roundish milkweed leaf beetle (*Labdomera clivocollis*) does not sequester cardenolides as it feeds on the leaves but derives protection merely by having black and red-colored elytra, another example of Batesian mimicry (fig. 245).

Fig. 244. Aside from eating milkweed, most of the life of adult four-eyed milkweed beetles is spent in reproductive activity.

Common Milkweed

Fig. 245. A leaf beetle (*Labdomera clivocollis*) also feeds on milkweed but does not sequester its poisonous compounds. It is protected from predation by birds by its deceitful "warning coloration" pattern.

A third beetle, this one a seed-eating weevil (*Rhyssomatus lineaicollis*), uses a different strategy to avoid predators. It is small and dark gray (fig. 246), and when it senses danger, it drops quickly to the ground, becoming easily camouflaged against the soil. In spring, female adults feed on young leaves (fig. 247) and then lay their eggs in a vertical slit that they make near the base of the milkweed stem (fig. 248). The larvae feed on the stem tissue in spring and early summer. In late summer, a second wave of larvae, developing from eggs laid in the outer part of the follicle, consume immature seeds within the fruits.

Fig. 246. (Left) The milkweed weevil (*Rhyssomatus lineaicollis*) is cryptically colored and defends itself by falling quickly to the ground when threatened.
Fig. 247. (Right) The feeding of milkweed weevils leaves a distinctive pattern of holes in the young leaves. Like some other milkweed-feeding insects (e.g., monarch caterpillars), they first sever the veins in order to stem the flow of sap to the part of the leaf on which they then feed.

Fig. 248. Female milkweed weevils make a slit near the base of the milkweed plant's stem in which they lay their eggs in the late spring.

Dogbane Family

Two bugs, the large and small milkweed bugs (*Oncopeltus fasciatus* [fig. 249] and *Lygaeus kalmia* [fig. 250] respectively), also utilize a contrasting red and black pattern to deter predators. In spite of

Fig. 249. In this image, a mature adult of the large milkweed bug (*Oncopeltus fasciatus*) (on the left) advertises its unpalatability with its black and red colors. The immature nymphs sport the same colors. On the lower right of the milkweed follicle, two recently emerged adults have not yet developed their warning coloration. The nymphal exoskeletons from which they emerged can be seen still clinging to the lower edge of the milkweed fruit. Note the rough surface of this common milkweed follicle, which differentiates it from that of the similar, but smooth, follicle of purple milkweed.

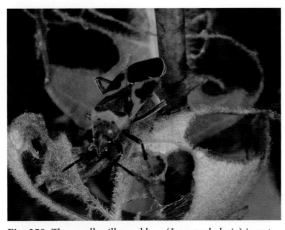

Fig. 250. The small milkweed bug (*Lygaeus kalmia*) is not very different in size from the large milkweed bug, but the different pattern of its black and red coloration can be seen by comparing the bug in this image with that of the large milkweed bug in fig. 249.

their common names, there is not much difference in the size of the bugs; rather they are distinguished by the pattern of the red and black markings. Both the nymphs and the adults feed on the seeds of milkweed plants by inserting their long proboscises into them, secreting enzymes to liquefy the seeds' interior and sucking out the contents. The seeds contain three times the concentration of cardenolides as the leaves, and this is sequestered by these seed bugs.

Five of the insects—the two principal caterpillars, the four-eyed beetle, the leaf beetle, and the weevil—utilize the same technique to avoid consuming excess sap, that of trenching, or cutting veins that supply an area of the leaf before feeding on it. Once this is done, even slugs have been observed feeding on the relatively toxin-free sections of leaf.

Rarely noticed is the larva of a fly (*Liriomyza asclepiadis*) that spends its entire larval stage as a leaf miner, feeding on leaf tissue between the upper and lower surfaces of a milkweed leaf. The only evidence of its presence is a discolored blotch on the leaf. Observant wasps may parasitize a larva living within a leaf by ovipositing eggs within its body. The fact that these larvae do not sequester cardenolides is of limited significance, as they are hidden from the sight of most predators. The wasp larvae that hatch from the eggs will consume the fly larva internally, causing it to die just before the wasp larva is ready to pupate. The larvae of leaf miners not parasitized emerge from the leaf and drop to the ground where they pupate in the soil.

Over eons, the same genetic metabolic modification noted previously to have evolved in monarch butterflies also occurs in the leaf beetle, the leaf-mining fly, and the two seed bugs.

Rounding out the guild of milkweed feeders are three species of aphids, all of which sequester cardenolides. These species avoid competition by being active at different times of the year. The most easily observed are the brightly colored *Aphis nerii* with their striking yellow-orange bodies and black legs (fig. 251). The aphids cluster together, sucking the sap from stems and inflorescences of common milkweed and almost any other member of the Apocynaceae. These aphids originate in the South, are blown northward, and do not arrive here until the latter part of the summer. They are preyed upon by ladybird beetles (fig. 252).

Myzocallis asclepiadis aphids feed exclusively on *Asclepias syriaca*. Because they are generally greenish

Common Milkweed

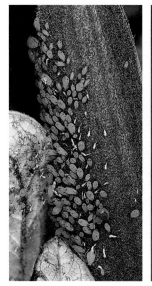

Fig. 251. (Left) The most obvious of the three species of aphids that feed on milkweed is the orange and black oleander aphid (*Aphis nerii*), seen here imbibing phloem from a fruit of swamp milkweed (swamp milkweed has a low concentration of cardenolides). Small droplets of white sap can be seen where the fruit has been pierced.

Fig. 252. (Right) Many species of ladybug visit milkweed plants to prey on the aphids.

and feed on the undersides of the leaves, they are seldom seen. Like other aphids, *M. asclepiadis* excretes "honeydew," a liquid waste product relished by ants. In this case, the ants don't take the honeydew directly from the aphids, but instead crawl about on the leaves beneath those being fed upon and lap up the droplets of honeydew excreted by the aphids above. Mid-July is the time to look for these aphids.

The third aphid, also greenish, is *Aphis asclepiadis*. In a classic mutualistic relationship, ants "tend" the aphids by stroking them in order to stimulate them to emit droplets of honeydew. The honeydew not only is a food source for the ants but, because it is laced with cardenolides, also provides the ants that consume the honeydew with some protection from predators. In return, the ants vigorously defend the aphids from predaceous insects such as ladybird beetles and their larvae, as well as any other visitor. They may be observed in mid-June.

For more than 100 years the amazing story of monarch migration was unknown. Some people thought that monarchs overwintered in the southern United States; others believed that the same monarchs seen in summer spent the winter at some undetermined location and then traveled all the way back to the northern United States and Canada. It was through the efforts of a Citizen Science project begun in the 1950s by Fred and Nora Urquhart that this mystery was finally solved. The Urquharts trained volunteers to capture and tag monarch butterflies by affixing small adhesive tags to their wings. On each tag the finders recorded the date and place of capture and sent that information to an address on the tag. After years of such efforts, a pattern began to emerge: monarch butterflies migrated south in the early fall, funneling through Texas into Mexico—but where in Mexico? That part of the question was not answered until 1972, when a citizen scientist in Mexico wrote to Fred Urquhart stating that he had seen thousands of monarchs in the mountains near Mexico City—including two butterflies that carried tags from previous captures in the northern United States. Urquhart, himself, traveled to Mexico in 1975 and located the monarch roosting area. The exciting discovery was the cover story of an issue of National Geographic early the following year. The exact locality, however, was kept secret, even from other legitimate butterfly researchers. One of these, Lincoln Brower, set out with another lepidopterist in January of 1977 to search for the area themselves. They not only found the monarch roosting sites, but they also encountered the Urquharts working at the site. Despite their disagreements, both scientists continued their research projects on monarchs for many years and contributed much to the understanding of monarch life history. It's now known that there are 12 large sites and about 30 smaller ones, all in close proximity in Mexico.

The migration of monarchs is fascinating and aspects of it are still not entirely understood. In March, as the days lengthen and the temperatures warm in the mountains of Mexico, monarch butterflies become reproductively mature and instinctively begin to mate. They then begin their long journey north, with most getting only as far as the southern part of the United States. When the first wave of adult monarchs return from their wintering grounds in Mexico (considering only those monarchs that originated in the Northeast at the end of the previous summer, not those from the western United States), the fertilized females seek out young milkweed plants in the southern part of

Dogbane Family

the United States on which to lay their eggs, releasing some of the stored sperm from the male as each egg is laid. Some monarchs, their reserves depleted, will oviposit on milkweed plants while still in Mexico, and others will die en route, thus not all the returning Mexican generation will complete the journey back to the States.

Although monarchs are thought to have originated in the tropics, they are now mostly temperate zone creatures. Mexico has a large number of milkweed species, but they are widely scattered, perhaps compelling monarchs to fly northward in search of food much like the tropical bird migrants that head to the United States and Canada each spring to find enough food to feed their young.

In most cases the returning monarchs lay their fertilized eggs on the still young plants of a south-central species known as Ozark (or spider) milkweed (*Asclepias viridis*) (see fig. 273 in the sidebar). They then continue to fly north, nectaring on a variety of early blooming flowers and continuing to lay eggs on milkweed until they have finally exhausted their reserves and die. With luck, the eggs laid by the returning migrants will hatch into caterpillars, feed, metamorphose, and continue northward as adult butterflies. Females with still unfertilized eggs are eagerly sought as mates by migrating male monarchs. (Male monarchs may be discerned from females by a noticeable black scent pocket on one of the veins of each hind wing [fig. 253]. Males have scent glands on the ends of their abdomens that they curl upward to transfer some of the scent to the pockets on the wings where they are absorbed by special scales. The liquid acts as a seductive perfume—a pheromone—to induce female monarchs to land for copulation).

The process of egg-laying and metamorphosis is repeated by two to four additional generations, each living for about a month, with some of the adult butterflies remaining where they emerge and others traveling further north. By late August to early September the last generation (the fourth or fifth) of the season will have begun its long journey south to the wintering grounds in Mexico, cued by changing day length and by the decline in milkweed. Having never been to Mexico, the monarch generation's ability to perform this feat of navigation is all the more incredible. A discussion of the possible methods of navigation to their winter destination lies beyond the scope of this essay, but for detailed information about monarchs and their life cycle, I highly recommend a book titled *Monarchs and Milkweeds: A Migrating Butterfly, A Poisonous Plant, and Their Remarkable Story of Coevolution* by Anurag Agrawal, a professor at Cornell University (see reference section for this chapter). The exciting part of this story is that there are still mysteries to be solved.

The final monarch generation of the season, maturing slowly due to low levels of juvenile hormone, lives for about eight months, flying up to thousands of miles to reach the mountainous region west of Mexico City where they convene by the hundreds of millions to spend the winter. The journey may take more than two months, depending on weather and wind conditions. Monarchs fly only during daylight and often cluster together on trees at night. During this time, it is important that migrating monarchs have an

Fig. 253. A monarch perches on an inflorescence of butterfly-weed (*Asclepias tuberosa*); it is a male, determined by the presence of two black spots on the veins of the hind wings. These spots function as scent pockets and emit aromas that are effective in attracting females of the species.

Common Milkweed

ample supply of nectar from flowers along their route. The butterflies use some of this nectar as an immediate source of fuel and convert the rest into lipids that are stored in their bodies to sustain them throughout the ensuing winter months. Many members of the aster family are in bloom at this time of year, including goldenrods and asters, and monarchs serve as pollinators of many of these species (fig. 254). A drought in the southern portion of the migration route may result in fewer plants in flower, thereby contributing to a decline in the numbers of butterflies that reach their winter roosting area.

Once arriving in the mountains of Mexico, the orange-winged monarchs cloak the fir trees, giving them the appearance of autumnal colors. The butterflies roost only on the middle sections of a particular species of evergreen fir tree—*Abies religiosa* (called *oyamel* in Spanish). The trees selected grow within a limited altitudinal range, where the temperatures generally remain in the 41°F–54°F range. There, the butterflies have the best chance of survival; freezing temperatures would kill them and warmer temperatures (55°F and above) would cause them to become active and expend the precious reserves needed to complete their return journey back to the southern United States. Each monarch butterfly weighs only 1 gram—the equivalent of a dollar bill, but their combined weight on a single tree may exceed 80 pounds. The butterflies spend about four months resting in a state of reproductive diapause, during which time their reproductive organs mature.

Generally, during this diapause phase, the poisonous monarchs are safe from predators since only a few bird species have evolved strategies that permit them to eat the cardenolide-infused butterflies. Of those studied, black-headed grosbeaks, whose winter range is in Mexico, have developed a resistance to cardenolides and eat the entire butterfly body (after removing the wings) without ill effect. Nevertheless, there is a limit to how much cardenolide they can consume, and they generally feast on monarchs for about a week and then have to abstain for a period to allow the level of toxins in their bodies to subside.

Another bird that preys on monarchs at the wintering site is the black-backed oriole, whose method of feeding differs from that of the grosbeak. Since orioles are more sensitive to cardenolides, they employ a method of slicing open the monarchs' bodies with their beaks in order to eat the insides while avoiding the more toxic cardenolide-laden cuticle (exoskeleton). This more careful method of avoiding the most toxic parts of the butterflies allows the oriole to participate in the butterfly bounty as well. If the birds detect higher than normal levels of the toxins in a particular insect, they discard it. Orioles selectively eat more males, which are lower in cardenolides. In addition to birds, mice are also known to prey upon the monarchs.

Initially, monarchs locate potential host plants by sight and then confirm that they have found an appropriate species by "tasting" the plant with special sensory organs on their legs and antennae. In some cases, they scratch the leaf surface with sharp spines on their legs to access the interior. Surprisingly, milkweeds have some chemicals that stimulate female monarchs to lay their eggs on the plants. This would seem counterproductive, but the hypothesis is that these

Fig. 254. This female monarch (note the absence of a black spot on the veins of the hind wings), nectars here on the autumn-blooming New England aster (*Symphyotrichum novae-angliae*) to fuel itself for its long journey to the mountainous fir forests of Mexico.

Dogbane Family

chemicals serve some other important (but undocumented) purpose for the plant. If the plant is verified as a milkweed, the butterfly then examines it to ascertain whether monarch eggs are already present. If not, she will deposit a drop of "glue" on the underside of a leaf and then oviposit on that drop. The pale yellow eggs are tiny but wondrously ornamented when viewed through a magnifying lens (see fig. 233). Each has 25 vertical ridges and about 34 horizontal rows. Monarchs generally lay only one egg per leaf—often only one per plant—unless there are few milkweed plants available, in which case they may lay several eggs on a single plant. The young caterpillars are cannibalistic, and the first one to hatch out on a plant avoids having to share the food source by eating any later-hatching caterpillars. To deter another female monarch from depositing an egg on the same plant, the first egg-laying female monarch will mark the leaf with a special scent that warns any subsequent monarchs not to deposit an egg there.

The caterpillars that do survive the first few days of life become voracious feeding machines, increasing in size by 2000 times, which necessitates shedding their outer skin five times during their 10–12 days as a caterpillar. When the caterpillar is ready to form a chrysalis, it often leaves the host plant in search of another place to pupate such as the twig or leaf of another plant. It excretes all waste, climbs its selected plant, and attaches itself, hanging by its hind pair of claspers. From a gland on its lower lip the caterpillar spins a small silken pad to which it attaches its hind end and begins to shrug off its skin for the final time (fig. 255). A green jadelike pendant is revealed, embellished with gold and black markings (fig. 256). It contains the pupa during the period of metamorphosis. The chrysalis stage lasts for about 8–14 days depending on the weather. Toward the end of that time, the green chrysalis becomes clear, and the orange and black wing pattern of the developing butterfly can be seen through it (fig. 257). On a sunny day, the mature butterfly will split open the chrysalis and emerge within a matter of minutes (fig. 258). Before it can fly, the newly emerged butterfly must hang in place until it has pumped fluid from its enlarged abdomen into its crumpled, heretofore unexpanded, wings to stiffen the supporting network of veins (fig. 259). Hanging exposed and unable to fly, the first few hours are a perilous time for the new butterfly (fig. 260). During this period the monarch excretes a drop of red liquid, wastes which have developed during the pupal stage (see fig. 260). In the past, finding these spots of "bloody rain" alarmed people who thought that they were caused by witches.

Fig. 255. (Left) A monarch caterpillar has affixed itself to a silken patch produced from glands on its mouth. It hangs upside down from the twig as it transforms from caterpillar to chrysalis. **Fig. 256.** (Middle) The monarch chrysalis is a jewel-like structure of jade green adorned with a semicircle of gold and black dots and dashes near its top and a few scattered gold dots near the bottom. Inside the chrysalis the substances of the caterpillar's body are broken down and utilized in the making of the adult butterfly. **Fig. 257.** (Right) Toward the end of the period of metamorphosis, which may take from 7 to 14 days, the chrysalis becomes transparent, and the orange and black wings of the adult butterfly become clearly visible.

Common Milkweed

Fig. 258. (Left) On the final day—a sunny one—the chrysalis splits open and the butterfly begins to wriggle its way out. This takes only a few minutes. **Fig. 259.** (Middle) Once the monarch is free of its enclosure, its crumpled wings are limp and useless for flight. Fluids stored in its enlarged body are pumped into the veins of the wings to expand and stiffen them. **Fig. 260.** (Right) The butterfly must then hang in place, occasionally spreading its wings until they dry. This may take a few hours and is a very hazardous time for the immobile butterfly. On the leaf behind the monarch is a drop of red liquid excreted by the butterfly after it emerged.

While members of the genus *Asclepias* are native to the New World (North, Central, and South America), some species in the genus have been introduced into Spain, Australia, New Zealand, and Hawaii, where they now serve as a food source for monarch butterflies that have also been introduced to these areas. In such cases the introduced monarchs are nonmigratory. There are many other species in the milkweed subfamily, Asclepiadoideae, which are native to other parts of the world, especially in Africa, where the genus *Gomphocarpus* is predominant. In such regions, similar guilds of insects have evolved the ability to feed on these plants despite their toxic sap. Considering only butterflies, there are more than 170 species of "milkweed butterflies" documented worldwide.

Fig. 261. Many milkweeds are attractive to butterflies because they provide a copious source of nectar. Here a great spangled fritillary (*Speyeria cybele*) has placed its proboscis deep into one of the nectar-holding hoods of the flower of common milkweed.

Dogbane Family

Fig. 262. (Left) Moths also seek nectar from milkweed flowers, both by day and by night. Here a diurnal snowberry clearwing (*Hemaris diffinis*) hovers over purple milkweed (*Asclepias purpurescens*) with its proboscis inserted into the flowers. **Fig. 263.** (Right) Bumblebees are the most effective pollinators of most milkweeds because their strong, spiny legs easily hook the pollinia from the anthers of the flowers as the bees seek nectar. They then transport the pollinia to other milkweed plants. Here, the bee's long proboscis is reaching down into a hood of a swamp milkweed flower (*Asclepias incarnata*) to access nectar. Note how one of its legs has slipped down between the hoods on the right side. It may be withdrawn carrying a pollinarium.

The interactions between milkweed and insects extend far beyond the guild of those that feed on the plant. Many butterflies (fig. 261), moths (fig. 262), bees (fig. 263), wasps, and flies visit the flowers for nectar. The presence of so many insects makes the milkweed plant an attractive environment for predators and parasitoids of those herbivores as well. Milkweed plants provide a fertile hunting ground for predators such as mantids (fig. 264) and ambush bugs (fig. 265), parasitoids (e.g., flies [fig. 266] and wasps that lay eggs on or in the larvae of other insects), and spiders. Many of the above lie in wait, camouflaged by the plant until a victim chances to visit. Crab spiders (*Misumena vatia*), are noted for lurking in flower clusters (fig. 267).

Fig. 264. Monarchs making the long journey to Mexico encounter many risks—from stormy weather to lack of flowers from which to obtain nectar. However, even when there are ample flowers, monarchs may fall prey to predators that take advantage of a flower's ability to draw insects to it. Predaceous insects, such as this alien mantid species, lie in wait for nectaring monarchs and other butterflies. This monarch's flight to Mexico was canceled at its stopover in Cape May, New Jersey.

Common Milkweed

Fig. 265. (Left) Many insect predators hide among the flowers of milkweed species waiting to pounce on the butterflies that are attracted to the flowers. Here an orange sulphur (*Colias eurytheme*) has been captured by an ambush bug (the light brown insect on the butterfly-weed flower just above the dead butterfly). The ambush bug is in the process of liquifying the butterfly's body and sucking the juices through its straw-like proboscis. **Fig. 266.** (Right) Monarch caterpillars are sometimes parasitized by adult wasps or flies (as in this case) that lay their eggs in or on the body of the caterpillar. The larvae that hatch from the eggs feed on the interior of the caterpillar, but they don't eat its vital organs until the parasitoids are ready to pupate. They then consume the remainder of the caterpillar's organs, killing it, and emerge from the caterpillar, dropping to the soil to pupate. In this case, a larva emerged just as the monarch caterpillar was shedding its skin to transform into a chrysalis.

These spiders often lay their eggs on the underside of milkweed leaves, guarding them until they hatch. In a study comparing the success of prey capture by crab spiders on three common species (pasture rose, goldenrod, and milkweed), milkweed was found to be the most productive plant utilized by the spiders. It exceeded goldenrod as a provider of prey because it continued to attract insect visitors (moths) even at night. In fact, up to 25% of milkweed flowers are pollinated at night, primarily by moths (see fig. 240). These complex relationships make this common weed worthy of extensive study.

Not only is milkweed the linchpin of a vast ecological community, but it also has played many roles over a long period of use by humans. Perhaps the best known of these was the use of the milkweed seed's silken parachute (the coma) (fig. 268) as a substitution for the silky seed fibers of kapok in life vests during World War II. The U.S. supply of kapok (*Ceiba pentandra*) from Java was cut off during the war, and

Fig. 267. Crab spiders (*Misumena vatia*) frequent the inflorescences of many flowers, including milkweeds, which provide fertile hunting grounds for them. I've seen crab spiders with prey much larger than themselves, including butterflies and bumblebees.

Dogbane Family

the Department of Agriculture developed a program to collect and process the hollow, wax-coated floss that served to carry milkweed seeds on the wind. Farmers were paid to plant milkweed seed, and children were urged to do their part for the war effort by collecting the seeds of the resultant plants (earning 20 cents per large bag), which were then sent to a processing plant in Michigan. In 1944 alone, children collected more than 150,000 pounds of milkweed silk. While not as buoyant as kapok-filled life vests, the milkweed floss vests were capable of keeping a 150-pound man afloat in saltwater for 48–72 hours. The floss also served well as an insulating lining in flying suits. Processing milkweed floss was much costlier than kapok fiber, so the end of the war signaled the end of this industry.

Milkweed floss is currently being used as a stuffing for hypoallergenic pillows and winter coats, and in the past was used to make a very fine, silklike fabric in France. The hollow fibers are also an effective absorbent and have been used in cleaning up oil spills in the sea or other waterways. Like thistle down, milkweed floss is used by goldfinches to line their nests.

Native Americans used the tough milkweed stem fibers to make cordage, much as they did with the fibers of the related dogbane (*Apocynum* spp.). They sweetened foods with the copious, sweet nectar of milkweed flowers, and they used parts of the plant as food, eating the young shoots, fruits, and flower buds after carefully processing them in several changes of boiling water. Medicinal uses included potions used as laxatives, and for treating warts, venereal disease, and dropsy (an old name for edema, often caused by congestive heart failure). Their use for the aforementioned conditions is debatable, but there is a basis for the last ailment. While it is dangerous to consume the toxic cardenolides in milkweed, as with many toxins, prescribed in the proper dosage, they have been used effectively in medicine, particularly for increasing the strength of heart contractions in the treatment of congestive heart failure.

Early in the twentieth century, Thomas Edison investigated the use of milkweed sap as a rubber substitute, as did the Germans later during World War II; the sap proved unsuccessful as a substitute. Also, milkweed seed oil has been found to be an effective lubricant. But perhaps the best-known use of some species of milkweed is in the horticultural realm. In the Northeast, the orange flowers of butterflyweed (*Asclepias tuberosa*), and the magenta-flowered swamp milkweed (*Asclepias incarnata*) are favorites of gardeners.

Fig. 268. The coma (also called the floss) of the seeds of common milkweed (*Asclepias syriaca*) played an important role in World War II. It served as a substitute for kapok fibers that were formerly used to keep life vests afloat. Kapok was difficult to obtain during the war, so the Department of Agriculture developed a program whereby children were paid to collect milkweed fruits. The "pods" were then sent to a processing plant that separated the silken fibers from the seeds.

Common milkweed is a valuable component of the North American continental ecosystem, but most studies show that population levels of milkweed are declining. In the Midwest, a decline as high as 50% has been noted in some areas. As a consequence, monarch butterflies are declining as well. In 2008, the commission for Environmental Cooperation (comprising

Common Milkweed

members from the United States, Mexico, and Canada) signed a treaty, the North American Monarch Conservation Plan, citing declines in milkweed across the continent and proposing to remedy this situation. Reasons cited for the decrease include clearing of land for development and agriculture, the use of genetically modified (GMO) crops able to withstand the widespread broadcasting of insecticides and glyphosate herbicides (which kill milkweeds and other non-GMO plants), and a decline in pollinators. Professor Anurag Agrawal disagrees that milkweed populations are declining. Based on his own observations, there is still a vast amount of milkweed within its native range. Agrawal feels that the reasons for a decline in monarchs may lie elsewhere, most likely at the stage when the last generation born in North America is migrating south. Their numbers do not match up with those of the overwintering population that reaches Mexico; thus, it appears that there is a decline in numbers of butterflies en route.

Monarch populations have always fluctuated, but there has been a long-term downward trend. Possible causes may include the aforementioned U.S. droughts, illegal logging at the Mexican wintering site, occasional freezing temperatures at those sites, or an excess of ecotourists traveling to the site to witness the magnificent spectacle of millions of roosting monarchs (others say that the influx of tourists reinforces the value of preserving this phenomenon). For unknown reasons, the proportion of female monarchs to males in Mexico is lower than in former years.

Many other possible factors may contribute to the decline in monarchs. Collisions with cars kill millions of butterflies each year, and an increase in highways—and the cars using them—has resulted in a parallel increase in butterfly deaths. In an attempt to prevent a loss of trees during the widespread gypsy moth outbreaks in the 1970s and beyond, parasitoids were introduced to attack the caterpillars of the forest-eating gypsy moths. They have since been found to parasitize monarchs as well, causing death at the larval stage. Even disturbance in the electromagnetic field has been mentioned as a possible factor in lowering monarch numbers. Clearly, the cause of monarch decline is difficult to pinpoint, and perhaps a combination of these many factors will yield the best answer. Only time will allow this multiplex puzzle to be solved.

Once overwintering monarchs begin their return to the States, they must find young milkweed plants on which to lay their eggs, but they also need other plants already in flower from which to take nectar. Our changing climate has resulted in drought in several parts of the monarch migration route over a number of recent years. As we have pointed out, drought at this time, or during the fall migration, can have a negative effect on butterfly survival.

Meanwhile, a monarch conservation organization, Monarch Watch (as well as the North American Monarch Conservation Plan), recommends planting milkweed, both in your garden and along roadsides, to ensure there are available host plants throughout the range of the monarch. Planting other flowering plants is equally important so that adult butterflies have an adequate source of nectar throughout their time in North America. While I don't think that it can hurt to add milkweeds to the garden, and I have added four milkweed species to our own perennial garden over the past years, I have since had to remove two of them: common milkweed and poke milkweed, both of which overstepped their intended bounds in short order. The remaining species, swamp milkweed and butterfly-weed, add color to the garden in summer and attract many species of butterflies, including monarchs, and other pollinators. As a bonus, they are not browsed by rabbits or deer.

Dogbane Family

Some other milkweed species of the Northeast

Although common milkweed is the most frequently encountered milkweed species in the northeastern part of the continent, several other species may be found as well. Many of them share the same pollinators and herbivores that visit common milkweed. For a look at other milkweeds found in the Northeast, I provide summaries and photos of some you might see. There are several other members of the genus native to the Northeast and Midwest; only those I have personally encountered are mentioned here.

Swamp milkweed (*Asclepias incarnata*) (fig. 269) most often grows in wetter areas than common milkweed, but it does well when planted in normal garden soil. Its flowers are smaller than those of common milkweed and a bright magenta-pink; it also differs from common milkweed in that its leaves and pods are narrower.

Butterfly-weed (*A. tuberosa*), with its bright orange flowers (varying from yellow-orange to red-orange) (fig. 270), inhabits drier habitats with well-drained soils. Its leaves are alternate on the stem, and it lacks a milky sap. That it is one of our better-known milkweeds is due to its popularity as a garden plant and its superior ability to attract butterflies (see fig. 261). It is a preferred host for the orange and black caterpillar of the unexpected

Fig. 269. (Left) Swamp milkweed (*Asclepias incarnata*) has smaller, brighter pink flowers than common milkweed. Both its leaves and its fruits are narrower as well. It most commonly grows in wet areas, but I find that it does well in the average soil of my perennial garden.

Fig. 270. (Left) Butterfly-weed (*Asclepias tuberosa*) thrives in sunny situations with poor, sandy soil, but also does well in typical garden soil and is a popular plant with gardeners. It is fed upon preferentially by the caterpillars of the unexpected cycnia moth (*Cycnia inopinatus*), which also feed on dogbane. **Fig. 271.** (Right) A lesser known orange milkweed is the lance-leaved milkweed (*Asclepias lanceolata*). It has small inflorescences of just 7–10 flowers, lance-shaped leaves, and grows in swamps, bogs, and brackish marshes.

Common Milkweed

Fig. 272. (Left) Green milkweed (*Asclepias viridiflora*) is sometimes known as green comet milkweed because its flower petals are so strongly reflexed backward and its hornless hoods so tightly appressed to the gynostegium that the flowers appear to be narrow, shooting comets. Green milkweed grows as scattered individuals in sunny openings on sandy or limy soil. It is an indicator of a habitat that may harbor other interesting species. **Fig. 273.** (Right) Ozark, or spider, milkweed (*Asclepias viridis*) is common in south-central states and is often the first milkweed encountered by monarchs on their northward migration from Mexico to the United States. Its flowers are unusual for milkweed flowers in that the petals are not reflexed, but spread open, making the flowers appear more bowl-like.

cycnia moth (*Cycnia inopinatus*), which also feeds on dogbane.

A lesser known orange-flowered milkweed is the few-flowered milkweed (*A. lanceolata*); its flower color may vary from orange to reddish purple (fig. 271). As its common name suggests, few-flowered milkweed has umbels consisting of only 7–10 flowers. It inhabits swamps and bogs as well as brackish marshes.

Green milkweed (*A. viridiflora*) has green flowers with strongly reflexed petals and is sometimes referred to as green comet (fig. 272). It is a less common plant that grows in dry prairies and sandy woodlands.

A second green-flowered species, Ozark (or spider) milkweed (*A. viridis*) has flowers unlike those of other milkweeds in that the petals are upright and spreading with narrow purple hoods (fig. 273). Though native to parts of the Midwest, its range extends south into Florida

Fig. 274. Purple milkweed (*Asclepias purpurescens*) is shorter than common milkweed, and its flower heads are grouped more at the apex of the plant. The flowers are a deep magenta-purple unlike the dull pink flowers of common milkweed. Purple milkweed's fruits are smooth rather than warty as in those of common milkweed. It grows in meadows and occasionally in open woodlands.

continued overleaf

Dogbane Family

Fig. 275. Blunt-leaved milkweed (*Asclepias amplexicaulis*) has flowers that vary from a dull purplish-green to a striking magenta, as seen in this image. Its leaves have distinctly wavy margins. It grows in dry, sandy soils.

Fig. 276. Whorled milkweed (*Asclepias verticillata*) is named for its whorls of delicate, linear leaves that have rolled margins. It is a small plant with tiny (5–6 mm diameter) white flowers. It is another plant of dry, sandy soils in open, sunny areas. Whorled milkweed rapidly spreads in areas with bare soil and can become aggressive.

and Texas. This early-appearing species is an important food plant for the caterpillars of the monarchs that return from Mexico in early spring.

Purple milkweed (*A. purpurescens*) (fig. 274) resembles common milkweed but is shorter and has most of its flower umbels at the apex of the plant. The flowers are a deeper, more purple color, and its fruits are smooth rather than warty. It prefers dry habitats.

Blunt-leaved milkweed (*A. amplexicaulis*), also found in open dry habitats, has flowers that range from a deep magenta to drab greenish-purple (fig. 275). Its leaves are distinctively wavy along the margins.

Whorled milkweed (*A. verticillata*), named for its whorls of very narrow leaves, is a small species with umbels of diminutive white flowers (fig. 276). The leaves of this species are fed on by fuzzy gray caterpillars of the delicate cycnia moth (*Cycnia tenera*) (fig. 277), which also feed on dogbane.

Poke (tall) milkweed (*A. exaltata*), unlike most other milkweeds, can grow in more shaded situations. It is one of the tallest of the local milkweeds and has leaves somewhat reminiscent of those of pokeweed (*Phytolacca americana*). Its drooping, green and white flower clusters remind me of exploding fireworks (fig. 278).

Another *Asclepias* that can tolerate more shade than most other milkweeds is the four-leaved milkweed

Fig. 277. This fuzzy gray caterpillar of the delicate cycnia tiger moth (*Cycnia tenera*) generally feeds on milkweed's relative, dogbane (*Apocynum* spp.) as seen here, but it also feeds on the leaves of whorled milkweed.

Fig. 278. The white and green flowers of poke (or tall) milkweed (*Asclepias exaltata*) hang in dangling clusters like exploding fireworks. It is a milkweed that can tolerate some shade including forest edges and openings. I have found it to be too aggressive in my partially shady garden.

Common Milkweed

Fig. 279. Another milkweed that tolerates shade, four-leaved milkweed (*Asclepias quadrifolia*), has most of its leaves arranged in fours around the stem's nodes. The flowers are white (sometimes tinged with pink or lavender) and occur in umbels at the top of the stem. It is most commonly a forest species.

Fig. 280. Although the flowers of red milkweed (*Asclepias rubra*) are sometimes a dull red, they are more frequently deep pink to purple, much like those of swamp milkweed. It has a limited range from southern New Jersey to eastern Virginia in wet to moist habitats.

(*A. quadrifolia*), a white-flowered species with only a few flower clusters at its apex, and with leaves in whorls of four (or sometimes oppositely paired) on the stem (fig. 279).

Red milkweed (*A. rubra*) (fig. 280) is somewhat of a misnomer since its flowers are no redder than those of swamp milkweed, purple milkweed, or some plants of blunt-leaved milkweed. It has a limited range from southern New Jersey to eastern Virginia.

Tropical milkweed (*Asclepias curassavica*) is a native of South America and is gaining in popularity as a garden plant throughout North America due to its bright red-orange and yellow flowers (fig. 281), which excel in attracting butterflies. In the southern parts of the United States, it may naturalize and overwinter, but in our part of the country, it is grown as an annual. Controversy about the appropriateness of introducing tropical milkweed has arisen among native plant advocates. As with butterfly-weed, the flowers act as a butterfly magnet for many species, including monarchs. It serves both as a source of nectar (necessary for monarchs fueling up for their long southward migration), and as a plant on which they can lay their eggs. Countering those two positive aspects, some say that in places where tropical milkweed can thrive year-round (e.g., Florida), its presence may induce monarchs to remain there rather than continuing on to Mexico, thus disrupting their natural breeding cycle. In addition, the presence of old leaves on plants that have been visited by many butterflies allows for the build-up and dispersal of a fungal infection that negatively affects monarchs. In horticultural zone 7 and northward, this is not a problem as the plants are naturally killed back by frost in the fall. A work-around compromise has been suggested for gardeners wishing to grow tropical milkweed in the South. Since tropical milkweed regrows readily when cut back, gardeners are urged to cut the plants back a few times a year and dispose of the cut portions to ensure that monarchs are not infected by fungal spores on the overvisited parts of the old plants. This practice would also discourage monarchs from remaining in the southern part of the country rather than following their natural pattern of migrating to Mexico to overwinter. Moreover, since the level of cardenolides in tropical milkweed is generally higher than in most other milkweeds, monarchs that sense that they are infected with a fungus may preferentially choose to lay their eggs on this species, ensuring that their larvae would derive more protection from ingesting these more potent compounds.

Fig. 281. Tropical milkweed (*Asclepias curassavica*) is native to the South American tropics, but it is now found throughout tropical and subtropical regions of the world and is planted as a garden annual in temperate areas. Its brilliant red-orange and yellow flowers, which bloom over an extended period, have made it a favorite worldwide. It is known for attracting many species of butterflies as well as hummingbirds.

Figwort Family

Fig. 282. The flowering stalks of common mullein may reach 2 meters in height, making it one of our tallest naturalized wildflowers. The plants stand out in any landscape.

Common Mullein

Fig. 283. Common mullein (*Verbascum thapsus*) is a pioneer invader of disturbed soils in pastures and barnyards.

Common Mullein
Verbascum thapsus L.
Figwort Family (Scrophulariaceae)

Mullein plants are a common sight along roadsides and in recently disturbed areas. They tower over most other weedy vegetation and are easily recognized by their tall flowering stalks (fig. 282). Yet, there are no New World mullein species. Common mullein and its relatives in the Northeast were all introduced from Eurasia.

Habitat: Common mullein can invade almost any vegetation type, from abandoned farmyards (fig. 283) and pastures to meadows and prairies, and can even become established in forest openings (fig. 284) as long as there is an area of bare soil and an opening in the canopy (usually caused by some form of disturbance such as logging or fire).

Range: *Verbascum thapsus*, a native of Europe, is found as an introduced and naturalized plant in all 50 states, being most prevalent in the Northeast, the West, and the easternmost states of the Midwest. It also occurs in all provinces of Canada (including Newfoundland but not Labrador), but it is not found in the colder, more northern Canadian territories.

Scrophulariaceae is a family that has been radically restructured based on molecular evidence that has shown many of the genera previously included in the family to be more closely allied with a number of other families. The result is that the family is notably smaller than it was at the time the *Gleason and Cronquist Manual of Vascular Plants of the Northeastern United States and Adjacent Canada* was revised in 1991. At that time Scrophulariaceae included 37 genera within the region covered by the *Manual* (the same region covered by this book); today that number has been reduced to 4 genera. Nevertheless, the family remains large and has a worldwide distribution comprising 59 genera and nearly 2000 species. Familiar former "scrophs" now excluded from the family include *Mimulus* (monkey-flower), now placed in the formerly monotypic lopseed family (Phrymaceae); and *Chelone* (turtlehead), *Penstemon* (beard-tongue), and *Veronica* (speedwell), which are all now in the plantain family (Plantaginaceae). Genera having a hemiparasitic lifestyle were transferred to the Orobanchaceae.

Thus, the great majority of the genera that moved from the Scrophulariaceae to other families now reside in either the Plantaginaceae, a family we know primarily for the weedy *Plantago* (plantain) species that grow in our lawns, or the Orobanchaceae,

Figwort Family

a family that formerly consisted of strictly parasitic plants (holoparasites), e.g., beechdrops (*Epifagus virginiana*, included in this book) and squaw-root (*Conopholis americana*).

The name for *Scrophularia*, the genus that is the basis for the family name, Scrophulariaceae, is derived from the Latin word *scrophula*—an inflammation and swelling of the lymph nodes in the neck caused by a bacterial infection, most commonly a form of tuberculosis. Symptoms of scrophula in the past were treated with plants in the genus *Scrophularia* (figwort).

The genus *Verbascum* has as its center of diversity, Eurasia—from the Mediterranean to central Asia—a region of mostly dry, rocky soils. Thus, most species of *Verbascum* do well in arid, and even poor, soils wherever they have been introduced. I have, in fact, seen common mullein growing from crevices in trees and rocks (fig. 295). *Verbascum* is the ancient Latin name for plants in this genus, perhaps derived from *barbascum*, referring to a beard (*barba*), and indicating the dense hairs present on the leaves of many species (fig. 286); the epithet, *thapsus*, was bestowed upon the species to commemorate the ancient Roman port city of Thapsus, now in present-day Tunisia. Wherever common mullein has been introduced, it grows almost exclusively in recently disturbed sites and rarely persists for more than a few years as natural succession proceeds. Thus, the species has just a small window of time in which to establish itself. Its seeds, however, are long-lived and may remain viable in the soil for many decades until another disturbance takes place that affords the seeds access to the light necessary to stimulate germination.

In addition to the probable accidental introduction of *Verbascum* seeds to North America, there were also deliberate introductions because of the plant's long history of medicinal use. Preparations made from the plant were used to alleviate chest pain and coughs; the leaves were smoked for the same purpose as well as for stimulatory effects. Other folk uses for the leaves have prompted the names "Quaker rouge" for its effect of causing a blush when rubbed on the cheeks of young girls who were not allowed to use makeup, and "hiker's toilet paper" for obvious reasons.

There is also evidence of common mullein being used on a small scale as a fish poison in the Blue Ridge Mountains during the mid-1700s. Such toxins "stunned" the fish and made them easy to collect as they floated on the surface. By 1839, common mullein had spread as far west as Michigan, most likely by settlers taking the seeds with them as they moved west, or via long-distance movement of soil and road materials that harbored the mullein seeds; in 1880 mullein was reported as naturalized in California.

Fig. 284. (Left) Even a recently disturbed area of forest, as may be caused by a treefall, can be quickly colonized by common mullein. Such colonization results primarily from the long-lasting seedbank in the soil rather than by dispersal from other areas. **Fig. 285.** (Right) The native Eurasian habitat of common mullein is generally dry and rocky; thus, mullein is tolerant of difficult growing conditions—as shown here where a plant has grown from a seed that had lodged in a rock crevice.

Common Mullein

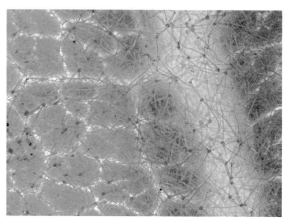

Fig. 288. The leaf hairs on both the upper and lower surfaces of the leaves of common mullein are branched or starlike (stellate), best seen through the microscope. The hairs tend to shed water from the leaves.

Fig. 286. (Left) The dense, downy hairs that cover the vegetative parts of common mullein and some other species in the genus were probably the feature for which the genus was named, *Verbascum* being a corruption of *barbascum*, referring to a beard. **Fig. 287.** (Right) Mullein plants are typically biennial, producing a large basal rosette of leaves during the first year and storing carbohydrates in their long taproots that will serve as the energy source for the growth of a flowering stalk the following year (see fig. 285).

As previously noted, the seeds of common mullein require patches of bare soil in order to germinate, and sunlight is an important component in their success. Once germinated, plants that do not develop a critical minimum size of rosette (at least 6 cm in diameter) during their first growing season are unlikely to survive the winter. Thus, the earlier a seed germinates and produces a rosette, the more likely the rosette will reach a size that allows it to overwinter. However, even when conditions are favorable and plants survive the winter to produce flowers and seeds in the second year, success of the population is short-lived. A population that colonizes a site with bare soil is generally ephemeral because the seeds of other pioneer plants will have also taken advantage of the open site and are also likely to become established, thus the bare spot soon becomes covered with vegetation. Therefore, the seeds produced by the first mullein plants will not fall onto the bare soil environment ideal for germination and must lie in wait until conditions again become favorable for their germination.

Verbascum thapsus is typically a biennial plant producing a basal rosette of large (up to 30 cm long), softly hairy, pale green leaves (fig. 287) and a deep taproot during the first growing season. The leaf hairs are branched or stellate (starlike) (fig. 288), giving the leaf a soft texture that is responsible for one of the plant's alternate common names, flannel leaf (mullein, is derived from the Latin *mollis*, meaning soft). The hairs help protect the leaves from strong sunlight and also prevent excessive water loss by transpiration. They can hold droplets of dew on the leaves well into the morning after a cool, damp night (fig. 289).

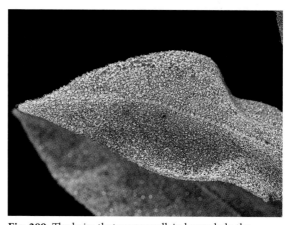

Fig. 289. The hairs that cover mullein leaves help the plant to conserve moisture in its dry, sunny environment by inhibiting transpiration and capturing and holding morning dew.

Figwort Family

Fig. 290. (Left) If the rosettes of common mullein are large enough, they can withstand harsh winters, even surviving burial by snow. This rosette is just being dusted with snow at the beginning of a snowstorm.
Fig. 291. (Right) The inflorescence of common mullein is generally spikelike, but sometimes it branches, especially if the apical meristem (growing tip) has been damaged.

The plant undergoes a period of dormancy, often covered by snow, during its first winter (fig. 290) before it produces an elongate (up to 2 m), dense flowering stalk during its second year. The spike is usually simple, but occasionally branches (fig. 291), especially if it has been damaged early in its development. The leaves that arise from the lower part of the stalk are decurrent on the stalk, meaning that their bases extend down along the stalk beneath the free portion of the leaf, giving the stalk a winged appearance (fig. 292).

Mullein stalks begin flowering at the bottom with small clusters of large (up to 2.5 cm across), slightly bilaterally symmetric, yellow flowers (rarely, the flowers are white). Flowering begins in late June and peaks in August. The small flower clusters spiral upward around the stalk, with the older flowers forming fruits as the younger ones begin to bloom. Some inflorescences will have a few clusters blooming simultaneously, but, in general, only a small number of flowers are in bloom at any one time. Each flower opens for only a day and closes before evening. Closer inspection of the flower reveals its beauty. The flower has a five-lobed corolla with the two-lobed lower portion slightly larger than three-lobed upper part. The stamens are adnate (attached) to the corolla lobes and are dimorphic (of two types). The filaments of the three upper stamens are densely covered with whitish to yellowish hairs and have short anthers; the lower two stamens have longer, nearly hairless filaments and linear anthers that extend down the filament for a short distance (fig. 293); the pollen of all five anthers is bright orange. The style, having a capitate stigma,

Fig. 292. The bases of the leaves of common mullein extend down the flower stalk to the leaf below, giving the stalk a winged appearance.

Common Mullein

Fig. 293. (Left) Flowers of common mullein have stamens of two types: the filaments of the three upper stamens are covered with yellowish hairs, while the filaments of the two lower stamens are hairless or have only a few hairs. The anthers differ as well; those of the lower two stamens are linear and longer than those of the upper three. The styles persist after the corollas have fallen (seen here as thin, dark brown, hairlike structures protruding from the closed calic§es beneath the blooming flowers). **Fig. 294.** (Right) Bumblebees are among the most successful pollinators of common mullein. Note the three syrphid flies (also known as hover flies or flower flies) on or near the flower. They feed on the pollen or sip the scant nectar at the base of the tubular portion of the corolla; occasionally, the flies serve as minor pollinators.

persists long after the corolla has fallen (see fig. 293). The plant dies after flowering.

Many insects visit the flowers, primarily to collect pollen (the flowers produce only small amounts of nectar). Bees are the most effective pollinators, with bumblebees (fig. 294) and honeybees playing the most important role, but smaller bees (fig. 295) and syrphid flies (fig. 296) actively collect pollen and inadvertently

Fig. 295. Smaller bees, such as this halictid bee in the genus *Lasioglossum*, collect mullein's orange pollen on their abdomens and legs, pollinating mullein as they do so.

Fig. 296. Many species of syrphid flies (this one is *Toxomerus marginatus*) visit the flowers of mullein. They are too small to be very effective at pollinating the flowers.

Figwort Family

pollinate the flowers as they do. As the flowers close at the end of the day, the anthers come into contact with the stigma, and if pollination by outcrossing has not already occurred through insect visitation, the flowers will self-pollinate. A study of insects that visited the flowers of common mullein growing in a large population showed that more insects visited the taller flower stalks. However, since the taller plants also had more open flowers, it was difficult to determine whether the primary attraction was the height of the plant or the larger number of flowers.

Another insect, this one detrimental to the plant, is commonly found on mullein and is in fact a specialist on that species. It is thus called mullein weevil. Mullein weevil (*Rhinusa tetra*) is an insect probably introduced accidentally from Europe along with the plant. This tiny (2–4 mm), furry brown weevil takes shelter in the rosettes of mullein early in the second year and in summer lays its eggs in the developing seed capsules, where the larvae feed on the seeds and then pupate within the capsules. Downy woodpeckers are aware of this food source and have been observed pecking at the fruits to feed on the pupae inside.

The fruits that develop from the superior ovaries are two-celled dry capsules that split into two valves (fig. 297) before releasing their seeds. Seeds are merely shaken out by the wind or by the jostling of a passing animal; most fall within a meter of the plant, but some are carried up to 11 meters away if the wind is strong. *Verbascum* species are prolific seed producers. Katherine Gross, in a Michigan study, found that the average number of seeds produced by a plant of *V. thapsus* was 180,000 (a higher number than was found in other studies). The small seeds easily sift into the soil in the autumn, but rain during the following spring may wash the soil away, exposing the seeds to the light necessary for germination.

I was intrigued by the seeds when I viewed them through my hand lens. Although small (50–100 μm in length), they are odd in appearance, light brown and having a columnar form vertically ridged with "squiggly" lines (fig. 298). When I arrived home, I put the seeds under the dissecting microscope and noticed small, black protuberances on them (see fig. 298). I found nothing in the literature that described these spots and, on a hunch, I sent some seeds to a mycologist to see if they might be fungi. Dr. Sabine Huhndorf, an ascomycete specialist, identified the black bodies as belonging to the fungal genus *Phoma*, a large genus that parasitizes many different hosts.[1] Three species of *Phoma* have been described from *Verbascum*, two collected on stems and the other from an unknown plant part. Thus, this is possibly the first report of *Phoma* being found on the seeds of *Verbascum*.

In a famous seed viability experiment initiated in 1879 at the Michigan Agricultural College, Dr.

---

1 Sabine Huhndorf, e-mail message to author, January 15, 2013.

Fig. 297. The fruits of mullein are two-celled capsules densely packed with seeds. Once they have split open, the many seeds are shaken out of the capsules by the wind. The fruits often contain weevil larvae that consume many of the seeds.

Fig. 298. Common mullein seeds are tiny, measuring only 50–100 μm in length. When photographed through a dissecting microscope, their intricately patterned surface is evident. The round, black areas are a fungus in the large genus *Phoma*.

Common Mullein

William Beal filled 20 pint bottles with seeds representing 23 species of plants (50 seeds each) with the object of storing them buried in the ground and retrieving one bottle of each species every five years to check the seeds for viability. Included in the experiment were two species of *Verbascum*: common mullein (*V. thapsus*) and moth mullein (*V. blattaria*). The five-year experimental cycle was repeated by Dr. Beal until his retirement in 1920. The project was continued by Dr. H. T. Darlington, Beal's colleague. Beginning in 1920, Darlington lengthened the excavation intervals to 10 years in order to extend the experiment, and subsequent overseers of the project decided in 1990 not to excavate that decade's cohort, but to lengthen the intervals to 20 years. After the fifteenth series of bottles was unearthed in 2000, the remaining five lots could potentially carry the experiment to the year 2100.

In 2000, the 120th year of the experiment, only two species yielded viable seeds: *Verbascum blattaria*, with 23 seeds germinating (43% of those tested) and *Malva rotundifolia*, with one seed successfully germinating (2%). *Verbascum blattaria* was the only species with seeds that had germinated in all 15 trials. One seed of *Verbascum thapsus* had germinated in 1980, the 100th year of the project, after several previously excavated lots of that species had produced no viable seeds. Thus, to date, the experiment has documented that seeds of both *Verbascum blattaria* and *Malva rotundifolia* are capable of germination after 120 years of dormancy and that the resultant plants yield viable seeds. As seen in the case of the one successfully germinated seed of *V. thapsus* in 1980, seeds from a cohort that have not germinated in a series of previous excavations may potentially germinate in successive attempts, so one cannot say that only two species retain the potential to germinate in 2020 or beyond. Astoundingly, Gross and Werner report that *Verbascum* seeds recovered from soil samples archaeologically dated from 1300 C.E. have proven to have retained their viability.

When comparisons are made of *V. thapsus* growing as aliens in the United States (particularly at high altitudes in the arid western states) with the same species growing in its native range in Europe, the introduced plants are found to form larger, more permanent populations comprising larger individuals. The reason for this is hypothesized to be that the introduced plants thrive because they have escaped from their natural predators (e.g., caterpillars and other insects that feed on the plants in their native range). Because there is some concern about common mullein outcompeting species in the native plant communities in these high-altitude regions, consideration is being given to introducing native European pests of mullein into those areas of the United States.

A small number of other *Verbascum* species may be encountered growing in the Northeast, all, of course, introduced from Eurasia. Moth mullein (*V. blattaria*) (fig. 299) is the second-most common mullein encountered in our region. Its specific epithet is derived from its presumed use as a cockroach repellent (*blatta* is Latin for cockroach). Like common mullein, moth mullein is a biennial with a similar life history, but unlike common mullein, moth mullein prefers rich soils, albeit in waste places. It differs from common mullein in many other respects. It is an overall smaller, more delicate plant with a flowering stem that rarely exceeds a meter in length. The sessile stem

Fig. 299. Moth mullein (*V. blattaria*) is also a common species in the Northeast. It is a more delicate plant having flowers spaced singly along the flowering stalk. The flowers may be yellow or, more rarely, white.

Figwort Family

Fig. 300. The leaves of moth mullein are also quite different from those of common mullein in that they are smaller, toothed, hairless, and not winged (decurrent) on the stem. The basal leaves are hairless and sometimes wither and disappear before the plant flowers.

Fig. 301. Whether yellow or white, the flowers of moth mullein have a purple spot at the base of each of the five petals and purple hairs on the staminal filaments. The glandular hairs of the calyx are just visible on the upper pentagonal-shaped bud.

leaves are hairless (fig. 300) and range from 2 to 6 cm long; they are toothed or sometimes even pinnately lobed but do not form winged bases along the stem (as seen on *Verbascum thapsus* in fig. 292). The leaves of the basal rosette are longer than the stem leaves, up to 35 cm (these basal leaves sometimes wither and disappear before the plants are in flower). The flowers of moth mullein are borne singly on short pedicels, each with a small bract beneath, and spread from the raceme in a loose manner. The calyx is covered with simple glandular-tipped hairs (as are the pedicel and stem). Corollas are typically yellow or sometimes white with a purple splotch at the base of each petal, with both yellow- and white-flowered plants occasionally growing at the same site. As in common mullein, the stamens are fused to the five petals and are of two types: the upper three are smaller and have smaller, kidney-shaped anthers, while the anthers of the lower two stamens are obliquely inserted and extend slightly down the filament. All the staminal filaments are densely covered with violet hairs that cover the entire filament in the upper three stamens and about half the filament in the lower two (fig. 301). The stigmas persist in the fruit, which is a more or less flattened sphere measuring about 5–8 mm (fig. 302). A species similar in appearance to moth mullein is

Fig. 302. The styles of moth mullein's flowers are persistent, remaining attached even into the fruiting stage.

Common Mullein

an occasional escape from gardens—*V. phoeniceum* (purple mullein). It is easily distinguished from moth mullein by its purple flowers (fig. 303) and its sparsely hairy leaves.

Verbascum lychnitis, although commonly called white mullein, generally has pale yellow flowers. It has in common with *V. thapsus*, a densely congested inflorescence and branched hairs on its stem and lower side of the foliage, but is distinguished from common mullein by having a panicled inflorescence (fig. 304), the source of its specific epithet—*lychnitis*, meaning lamplike—a reference to the overall candelabra-shaped form of the inflorescence); it also is distinguished by having leaves that do not descend down the stem. The stamens are all alike, with the filaments covered with whitish hairs, except that the two lower ones have

Fig. 304. White mullein (*Verbascum lychnitis*), despite its name, usually has yellow flowers. The inflorescence is very much branched. It is seen here in its native range, above the village of Molines in the French Alps.

Fig. 303. Purple mullein (*Verbascum phoeniceum*) is often planted in gardens for its showy flowers. It resembles moth mullein in many respects except for flower color and the sparse hairs on the basal leaves (not visible in this image).

hairs along only one side of the filaments and lack hairs near the tips (fig. 305). The anthers are transverse and kidney-shaped, and the stigma is capitate.

Similar to *V. lychnitis*, but with either a spikelike or a panicled inflorescence, is black mullein (*V. nigrum*), so called because its stellate hairs are blackish. It is occasionally reported from the northern half of our range. It prefers moist soils within its native European range. The yellow flowers differ from those of *V. lychnitis* in that they have a purple spot at the base of each petal, and the stamens are all alike, having filaments covered by conspicuous purple hairs (fig. 306).

Another species similar in overall gestalt to common mullein is clasping mullein (*V. phlomoides*—from the Latin, meaning resembling *Phlomis*, a genus in the mint family that includes many species with soft leaves and yellow flowers). Clasping mullein may be

Figwort Family

Fig. 305. (Left) The filaments of the upper three stamens of white mullein are covered with whitish hairs, looking like mini bottle brushes, but those on the lower two stamens are along only one side of the filaments, more like a toothbrush, and extend only to a point shortly beneath the anthers (the anther of the middle stamen is hidden behind another). Note that the style of the green ovary at the center of the flower appears to have been broken off. **Fig. 306.** (Right) Black mullein (*Verbascum nigrum*) is similar to white mullein, but its yellow flowers have a purple spot at the base of each petal, and all of its stamens are alike, with purple hairs on the filaments.

differentiated from common mullein by its toothed, sessile leaves covered with stellate hairs (and, if decurrent, then only partially down to the next node—giving the "clasping" appearance) (fig. 307). In its early stage of development, the flowering stalk of clasping mullein is much like that of common mullein, but as it matures, it elongates, exposing gaps between the flower clusters (fig. 308). The flowers are yellow or white, larger than those of common mullein and with slightly ruffled petal margins. The stamens are dimorphic—the three upper ones having anthers that are transverse and kidney-shaped and filaments

Fig. 307. Clasping mullein (*Verbascum phlomoides*) is encountered frequently in the Northeast. The leaves are toothed and usually do not extend down the stem, but if they do, they do so for only a short distance.

Common Mullein

Fig. 308. At first glance, clasping mullein looks like common mullein, with a single tall stalk of yellow flowers, but as the stalk elongates, gaps are left between the flower clusters.

entirely covered with yellowish hairs; the two lower ones without hairs on the filaments (fig. 309) and with oblique anthers that are incurved and extend down the widened tip of the filament. The stigmas are spatula-shaped and also extend down the style (see fig. 309).

Verbascum densiflorum, a yellow-flowered species with dense flowering spikes, is reportedly locally abundant in parts of the Midwest, but I have never encountered it. Some of the above species will hybridize with the others; the seeds produced are usually, but not always, sterile. Although all are weedy plants that colonize disturbed soils, none are currently considered invasive. Nevertheless, in some of our western national parks, *V. thapsus* is one of the few non-native plant species to have established itself at high altitudes, and its ability to produce plants with varying physical characteristics when exposed to different environments makes it a species worthy of careful monitoring. It is definitely not a welcome addition to the western flora. Such plants may be present in the environment for a long time before they are triggered by some indeterminate factor to increase suddenly and exponentially and then be considered invasive.

Fig. 309. The yellow flowers of clasping mullein are showier and larger than those of common mullein and have petals with slightly ruffled margins. Note the spatula-shaped tip of the style with a stigma that extends for a short distance along the sides of the style.

Evening-primrose Family

Fig. 310. The reflexed sepals and large, X-shaped stigmas are evident on these newly opened flowers of evening-primrose.

Evening-primrose

Fig. 311. The woody, four-parted capsules of evening-primrose may remain on the plant throughout winter.

Evening-primrose
Oenothera biennis L.
Evening-primrose Family (Onagraceae)

Evening-primrose (often referred to as common evening-primrose) is sometimes mistakenly thought to be an alien weed because of its preference for growing along roadsides, in abandoned pastures, and in other "waste" places; however, evening-primrose is one of our true native wildflowers. As its name suggests, the flowers open in the evening—timed for the nocturnal activity of its principal pollinators, sphinx (or hawk) moths. It's easily recognized by its four-petaled flowers, each bearing a large X-shaped stigma (fig. 310), and its four-parted brown capsules that remain in winter (fig. 311).

Habitat: Sunny places along roadsides and railways; old fields; areas with poor, sandy soil; and other places where the soil has been recently disturbed.

Range: Varieties of common evening-primrose are found throughout the continental United States with the exception of Idaho, Wyoming, Utah, Arizona, and Colorado. It grows in most of Canada, except for Labrador, Nunavut, the Yukon, and the Northwest Territory. Evening-primrose's probable origin is Mexico and Central America, from where it dispersed widely throughout the Americas. It now has been introduced throughout Europe and Asia.

> When once the sun sinks in the west,
> And dewdrops pearl the evening's breast;
> Almost as pale as moonbeams are,
> Or its companionable star,
> The evening primrose opes anew
> Its delicate blossoms to the dew;
> And, hermit-like, shunning the light,
> Wastes its fair bloom upon the night,
> Who, blindfold to its fond caresses,
> Knows not the beauty it possesses;
> Thus it blooms on while night is by;
> When day looks out with open eye,
> Bashed at the gaze it cannot shun,
> It faints and withers and is gone.

This lovely poem, included in the 1835 publication *The Rural Muse*, by English poet John Clare, recognizes the fleeting beauty of evening-primrose, a plant too often underappreciated or dismissed as "just a weed" by people who never see it in its nocturnal beauty.

When the flowering stalk of evening-primrose elongates, it may reach 5–6 feet in height. The inflorescence of evening-primrose produces many tubular yellow flowers, but only a few open each evening, each flower remaining open for only one night. The flowers close at dawn unless the day is overcast, in which case flowers may remain open into the day. This sequence favors visitation by nocturnal moths over diurnal insects. The flowers exhibit a typical

Evening-primrose Family

moth-pollination syndrome: opening in evening; having a light-colored, tubular corolla; and producing a fragrant aroma. There is also a strong UV pattern that is unperceivable to the human eye but strongly visible to insects. It indicates where the moth must place its proboscis (tongue) to reach the nectar. Four long sepals encase the bud, held tightly closed by tangled hairs at their margins. The expansion of the flowers forces the sepals to "unzip" and abruptly release the four, lobed petals. Even before the flower is open, the anthers have dehisced (opened) to release their pollen (fig. 312). The flowers open quite rapidly (often in just a few minutes), a phenomenon that is magical to watch. Nocturnal flowers are often stimulated to open by the increase in relative humidity that results when cooler evening temperatures cause condensation to form on the vegetation. The opposite is true of flowers that open in morning, when the warming sun dries the dew from the flower buds.

Large-bodied hawk moths are attracted to the flowers by a sweet aroma that wafts into the nighttime air and triggers a search behavior in sphinx (hawk) moths. A principal constituent of the aroma is linalool, a chemical present in the floral fragrance of many flowers from diverse families, but particularly prevalent in night-blooming flowers such as jasmine. Linalool is widely used in the flavor and fragrance industry as a component of foods, cosmetics, and perfumes (including the well-known Yves Saint Laurent "Opium"). A moth will hover in front of the flowers and extend its proboscis down the long tubular portion of the corolla (fig. 313) to reach the nectar at its base. In doing so, the moth's hairy body will become covered with pollen that is then easily transferred to the large, cross-shaped stigma of another flower (or sometimes to that of the same flower) (fig. 314). The best known of these hawk moths is the evening primrose moth (*Schinia florida*), the female of which will often lay her eggs in the flower. There the caterpillars hatch out and feed on the flowers and seed capsules until they are ready to pupate. When the adult emerges the following summer, the cycle begins again. The primrose moth is active at night but is sometimes caught napping in the flowers during the day. Twice my attention was drawn to a flower of evening-primrose because of its unusual yellow and pink colors. On closer inspection I could see that the pink color belonged to a moth that had tucked itself, headfirst, into the flower until nightfall (fig. 315). Hawk moths visit the flowers of many other night-blooming flowers (see the essay on Jimsonweed in this book), including other species of *Oenothera* and those of plants that were formerly placed in the genus *Gaura* (fig. 316). On overcast mornings, many other insects take advantage of the still-open evening-primrose flowers by feeding on the remaining pollen (fig. 317). They may also eat the buds, leaves, immature fruits, and seeds (fig. 318).

The flowers of Onagraceae have specialized pollen that produces sticky threads known as viscin threads. Attached by these viscin threads, the pollen grains stick together to form masses of pollen that may exceed 100 grains per clump. It is therefore likely that

Fig. 312. The stigmatic lobes have not yet spread apart in this opening flower, yet the anthers have already released their pollen. Note how the pollen adheres in clumps to the sticky threads.

Fig. 313. A lateral view of partially open evening-primrose flowers, each flower showing the long light yellow-green corolla tube that arises from a short green ovary in the leaf axils of the plant.

Evening-primrose

Fig. 314. The four petals of evening-primrose are bilobed. Note the large, protruding, X-shaped stigma of the flower, which moths must brush against as they probe deeply into the corolla to get nectar.

Fig. 315. A yellow and pink evening-primrose moth, a type of nocturnal hawk moth, spends the day resting within a flower. It may have already laid its eggs in the flower.

these clusters of pollen on sticky threads are left on the protruding stigma of the same flower, contributing to the self-pollination of *Oenothera*. Because of evening-primrose's efficient method of pollen delivery, the pollination of the flowers and subsequent fertilization of the ovules is more or less ensured. Accordingly, *Oenothera* maintains a low pollen to ovule ratio (i.e., the anthers produce relatively few pollen grains in comparison to the number of ovules to be fertilized). There is no need for the plant to produce an excessive number of pollen grains when fertilization is almost assured to take place. Based on similar outcomes in fertilization, but via a different method of mass pollen delivery, low pollen/ovule ratios are also found in the flowers of milkweeds and orchids.

The story of evening-primrose reproduction is complicated and unusual. Rather than having pairs of chromosomes as found in the majority of other organisms, the 14 chromosomes of most *Oenothera* egg cells are arranged in a ring of single chromosomes, attached end to end, in which those from the female parent alternate with those of male origin. When the chromosomes separate to go into different reproductive cells, as happens in most other plants, the adjacent chromosomes of evening primrose disjoin and go into different daughter cells, but only with *like* chromosomes from the same parent. Thus, rather than pairing randomly as in most other organisms, all maternal chromosomes join with others of their kind, and all of paternal origin do the same. The result is that there are only two kinds of eggs and sperm, and each is identical in terms of its set of genes to the two reproductive cells that had initially united to form the plant. But each set of genes includes one gene (termed a lethal gene) that prevents it from existing with an identical

Fig. 316. (Far left) A species of hawk moth is covered with pollen from the anthers of the *Oenothera gaura* flower it has just visited.

Fig. 317. (Left) A small flower fly visits an evening-primrose flower that has remained open into an overcast morning. It is scavenging pollen from the anthers.

Evening-primrose Family

Fig. 318. The exit hole in this unopened flower bud is evidence that some unidentified insect was feeding in the bud and has emerged from it.

gene within the same gene set. Since many evening-primroses are self-pollinated, this means that there can be no AA or BB plants as could occur with a normal mixing of genes according to Mendel's theory of heredity (which states that there is a probability of one of the four offspring being AA, one BB, and two AB). In evening-primrose, *only* AB plants would result (the same as the parent genome). Thus, a balance of genes is kept, and plants in all subsequent generations will also be AB plants.

Outcrossing does occur via insect-mediated cross-pollination; thereby a normal mixture of genes and resultant hybrid vigor are not lost. Rarely a freakish plant is encountered in which the stem, inflorescence, or other plant parts have become misshapen or contorted (fig. 319). This abnormality, termed fasciation, results from a defect in the apical meristem (the terminal growth point), caused by a genetic mutation, a bacterium, a virus, a fungus, or some other cause. Fasciation is seen in plants of other families as well (see the essay on goldenrods).

Evening-primrose was first brought from Virginia to Padua, in northern Italy, in 1614. Padua is the site of one of the earliest botanical gardens, Orto Botanico di Padova, which was founded ca. 1545 and continues to this day as a functioning botanical garden. Recognized as the oldest botanical garden in continuous operation in the same location, it is designated as a UNESCO World Heritage Site (fig. 320). Like all botanical gardens of the time, the Orto Botanico was a teaching garden where medicinal plants could be studied by herbalists, physicians, and students at the affiliated University of Padua.

There are 145 known species of *Oenothera* (if members of the genus *Gaura* are included, as is presently done by the nomenclatural authorities, Angiosperm Phylogeny Group and Tropicos), but it is *O. biennis* that deserves the name "common" evening-primrose as it is by far the most widespread species. When Linnaeus chose the name *Oenothera* for this genus he used a name that Theophrastus, an early Greek scholar and botanist, had applied to another plant, most likely *Epilobium*, another member of the Onagraceae. Regarding the vernacular name, it *is* accurate that the flowers of evening-primrose open in the evening, but the plant is *not* related to primroses, which belong to another family, the Primulaceae. A hyphen inserted before the second part of a compound plant name is sometimes used to indicate that a plant is not a true member of the named group (e.g., trout-lily, marsh-marigold, rattlesnake-plantain—and evening-primrose). This is one of the optional spellings, others being "evening primrose" and "eveningprimrose." Evening-primrose overwinters as a perennial plant does, but as in annual plants, it blooms only once before dying. It is one of a group of unrelated plants (including mullein, Queen Anne's lace, teasel, and thistles) having a lifestyle termed biennial. The scientific epithet *biennis* is somewhat misleading. Evening-primrose produces a circular rosette of leaves during its first year of growth, the rosette of leaves, lying

Fig. 319. The abnormal growth of this plant has been caused by elongation of the apical meristem cells at the growing tip, perpendicular to the stem, causing the stem of this evening-primrose to become broad and flattened—a condition called fasciation.

Evening-primrose

flat on the ground, thus, escaping harm from the worst of winter's weather. The ecologist Katherine Gross followed several species of biennials, including evening-primrose, over a period of years. She and her coworkers discovered that flowering doesn't necessarily happen in the second year. It might not occur until the third or fourth year—or never—depending on the environmental conditions. Gross discovered that flowering was determined more by the diameter of the rosette of leaves produced the previous year than by the age of the plant. Plants having rosettes smaller than 3 cm in diameter at the end of their first season of growth never flower in the following year. The larger the rosette size, the greater the likelihood that the plant will flower during the subsequent year. Experimentally, all plants with rosettes measuring more than 14 cm at the end of the year produced a flowering stalk during the next growing season. By the end of the third year, all plants had either flowered or died.

Evening-primrose is best known for the reputed medicinal properties of its seed oil. The seeds are tiny, only 2–3 mm in length, and are scattered by gravity or wind when the capsules split open in late fall or winter. Cultivars have been developed with capsules that don't split open on their own, allowing the seeds to be harvested more efficiently. The seeds are a source of an omega-6 essential fatty acid (gamma-linolenic acid [GLA]), which is sold as an over-the-counter product known as evening-primrose oil (EPO, an essential oil). The term "essential" indicates that this fatty acid is something our bodies require. While EPO is not produced in our bodies, it can be obtained from foods, especially nuts and seeds. The oil has been used for treating a variety of conditions, from skin irritation to cancer. Popularity of EPO increased in the latter part of the twentieth century when many new uses were developed in China and other countries where *Oenothera biennis* has become naturalized. A report by the Mayo Clinic states that the efficacy of EPO for treating most diseases and other conditions has not been documented. Based on studies in laboratory animals, the few diseases for which there may be potential benefit from treatment with GLA are eczema, skin irritation, diabetes, and multiple sclerosis. Follow-up testing is needed to determine whether these effects will be shown to be beneficial to humans. Warnings on the product include risk of seizures, headache, and gastrointestinal problems, so be aware of these risks.

In addition to its professed medicinal properties, evening-primrose is also used as a food plant. The fleshy roots can be boiled and are said to taste like parsnips; the young seedpods may also be cooked and eaten. In addition, the flowers can be added to salads to provide a colorful touch.

Even hardy, "weedy" wildflowers are sensitive to human-caused pollution. In a Canadian study it was demonstrated that reproduction in a closely related species of evening-primrose (*Oenothera parviflora*) was negatively affected by simulated air pollution, specifically that mimicking pollution caused by acid rain. The study was carried out in Ontario, an area strongly affected by acid rain. Pollen germination, pollen tube growth, and stigma receptivity all significantly declined when the pH was low (more acidic). Thus, evening-primrose reproduction was negatively affected in three different ways. Such results provide a warning regarding the effect of pollution on wild populations of plants in our fields and forests.

Fig. 320. A view within the Orto Botanico, an Italian botanical garden founded on this site in 1545. In the background are the steeples of the Basilica of St. Anthony of Padua. It was here that evening-primrose was first introduced to Europe for its medicinal properties.

Gentian Family

Fig. 321. A spectacular plant of fringed gentian (*Gentianopsis crinita*) that produced more than 50 flowers during its month-long blooming period.

Fringed Gentian

Fig. 322. Yellow gentian (*Gentiana lutea*) is a tall, large-leaved plant with roots that yield compounds used in medicines and for flavoring alcoholic beverages. It is the tallest member of its genus. Seen here growing in the French Alps.

Fringed Gentian

Gentianopsis crinita (Froel.) Ma
Gentian Family (Gentianaceae)

The gentians are among the last native wildflowers of the Northeast to flower, blooming from late August into October. As such they provide a regal farewell to the flowering season. I have a particular fondness for blue and purple wildflowers. To my eye the fringed gentian (fig. 321) is the prettiest of our northeastern gentian species, but I find all gentians worthy of interest and, therefore, will include a few other species in this account.

Habitat: Open, sunny areas in wet, calcareous meadows.

Range: Most states along the eastern coast of the United States, from Maine to northern Georgia and inland to Tennessee; west from West Virginia to northern Illinois, northern Iowa, and North Dakota—*but* with very limited distribution in some of the states where it is found; in Canada fringed gentian occurs from southern Quebec west to southern Manitoba.

Gentians (in the broad sense of the gentian family) are found nearly worldwide in both the Northern Hemisphere (where blue-flowered species are prevalent) and the Southern Hemisphere (where red-flowered species predominate). It is most likely that the color difference may be attributed to the coevolution of the plants with their primary pollinators: bees in the Northern Hemisphere and birds in the Southern Hemisphere. White flowers predominate in New Zealand, and yellows and pinks may be found throughout various other regions.

The gentian family takes it name from its largest genus, *Gentiana*, and that in turn, from King Gentius (Genthios), king of Illryia, a region of the western Balkan Peninsula that corresponds roughly to present-day Albania and the countries that formerly composed Yugoslavia. Legend has it that it was Gentius (Illryia's last king before the region's defeat by the Romans in 167 BCE) who first discovered the medicinal properties of gentians (most likely from the tall, yellow European gentian, *Gentiana lutea* [fig. 322], a common plant long considered a "bitter herb," and used in the treatment of venomous animal bites, liver and

Gentian Family

Fig. 323. A couple enjoying Aperol Spritz aperitifs at a restaurant in Cinque Terre, Italy.

Fig. 324. Spotted gentian (*Gentiana punctata*) is another yellow-flowered European gentian used for its medicinal properties and as a bitter flavoring.

stomach ills, inflamed eyes, and skin sores). Additionally, the roots of yellow gentian are ground with other herbs to produce a number of aperitifs and digestives, most of which are mixed with other ingredients to make them more palatable. In Europe these beverages are much relished for their bitter flavor and curative properties. In the summer of 2015, the Aperol Spritz, made with yellow gentian, bitter orange, rhubarb, cinchona bark (quinine), and other botanicals mixed with seltzer water, was the most popular pre-dinner drink on tables in northern Italy (fig. 323). A few of the other aperitifs that use yellow gentian as a component are Appenzeller, Campari, Cinzano, and Suze.

The aforementioned "yellow gentian" is not the only European species with yellow flowers; spotted gentian (*Gentiana punctata*), with yellow corollas dotted with minute purple dots (fig. 324), is common above 1500 meters in the Alps and other European mountain ranges. As with *G. lutea*, the roots of this species contain bitter-tasting glycosides, namely gentiopicrin and amarogentin. The latter compound is ranked 71 on a bitterness scale (BitterDB) that ranges from 7 to 942 and is used to rate this quality in nearly 700 natural and synthetic compounds (for comparison, the bitterness of caffeine is 46 and of theobromine [a component of cacao, from which we derive chocolate] is 749—with larger numbers indicating greater bitterness). *Gentiana punctata* is harvested for medicinal purposes as well. Over-collecting of these and other gentian species for both medicinal and for horticultural use has caused a decline in their numbers, and some gentian species are now considered threatened in parts of their ranges.

In the first half of the twentieth century, a popular, but somewhat bitter, nonalcoholic beverage made with unspecified gentian root extracts mixed with wintergreen (*Gaultheria procumbens*) and other ingredients was sold in this country under the brand name Moxie. Moxie was formulated in Maine as a patent medicine (a "nerve tonic") during the 1880s and was purported to increase health and energy much like the claims for today's highly caffeinated energy drinks. As a result, the term "moxie" came into use to describe an abundance of vigor or "spunk" when describing a person. Moxie was consumed as a carbonated soft drink primarily in the Northeast, and in the 1920s Moxie outsold Coca-Cola on a *nationwide* basis. Today Moxie has a decreasing number of devotees, but various towns in Maine still hold annual Moxie festivals that attract tens of thousands of attendees, and Moxie advertising paraphernalia has become a popular collectible. Now hard to find outside of Maine and a few other localized New England areas, Moxie can be ordered online if you wish to try it. I found it somewhat similar to cream soda.

I am sometimes asked about the origin of a product called gentian violet (crystal violet) by people wondering if it is derived from a member of the gentian family. This water-soluble dye is derived *not* from gentians, but from coal tar. It is used primarily in medicine—to stain bacteria for examination and identification and for treating certain fungal infections

Fringed Gentian

such as oral thrush. Another esoteric use of this compound is based on its property of producing an image when in contact with the fats found in sweat. It is thus used to reveal latent fingerprints on the print-lifting tapes used in forensics. Gentian violet was given its common name because of its color, which resembles that of the flowers of some gentians.

I am not alone in my admiration of gentians. Various species of gentian have been painted, memorialized in poetry, depicted on stamps, and used in horticulture. Two well-known American poets, William Cullen Bryant and Robert Frost, wrote odes to the gentian. As previously noted, the North American gentians are late season bloomers, some species waiting even until October to produce their flowers. Bryant refers to this tardiness in his poem, "To the Fringed Gentian":

Fig. 325. Narrow-leaved gentian (*Gentiana linearis*) is appropriately named for its narrow leaves (visible on the middle plant). It is one of a number of gentians known as bottle (or closed) gentians because their flowers never quite open.

> Thou blossom bright with autumn dew,
> And coloured with the heaven's own blue,
>   That openest when the quiet light
>   Succeeds the keen and frosty night.
>
> Thou comest not when violets lean
> O'er wandering brooks and springs unseen,
>   Or columbines, in purple dressed,
>   Nod o'er the ground-bird's hidden nest.
>
> Thou waitest late and com'st alone,
> When woods are bare and birds are flown,
>   And frosts and shortening days portend
>   The aged year is near his end.
>
> Then doth thy sweet and quiet eye
> Look through its fringes to the sky,
>   Blue—blue—as if that sky let fall
>   A flower from its cerulean wall.
>
> I would that thus, when I shall see
> The hour of death draw near to me,
>   Hope, blossoming within my heart,
>   May look to heaven as I depart.

Many people seem to have a broad perception of the color blue that includes all shades of lavender, violet, and purple. I beg to differ with Bryant's comparison of the color of fringed gentian to that of the sky. Although the perception of color is somewhat subjective, it is generally acknowledged that in the northeastern United States most gentian flowers range from a deep purple to lavender or occasionally to pink (fig. 325)—and rarely to pure white (fig. 326). Those who consider gentians to be blue in color are perhaps thinking of the gentians of Europe and of other parts of the world, flowers that display a range of blues from the pale blue of *Gentiana sedifolia* (fig. 327) of Central and South American mountainous regions, to the bright, clear blue of *G. verna* (fig. 328), or the deep royal blue of *G. clusii* (fig. 329), both from calcareous mountain regions of Europe. It is the intensely blue species that are rock garden favorites—though often tricky to grow outside their normal habitat. Aside from being difficult to grow in garden situations, gentians are sometimes eaten by deer despite their bitter properties.

Fig. 326. In *Gentianopsis crinita*, occasional albino plants occur, such as this white-flowered fringed gentian found growing among hundreds of purple ones. This genus typically has four petals.

Gentian Family

Fig. 327. *Gentiana sedifolia* is a low cushion plant adapted for living at high altitudes in Central and South America (it is the only Andean gentian), where daytime temperatures are high, nighttime temperatures low, and sunlight and ultraviolet radiation strong. Its leaves resemble those of *Sedum*, thus the specific epithet, *sedifolia*. The toothed folds (pleats) between the five petals give the flowers the appearance of having ten petals.

Fig. 328. One of the shortest gentians, spring gentian (*Gentiana verna*), is widespread throughout Eurasia, most commonly in central and southeastern Europe and rarely in the British Isles. Its bright blue color is striking. The white flowers are those of *Dryas octopetala*, a member of the rose family.

Robert Frost also chose the fringed gentian as the topic of his poem, "The Quest of the Purple Fringed." After finally finding the fringed gentian in flower, Frost headed home feeling that summer was truly over and fall had begun now that the fringed gentian was in flower:

Fig. 329. *Gentiana clusii* grows in limestone and dolomitic mountains such as those near the Tre Cime formation in the Dolomites, where this photo was taken. The large flowers are deep blue and differ only slightly from those of *G. acaulis*, a species more common in more acidic soils.

> Then I arose and silent wandered home,
> And I for one
> Said that the fall might come and whirl of leaves,
> For summer was done.

Yet a third author, Emily Dickinson, compared herself to the gentian, a reference to her being a "late bloomer." In her poem "God Made a Little Gentian," she calls to mind the story of the ugly duckling that in time is transformed into a swan. In that poem, she depicts the color of our local gentians more accurately than Bryant:

> It tried to be a rose
> And failed, and all the summer laughed.
> But just before the snows
> There came a purple creature
> That ravished all the hill;
> And summer hid her forehead,
> And mockery was still.
> The frosts were her condition;
> The Tyrian would not come
> Until the North evoked it.
> "Creator! Shall I bloom?"

Dickinson's phrase, "The Tyrian would not come," refers to the color of the flower, which doesn't reveal itself until colder temperatures arrive ("Tyrian" was a name given to a purple color [actually more red-violet or magenta] derived from certain species of murex sea snails). Murex snails yield a rare and expensive,

Fringed Gentian

long-lasting dye described in historical accounts dating back to the fourth century BCE as being worth its weight in silver. During Roman times it was reserved for the dyeing of royal robes).

One last, and more personal, literary reference to gentians is that of Lucy Maud Montgomery, author of *Anne of Green Gables*, with whom I share an ancestor. "Maud" credits a poem titled "The Fringed Gentian" (part of a story that speaks of reaching lofty goals, and which was published in *Godey's Lady's Book* in 1884), as having been an inspiration for her writing career. In her autobiography she tells of clipping the poem and pasting it into the portfolio in which she kept her essays and letters. Seeing the poem each time she opened the portfolio motivated her to continue her literary efforts.

To what advantage is it to flower so late in the year—sometimes even after the leaves of trees and shrubs have begun to turn color (fig. 330) and just before the first killing frost? For one thing there is less competition for pollinators. Generally, only some of the asters and goldenrods are still in flower (fig. 331) providing sustenance for the season's remaining insects. Invariably, the principal pollinators of our northeastern gentians are bumblebees (*Bombus*). Bumblebees are large and densely covered with hair, which insulates them and allows them to be active even in the cooler temperatures of autumn (as well as in early spring). On sunny days bumblebees enter into the corolla (fig. 332) and probe the base of the flower where nectar is produced. When bees visit the flowers only their hindquarters (if even that) remain visible (fig. 333).

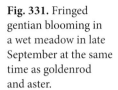

Fig. 331. Fringed gentian blooming in a wet meadow in late September at the same time as goldenrod and aster.

Fig. 332. A bumblebee (*Bombus* sp.) beginning its descent into the depths of a fringed gentian corolla in search of nectar.

Fig. 330. Our northeastern fringed gentian (*Gentianopsis crinita*) doesn't flower until late September into mid-October in the Northeast, by which time the leaves of many trees and shrubs, such as this lowbush blueberry (*Vaccinium angustifolium*) have turned red.

Fig. 333. Only the tail end of this bumblebee is visible as it backs out of the flower of fringed gentian. Note the pollen grains that are caught on the hairs of its abdomen and on its legs.

Gentian Family

Fig. 334. The margins of the sepals on this fringed gentian bud are so membranous that they are transparent. Note how the flower's petals are contorted so that each overlaps the adjacent petal to the right.

Fig. 335. (Left) The fruit of fringed gentian is a two-parted dry capsule that splits open to expose the seeds. As wind or other disturbance shakes the fruit, the seeds are dislodged and dispersed. Seeds germinate in spring, and in the first year they form rosettes of leaves. Flowers are not produced until the following fall. **Fig. 336.** (Right) Large carpenter bees (*Xylocopa* sp.) also visit the flowers of fringed gentian, but I have only observed them "robbing" the nectar by inserting their proboscises into the base of the flower rather than by entering the open corolla. By "stealing" nectar, they do not contact the reproductive parts and therefore are not effective pollinators.

Like some other species in this book (e.g., evening-primrose and Queen Anne's lace), fringed gentian has a biennial lifestyle. Only a small rosette of leaves is present during the first year, a period during which the leaves manufacture food and store it in the underground roots. During the plant's second year a branched leafy stem grows upward and produces several flowers. As with many of our spring ephemerals, the fall-flowering gentians open only on bright, sunny days, remaining closed in conditions when pollinators are less likely to be about—that is, overcast or rainy days and at night. Prior to opening, each petal overlaps the one to its right in a spiraled manner (fig. 334—note also the transparent margins of the calyx lobes seen in this image).

Once the flower's ovules have been fertilized, the ovary develops into a two-chambered capsule (fig. 335) that, after opening, remains upright on the plant throughout winter, the wind shaking its tiny seeds onto the surrounding soil. In spring, when sunlight increases and temperatures warm, the seeds germinate, developing into first-year rosettes by autumn.

Several times I've observed carpenter bees (*Xylocopa* spp.) visiting the flowers of fringed gentian, but in every case the bees, rather than entering the flower through the open corolla, instead inserted their proboscises from the outside of the flower into the base of the corolla (fig. 336) where the nectaries are located, thus failing to serve as effective pollinators of the flowers.

The name *Gentiana* was used for the largest genus within the gentian family as early as the first century CE by Pliny the Elder. The name *Gentiana* was later used in 1700 by the French botanist Joseph Pitton de Tournefort, and Linnaeus maintained the name for plants of that genus when he published his *Genera Plantarum* in 1737.

The gentians have a long and complicated taxonomic history. It has been only within my lifetime that the genus *Gentiana* was split, with several species segregated into a separate genus, *Gentianopsis*. I have simplified the various changes here to include only those species relevant to this chapter.

In 1951, a Chinese botanist, Yu-Chuan Ma, determined that because of morphological differences, many of the Asian gentians with fringed petals should be removed from *Gentiana* and placed into a newly created genus, *Gentianopsis*. As Ma did not

Fringed Gentian

have specimens of all New World fringed gentians for comparison, those he could not examine were left in the genus *Gentiana*. In 1957, the morphological diversity of those remaining *Gentiana* led John M. Gillett to segregate some of them further into a related genus, *Gentianella*. Then, in 1962, Dr. Hugh Iltis reclassified some species within the genus *Gentianella* (namely the fringed species, *G. crinita* and *G. procera*) as *Gentianopsis*. At that time, he also consolidated within the species *G. procera* many of the lesser *Gentianella* species whose differences he did not consider sufficient to delineate them as separate species. The possibility exists that these plants that exhibit a slightly different appearance are newly evolving species that have arisen since most of North America became free of glacial ice—about 10,000 years ago. Species documented to have evolved rapidly since that time include *Iris lacustris* and *Cirsium pitcheri*. Iltis also added a few more distinguishing characters of the genus *Gentianopsis* to those given by Ma: the flowers have a distinct gynophore (a stalk holding the female reproductive parts aloft), a large, bilobed stigma, and fringed corolla lobes. In addition, the nectaries of *Gentianopsis* are attached to the interior of the corolla (fig. 337) rather than to the ovary as in *Gentiana*.

Our two northeastern species of *Gentianopsis* are easily recognized by their delicately fringed petals. The fringes are thought to deter nectar thievery by crawling insects, such as ants, from entering the corolla. *Gentianopsis crinita* is more eastern in range and *G. procera* more Midwestern, extending even into the northern Rocky Mountains. There is some geographic overlap where the two species may intergrade. With some exceptions, *Gentianopsis crinita* is larger in all respects and grows in moister, more neutral soil; *G. procera* is shorter, has fewer, smaller, and often lighter colored flowers with longer pedicels, and requires a more alkaline soil. Iltis and others hypothesize that these two species originated from a single species that diverged into two populations that were geographically isolated by the Pleistocene glaciers, thus leading to speciation during that time.

To provide a quick introduction to some of our other gentian genera and discuss how they differ from fringed gentian, I'll begin with the pine barrens gentian (*Gentiana autumnalis*), a perennial plant of nutrient-poor and sandy pine barren areas from southern New Jersey to South Carolina. It is one of the last of the gentians to flower each year. One must search carefully for glimpses of purple among the weedy roadside vegetation (fig. 338); once sighted it is well worth pulling off the road to investigate these delicate flowers. Although it is typical for flowers of *Gentiana* to have five-lobed corollas (fig. 339), individuals with four or six petals (figs. 340–341) are sometimes encountered. The corollas of *Gentiana* have folds known as plicae, or pleats, between their lobes, easily seen in fig. 339.

Most of the flowers of pine barrens gentian are a deep blue-violet sprinkled with bright, yellow-green dots on the corolla lobes and stripes in the fused portion of the corolla. The flowers are open and thus

Fig. 337. A fringed gentian flower has been cut open longitudinally to show the interior. The ovary is elevated on a stalk (termed a gynophore), a characteristic of the genus *Gentianopsis*, and is topped by a bilobed stigma; the stamens are attached to the corolla. Small green nectaries are visible at the base of the corolla, below the stamens.

Fig. 338. A typical roadside habitat of pine barrens gentian (*Gentiana autumnalis*) in southern New Jersey.

Gentian Family

Fig. 339. Two normal flowers of pine barrens gentian with five petal lobes. Note the sprinkling of green dots and the green lines leading into the nectar-containing base of the flower. The "pleats" between the petal lobes of flowers in the genus *Gentiana* are evident here. The flower on the left has just opened, and its stamens are still pressed tightly against the pistil. In the right-hand flower, the stamens have spread away from the pistil, and the anthers are releasing pollen. The female stigma in the center has not yet opened its two lobes, thus preventing self-pollination.

accessible to visitation by many types of insects, including small bees (see figs. 340–341) and butterflies (fig. 342). One can see how pollen brushed onto a butterfly's wings could be transferred to the stigma of a subsequently visited flower. However, it has been reported that the most effective pollinators are bumblebees.

For comparison, let's look at one of the closed gentians (also called bottle gentians) in the same genus. *Gentiana linearis* is a tall plant of wet meadows with flowers that never truly open (fig. 343). Although "closed," the flowers do have small pleats between their small corolla lobes that can be seen by manually opening the corolla. Narrow-leaved gentian is a perennial plant found more often in the northern part of our range (northern New England, northern New York, northern Pennsylvania, the Upper Peninsula of Michigan, and isolated populations in New Jersey south to the mountains of Virginia and Tennessee, as well as in the eastern half of Canada (with the exception of the Maritime Islands and Nunavut). The flowers of the

Figs. 340 & 341. (Left and middle) These two images of an aberrant pine barrens gentian flower with six petal lobes show small, solitary bees that appear to be gleaning pollen from the withering anthers. The stigma is now in its receptive phase. **Fig. 342.** (Right) This swarthy skipper butterfly visited several flowers, inserting its proboscis into the openings at the base of the corolla to imbibe nectar. Note how its wings brush against the anthers of this flower.

Fig. 343. Narrow-leaved gentian (*Gentiana linearis*) growing in a wet meadow in the Adirondacks.

Fig. 344. (Right) A bumblebee is beginning to back out of a flower of narrow-leaved gentian, having taken nectar from the nectary at the base of the flower. Note that it also carries pollen in special structures on its hind legs. Bumblebees are strong enough to force the "closed" flowers open in order to get their reward.

Fringed Gentian

Fig. 345. Stiff gentian (*Gentianella quinquifolia*) is an upright, almost shrubby plant bearing many small lavender-pink flowers.

closed gentians appear as though they are in bud and about to open—but they never do. However, lest you fear that the closed flowers exclude pollinators, strong bumblebees are well adapted for forcing their way into the closed corolla lobes to drink nectar secreted at the base of the ovary and/or to collect pollen (fig. 344). In fact, in its collecting efforts, the bee sometimes completely disappears into the flowers as it does so. A furious buzzing sound may be heard during the time the bee is in the flower.

Yet another related genus is the small-flowered *Gentianella*, represented in the Northeast by the many-branched, stiff gentian (*Gentianella quinquefolia*), a denizen of woods and open wet places in portions of all states from Maine to Tennessee and northern Georgia (with the exception of Rhode Island), and in Ontario, Canada. Various sources list stiff gentian as an annual, a biennial, or a perennial; perhaps the latitude at which the species is growing determines the length of its reproductive cycle. The common name derives from its "stiff" habit (figs. 345–346), and the species is most easily recognized by its upright leafy stems that develop clusters of small, narrow (usually pinkish-lavender) flowers at their tops and in the axils of the upper leaves (see fig. 346). The flowers have no pleats between their lobes and are only barely open if the weather is sunny; they are most likely pollinated by long-tongued bees.

Take time to notice the arrangement of the leaves of stiff gentian, which arise from the stem in opposite pairs that alternate up the stem such that each pair is attached to the stem at a more or less right angle to the pair immediately below. This arrangement (termed decussate) prevents each pair of leaves from being shaded by the leaf pair above, thus maximizing the amount of sunlight striking the individual leaves.

Do extend your search for wildflowers into the fall so as not to miss this last, beautiful hurrah!

Fig. 346. Although the flowers of this stiff gentian plant appear closed when photographed on a cloudy day, they will open—but only slightly—when the sun appears.

Orchid Family

Fig. 347. Two immaculate inflorescences of white fringed orchid (*Platanthera blephariglottis*) glow in a wet meadow as evening descends.

Fringed Orchids

Fig. 348. The bright orange flowers of orange fringed orchid (sometimes called yellow fringed orchid) attract butterflies such as this spicebush swallowtail. The butterfly has already visited other flowers, as evidenced by the two yellow pollinia that are stuck to its eyes. Note how the spots on the forewing are blurred due to the butterfly's need to constantly flutter its wings in order to maintain its balance on the narrow lip of the flower.

Fringed Orchids

Platanthera spp.: White Fringed Orchid, *P. blephariglottis* (Willd.) Lindl.; Orange (sometimes called Yellow) Fringed Orchid, *P. ciliaris* (L.) Lindl.; Crested Yellow Orchid, *P. cristata* (Michx.) Lindl., which has orange flowers, not yellow; Greater Purple Fringed Orchid, *P. grandiflora* (Bigel.) Lindl.; Ragged Fringed Orchid, *P. lacera* (Michx.) D. Don; and Lesser Purple Fringed Orchid, *P. psycodes* (L.) Lindl.

Orchid Family (Orchidaceae)

Of the approximately 200 species of *Platanthera*, most are from temperate regions of the world, with 32 species found in North America, and 20 of those in the greater Northeast (some having just a toehold in the southern or western margin of the region). The genus name is derived from the Greek *platy*, meaning "wide," and *anthera*, meaning "anther," referring to the broad anthers with their widely separated pollen sacs. *Platanthera* species are known as bog orchids because of the boggy habitat of many of the species, or rein orchids for the imagined resemblance of some of the flowers to the reins of a horse. Of those in the Northeast, eight are commonly called fringed orchids because of the fimbriate margins of their lips (the large, lower petals); the six listed above are discussed here. Fringed orchids are much sought after by orchid aficionados who wish to admire their delicate beauty in the wild.

Habitat: Habitat preferences vary according to the species: White fringed orchid (fig. 347) is found in damp, acidic, frequently sandy soils but also inhabits sphagnum bogs; orange fringed orchid (fig. 348) grows in lightly wooded to sunny moist areas with acidic, peaty, or sometimes sandy soils; crested yellow orchid favors open areas in moist pine woodlands; greater purple fringed orchid can grow in full sun or shade in damp soils, usually under conifers; ragged fringed orchid is found in wet, acidic forests and meadows; and lesser purple fringed orchid grows in wet meadows, woodlands, and roadside ditches.

Range: Throughout the range of this book, from Maine to Minnesota and adjacent Canada, south through Virginia plus the southeastern corner of Missouri.

I must admit, I am a sucker for fringe—not on my clothing or lampshades, please—but when adorning flowers, it is something that draws me as though I were one of their pollinators (you might have noticed that fringed gentian and buckbean, with their beautiful purple- or white-fringed flowers, have also found a place in this book). The fact that these small flowers are embellished in such an extravagant manner piques

Orchid Family

not only my aesthetic sense, but my curiosity as well. Just as long eyelashes are said to attract members of the opposite sex in humans, the fimbriate margins on the lips of fringed orchids are believed to help entice their pollinators—primarily butterflies; the fringe is even visible in the opening buds (fig. 349).

During the second half of the nineteenth century, Charles Darwin studied the multiple methods of pollination found in the orchids of Europe. At the same time Asa Gray, the great American botanist, was studying the pollination of North American orchid species. What was fascinating to Darwin (and to other orchidologists since then) are the complex relationships between orchids and their pollinators. In orchids, flower form is intricately related to reproductive success. By examining a flower carefully, one can often deduce what type of pollinator would be necessary for the effective transfer of the male pollen to the female stigma of another flower. Because this mechanism is so specifically designed for each orchid/pollinator relationship, orchids don't often hybridize in nature. However, when the normal constraints of flowering time, location, and specific pollinators are removed, there are few barriers to prevent crossing one species with another. Since the nineteenth century, however, much of the research on orchid pollination has focused on the generally showier tropical species. Orchid growers have taken advantage of the ease with which species can be cross-pollinated manually and have bred thousands of hybrids through crossing and backcrossing orchids—even among different genera.

The reproductive parts of orchids are unique. The male and female parts are fused into a single structure called the column. Some more primitive orchids (e.g., the lady-slipper orchids in the genus *Cypripedium*) shed their pollen as single, somewhat viscid, grains, but in most species the pollen grains are grouped into waxy structures called pollinia. These pollinia, which in *Platanthera* are usually mealy in texture, are connected to a stalk (the caudicle), which in turn is attached to a very sticky, adhesive pad (the viscidium). This entire male structure, of which there are two in the fringed orchids (fig. 350), is the unit of pollen transfer. The stigmatic surface is located on the column just between or below the anthers. The only other plant group to have anything similar in terms of pollinia is the genus *Asclepias* (milkweeds), discussed elsewhere in this book.

The orchid species encountered in the Northeast, and, indeed, throughout most of the temperate regions of the world, are terrestrial, with exceptions occurring in warm temperate to subtropical regions such as Florida. This contrasts sharply with the majority of orchid species living in the tropics, where they are primarily epiphytic (from the Greek, meaning "on a plant"), living on the branches or trunks of trees high in the forest canopy (fig. 351). Growing as they do in the treetops, tropical orchids have access to greater light than they would as ground dwellers on the shaded forest floor. They do not take moisture or nutrients from the host tree as would a parasitic plant, but, conversely, when a number of epiphytes grow in close proximity, the host tree may produce supplementary roots from that branch that grow into the epiphytic mass and gather

Fig. 349. The fringed lip (labellum) of this orange fringed orchid is exposed as the bud begins to open.

Fig. 350. A pollinium of white fringed orchid clings to a finger by means of an adhesive pad (the viscidium). The mass of pollen grains (the pollinium) at the lower end is attached to the adhesive viscidium by a stalk (the caudicle). Together, the three structures constitute a dispersal unit.

Fringed Orchids

Fig. 351. (Left) The showy flowers of this *Cattleya violacea* orchid (*the* corsage orchid of the 1950s) brighten the dark rainforest canopy, where the plant lives as an epiphyte on the trunk of a tree in Amazonas, Brazil. Most orchids of tropical rainforests are epiphytic on other plants or even epilithic on rocky cliff faces, both substrates allowing the plants a more exposed position where they may receive more light.

Fig. 352. The long tubular spur of the flower of white fringed orchid holds nectar that can be accessed only by an insect with a long tongue (proboscis), such as a moth or butterfly. Here, one pollinium has been removed from an anther, but it did not adhere to the visiting insect, remaining hanging from the flower.

additional nutrients from the biomass. Epiphytic orchids have specialized roots with a unique layer—a spongy covering called velamen, which allows the orchid to take up moisture from the excessively humid air and nutrients for the tree from the debris that accumulates among the epiphytic masses that often develop on branches and trunks in the forest canopy.

Until the mid-1970s, most orchids that are now in the genus *Platanthera* were placed in the genus *Habenaria*, and even earlier many were included in the genus *Orchis*. *Platanthera* is distinguished from *Habenaria* by bearing the pollinia in broader anthers and by having a single stigmatic surface (*Habenaria* has a bilobed stigma); they also have a different type of root structure. Most (but not all) species of *Habenaria* are tropical.

Among our northeastern fringed orchids are species with white, orange, pink-to-purple, or greenish-white flowers. White fringed orchids appear almost perfectly adapted for moth pollination: their light color is easily seen at night, they are fragrant during the evening, and their floral nectar is held in a long tubular spur (fig. 352) accessible only to insects, such as nocturnal moths, that have a long proboscis (tongue). Indeed, in a 1976 study, several species of long-tongued hawk moths were documented visiting the flowers of *Platanthera blephariglottis* during the evening hours. Based on the numbers of capsules produced, white fringed orchids growing in only semi-open areas were pollinated with the same frequency as those in open areas, which were thus more visible. This finding may be viewed as an indication that the moths located the flowers primarily by fragrance rather than by visual means. In yet another study of this species, the flowers were found to be weakly fragrant both at night *and* during the day and were observed to be visited by several different species of

Fig. 353. Seen from the side, the flowers of white fringed orchid resemble small, flying white birds.

Orchid Family

Fig. 354. Bees also visit the flowers of fringed orchids, but other than some long-tongued bumblebees, they are ineffectual pollinators because they do not contact the anthers in such a way that the pollinia are removed. In this case, a small bee is investigating the anther of a white fringed orchid.

diurnal pollinators, particularly species of small sulphur (*Colias* spp.) and skipper butterflies. To my eye, the flowers resemble small flying birds when viewed from the side (fig. 353). Although I have seen white fringed orchids only in daylight hours, I did have a fleeting view of an unidentified black-colored swallowtail butterfly visiting the flowers. Various species of bees were also observed to visit the flowers by day (fig. 354), but even the proboscis of a bumblebee is not long enough to reach nectar held at the very base of the flower's tube. Bumblebees are considered only incidental pollinators of the flowers.

The pollinia of *Platanthera blephariglottis* (and some other species) are located to the right and left of the entrance to the nectary, arranged in such a way that they contact the eyes of long-tongued nectar seekers, including moths and butterflies, as they push their heads against the flowers to sip nectar. The sticky viscidia glue the pollen masses to the edges of the pollinator's eyes (see fig. 348)—which sounds terribly annoying! Within seconds the caudicle (stalk of the pollinium) begins to dry, resulting in a changed orientation of the pollinium, which becomes aligned in such a way that when the insect visits another flower the pollinium will easily come into contact with the stigma (located just above the opening to the nectar spur), thus depositing the pollen mass. Experiments have demonstrated that a pollinium can remain glued to a butterfly's eye for several days and that pollen can remain viable for up to five days after removal from a flower as well. The butterfly is, therefore, more likely to transfer the pollen to another plant rather than to the same one from which the pollen came.

Although this is a very precise method of ensuring pollination within a species, some closely related orchids might be accidentally cross-pollinated in this manner as well. For example, the bright orange-flowered *Platanthera ciliaris* resembles the white fringed orchid in almost all aspects but color. Both have large overlapping ranges extending south to Florida. They have the same habitat requirement: open to semishaded, acidic soil in bogs and other wet places, particularly in peaty sands and sphagnum mats. The presence of white fringed orchid serves as an indicator species of a healthy bog. Plants of white and orange species may grow side by side within the same geographical area, with their blooming periods overlapping by about 10 days, thus increasing the possibility of cross-pollination. The pollinia are positioned in the same way in both species, and thus a butterfly visiting one species (fig. 355), and then the other, may place a pollen mass from one species onto the stigma of the other species, giving rise to offspring having flowers with an intermediate color. Such orchids are often fertile and will produce plants having pale orange flowers. These hybrids, named *Platanthera* ×*bicolor* (Rafinesque) Luer (the use of the "×" [times sign] preceding the species name indicates that it is of hybrid origin). These offspring may further backcross with one of the parent species, resulting in the progeny having flowers

Fig. 355. Taller, more densely flowered inflorescences of orange fringed orchids are more successful in attracting butterflies. The butterflies generally begin by visiting the lower flowers and work their way upward.

Fringed Orchids

that are whiter or more orange, depending on the parentage. The hybrids make it difficult to recognize that these are true hybrids and not just variations in color of *P. ciliaris* or *P. blephariglottis*.

Unlike white fringed orchids, orange fringed orchids that grow in open, sunny areas are reproductively more successful than those that are lightly shaded. Their pollinators are butterflies that find their nectar sources visually, and the easier-to-see, open-growing plants receive the most attention. Size matters as well in orange fringed orchids; those with taller, more densely flowered inflorescences are showier, and thus more attractive to their butterfly pollinators (see fig. 355). The flowers are tightly arranged along the stem, appearing as a whorl when viewed from above (fig. 356). Conversely, the flowers on smaller, fewer-flowered inflorescences of white fringed orchids are visited just as often as those on larger inflorescences, attracting their pollinators primarily by aroma, not showy displays. Their moth pollinators visit only a few flowers per inflorescence before moving on to another plant.

The most common visitor to orange fringed orchids is the spicebush swallowtail butterfly (*Papilio troilus*), a convenient association considering that the host plant for the larvae of these swallowtails, the spicebush (*Lindera benzoin*), often grows nearby in wet forests. Spicebush swallowtails are large butterflies that must flutter their wings constantly in order to maintain their balance on the narrow orchid lips as they take nectar (see fig. 348). Other butterflies, diurnal moths, and bees (figs. 356–357) are also seen at the flowers. There has been a report of hummingbird visitation, but due to the hummingbird's long bill and tongue (which prevent its head from contacting the sticky viscidium) it is unlikely that it could be an effective pollinator.

Pollinators tend to visit the lower or middle flowers first and work their way upward, seldom visiting more than three or four flowers on any inflorescence. The lower flowers have greater amounts of nectar and remain receptive to pollination for a longer period of time than do the upper flowers. Reproductive success is high in both orchid species, with between 80% and 100% of the flowers producing fruits.

Neither *Platanthera blephariglottis* nor *P. ciliaris* is considered endangered or threatened on a national scale, but orange fringed orchid is considered endangered in Michigan, threatened in Florida, and extirpated in New Hampshire. Populations are declining throughout New England. Currently few populations survive in coastal plain areas of New England due to loss of appropriate habitat—both from development and from natural succession. One of the rare remaining Connecticut populations is located in a cemetery (fig. 358), demonstrating the value of smaller cemeteries for the secondary purpose of providing a refuge for species of plants and birds that have lost much of their original habitat. In the Connecticut example, volunteers, realizing the importance of preserving this beautiful orchid species, maintain the orchid areas by removing weeds, and by temporarily fencing

Fig. 356. A view from above shows the whorled arrangement of the flowers on the inflorescence of orange fringed orchids. The bumblebee has four pollinia glued to its head.

Fig. 357. The same bumblebee seen in fig. 356 visits (and perhaps will pollinate) a flower on another inflorescence of orange fringed orchid. It appears that one of the four pollinia is now missing, perhaps deposited on another flower. The bee's proboscis is not long enough to reach the full length of the nectar spur, but it pushes its head firmly against the flower to reach as deeply into the spur as possible.

Orchid Family

Fig. 358. This cemetery is the site of one of the few remaining populations of orange fringed orchids in southern New England. The undeveloped land of cemeteries often provides habitat for plants and birds that have lost their natural habitats due to plant succession or human caused development.

the plants to protect them from being mowed before they release their seeds.

Our other orange fringed orchid, known as the crested yellow orchid (*Platanthera cristata*) (fig. 359), is sometimes lighter in color (verging from yellow-orange to true orange) with a fringed lip the same length as that of the orange fringed orchid. The opening to its nectary is triangular (or sometimes keyhole-shaped) rather than round, and the spur is much shorter and straighter than the downward curving spur of *P. ciliaris*. The viscidia are placed closer together and project outward, and thus are better suited for pollination by bumblebees; however, crested yellow orchids sometimes receive visits from the same pollinators as the two earlier-discussed species, and hybridization does occur. Hybrids resulting from crested yellow orchids crossing with white fringed orchids are named *Platanthera ×canbyi* (Ames) Luer and those crossing with orange fringed orchid are *Platanthera ×channellii* Folsom, according to the treatment of *Platanthera* in *Flora of North America*. In some mixed populations of white fringed orchids and either of the two orange fringed orchids, the average rate of hybridization is less than 3%; however, in other populations having both orange and white species, hybrids can be more numerous than the parent species. The cause of these so-called hybrid swarms is unknown. On the dunes of eastern Long Island (New York), growing under pitch pine (*Pinus rigida*), are three populations of short-spurred, pale creamy *Platanthera* orchids that were described in 1992 as a distinct species: *P. pallida* P. M. Brown. Recognition of this population as a distinct species is still controversial among taxonomists, and many botanists believe the plants to be self-perpetuating hybrids or a pale form of the yellow crested orchid. The new species has not been accepted by the two authorities I have selected as standards for nomenclature in this book: Missouri Botanical Garden's *Tropicos* database (www.tropicos.org) and *Flora of North America*. Both presently treat *P. pallida* (pale fringed orchid) as a synonym of *P. cristata*, rather than as a distinct hybrid between the orange fringed species and the white fringed species. This decision is based on the presence of a spur that is much shorter than that of the flowers of either *P. blephariglottis* or *P. cristata*. It *is* acknowledged that the Long Island populations are unique and at least partially stabilized, but definitive taxonomic status will require further study.

There is a third orange-flowered northeastern species, *Platanthera integra* (Nutt.) A. Gray ex L. C. Beck, but since it is fringeless, it is not included here.

Similarity of shape and/or color is not necessarily an indication of shared pollinators. Minor differences in shape, size, and placement of the reproductive structures determine the true relationship between flowers and their pollinators. In the case of our two purple-flowered fringed orchids, *Platanthera psycodes* (lesser

Fig. 359. Similar in appearance to the orange fringed orchid is the crested yellow orchid (which is generally orange in color). The major difference seen is in the shorter length of the spur. Some flowers are light yellow-orange as seen here, and others are a true orange, like those of orange fringed orchid.

Fringed Orchids

purple fringed orchid) and *Platanthera grandiflora* (greater purple fringed orchid), the flowers and vegetative parts are so similar (figs. 360–361, respectively) that earlier botanists have at times classified them as varieties of the same species. The two species also share the same habitat requirements and an overlapping bloom period.

Differing from the orange- and white-flowered species, the lips of the two species of purple fringed orchids are three-lobed, with the side lobes of the lesser more often lying flat, or even curving backward, while those of the greater tend to curve forward. The narrow part of the lip (known as the claw) is sometimes white, especially in the greater purple fringed orchid. The lesser purple (*Platanthera psycodes*) is generally a smaller plant with smaller flowers than the greater purple (*P. grandiflora*), but the sizes of the two can overlap, and occasionally, depending on growing conditions, the "lesser" orchids can be larger than the "greater" orchids. There is also an overlap in range in the East, but *Platanthera grandiflora* is more common in northern forests and in uplands, while *P. psycodes* is more common in the southern Appalachians, but its range extends further westward and into adjacent Canada.

The principal difference in the flowers is found in the reproductive parts (figs. 362–363). The column of

Figs. 360 & 361. The habitat and blooming time of the lesser purple fringed orchid (left) overlap with those of the greater purple fringed orchid (right). Their flowers are similar: in the greater purple fringed, the "claw" of the lip is often longer and has a larger white area than that of the flowers of the lesser purple fringed. The offset of *peak* blooming time and the partially different ranges help to keep the two species distinct.

Figs. 362 & 363. The principal difference between the flowers of the two purple fringed orchids lies in the arrangement of the reproductive structures. The pollinia of the lesser purple fringed (fig. 362) are smaller, closer together, and hidden from the side by the wings of the column. There is a projection from above that extends partially between the anthers, causing the opening of the nectar tube to be more horizontal and the pollinia to appear like small dumbbells. The pollinia of the greater purple fringed (fig. 363) are spaced further apart with nothing blocking the round opening into the nectar tube. This discrepancy between the species results in each species having its own specific pollinators.

Orchid Family

the lesser purple has shorter (1.5–2 mm long) pollinia that are more closely spaced and angled downward when compared with the widely spaced (3 mm long) pollinia of the greater purple. Lesser purple fringed orchid pollinia are partially enclosed by the projecting wings of the column and are thus not visible from the side. Perhaps the most easily seen difference is in the shape of the opening to the nectary. That of the lesser purple is partially obstructed by a projection from the roof of the nectary, which may cause the opening to appear more or less dumbbell-shaped, while that of the greater purple is round.

These differences in column structure result in an overall difference in pollinators. The position of the wide-set, projecting pollinia of greater purple fringed orchid results in the transfer of their sticky viscidia to the outside edges of the eyes of large butterflies, most commonly the swallowtails, and large diurnal sphinx moths. Pollinia *can* adhere to nectar-imbibing bumblebees, but in many instances the bees have been found to remove the pollinia immediately from their bodies after withdrawing their proboscises from the flower.

The pollinia of the lesser purple fringed orchid are more closely set, thus, they generally dislodge onto the proboscises of smaller butterflies such as skippers and sulphurs. Because the proboscis of one of these smaller visitors is shorter than the nectar spur, and the head of the butterfly (or moth) does not come into contact with the flowers, thus the olive-colored pollinia are deposited close to the base of its tongue. In hawk moths and large swallowtails, which have longer tongues, the pollinia end up a bit further from the base of the tongue. Bumblebees also may carry pollinia to subsequently visited flowers. *Platanthera psycodes* has a sweeter fragrance that lasts into evening, thus making it a candidate for nocturnal pollination as well.

As already mentioned, the pollinia that adhere to a butterfly begin to change in position immediately, such that within two minutes, they are repositioned to deposit pollen on the stigma of another orchid. This interesting feature was noted by Charles Darwin, working on European *Platanthera* species in the mid-nineteenth century. As with the orange fringed and white fringed orchids, the mechanism is not foolproof, and hybrids between the two purple fringed orchids may have occurred.

The last of the northeastern fringed orchids to be discussed in this chapter is the greenish-white-flowered ragged fringed orchid (*Platanthera lacera*). A plant of wet forests, marshes, fields, and swamps (fig. 364), the ragged fringed orchid is the most common species of our northeastern fringed orchids and has the broadest range: from Manitoba south to Texas and in all states and provinces east of there other than Labrador.

As with the purple fringed orchids, the lip of ragged fringed orchid is three-lobed, and the plant is aptly named—all three lobes are deeply and unevenly fringed (fig. 365). The flower color ranges from pale green to white, rendering the flowers more visible to

Fig. 364. Ragged fringed orchid grows in swampy areas, as seen here. The light green to white color of its flowers makes them more obvious to its principal pollinators, nocturnal moths.

Fringed Orchids

Fig. 365. (Left) The fringe on the three-lobed lip of this orchid has a ragged appearance, the source of the species' common name. Pollinators are attracted by the sweet fragrance the flower emits in the evening.
Fig. 366. (Right) A crab spider lies in wait for the evening's visitors. It is well camouflaged in both color and form to blend in with the flowers of the ragged fringed orchid.

its nocturnal pollinators—moths. Flower buds begin to open in late afternoon but are not fully open until evening, when they release a pleasant aroma, an attractant to their evening visitors, members of the noctuid moth group. Although the ragged fringed orchid is self-compatible, an insect pollinator is needed to transfer the pale yellow pollinia to the stigma of another flower.

In a study carried out in Illinois, the most common moth visitors (in the wide-ranging genus *Anagrapha*) had proboscises that were shorter than the average length of the nectar spur, forcing the moths to push tightly against the reproductive parts in order to reach the nectar. The pollinia adhered to the bases of their proboscises. A related moth with a longer proboscis was an occasional visitor. Flowers remained fresh and capable of being pollinated for up to nine days, and a diurnal hawk moth called the hummingbird clearwing (*Hemaris thysbe*) was seen to visit the flowers during the daylight hours, as were some bees and flies (which did not carry pollinia). As with other flowers that are attractive to many kinds of insects, fringed orchids are often used as perches for crab spiders that lie in wait for prey (see fig. 366).

The hummingbird clearwing (also known to pollinate the small purple fringed orchid) may be one of the agents responsible for the interesting interspecies hybrid known as *Platanthera* ×*andrewsii* (Niles) Luer, a cross between *Platanthera psycodes* and *P. lacera*. In both orchids the pollinia adhere to the tongue of the hawk moth. In addition, small purple fringed orchid is still mildly fragrant in the evening, possibly attracting some of the nocturnal moth pollinators that are regular visitors to ragged fringed orchids. There are reports of another rare hybrid between ragged fringed orchid and the large purple fringed orchid: *Platanthera* ×*keenaii* P. M. Brown, but there are no specimens to voucher its existence and the name was never validly published.

Lepidoptera, both butterflies and moths, are pollinators of all the fringed orchids described here (only the crested yellow orchid has the bumblebee as its *primary* pollinator), and their visits can thus result in the production of hybrids between some of the species. All these lovely orchids are well worth seeking out in local swamps and bogs not only for their beauty, but for the enjoyment of watching their visitors as well.

Aster Family

Fig. 367. Showy goldenrod (*Solidago speciosa*) and New England aster (*Symphyotrichum novae-angliae*) blooming together in a field—an inspiration for "the two sisters" legend.

Goldenrods

Fig. 368. Early goldenrod (*Solidago juncea*) growing in the Ice Meadows along the Hudson River in New York's Adirondacks: photographed in flower on July 5th, earlier than usual.

Goldenrods
Solidago L. spp.
Aster Family (Asteraceae)

Habitat: Fields, roadsides, and other open sunny areas, with a few species growing in shady forests.

Range: Members of this genus are found throughout the range of this book and throughout the remaining United States, Canada, and Mexico. A limited number of species are native to South America and Eurasia.

Goldenrods, a large group of species, are closely related to asters (see the chapter on asters), with typical composite inflorescences having the appearance of individual flowers, but actually consisting of two types of flowers: disk flowers in the center and ray flowers around the perimeter. In almost all species of goldenrods, both types of flowers are yellow. Goldenrods and asters are both members of the Asteraceae (the aster or daisy family); they bloom at the same time of year (late summer into autumn), are often found growing side by side, and share many of the same insect visitors; they are often spoken of in the same breath when describing the flora of late summer as "the season of goldenrods and asters," *and* they can both be quite challenging to identify to species!

A Cherokee legend about these two wildflowers refers to them as "the two sisters," flowers that appeared on earth to represent the colors of the dyed doeskin dresses of two Cherokee girls, the only survivors of a massacre in their village. An old woman cast a magic spell on them to protect them from the feuding Indian tribe that had killed everyone else in their village. (A non-Indian version allows that the flowers represented the blue eyes of one sister and golden hair of the other, and they were turned into flowers in order to bring happiness to all.)

While observers may appreciate that the "two sisters" bring color to a fading summer landscape (fig. 367), goldenrods often fail to capture the admiration of many wildflower lovers. In fact, many of them go so far as to consider these flowers "boring," with few qualities to pique the imagination. It is only by taking a closer look that one can truly appreciate the various and interesting aspects of goldenrods. I have chosen to discuss the goldenrods as an entire group, singling out just a few species for special attention.

Aster Family

The sight of the yellow flowers of goldenrod is a sure sign that fall is on the way. While early goldenrod (*Solidago juncea*) (fig. 368) is found in flower in early August (and sometimes even in July), most goldenrod species flower in September, often continuing into October. This final burst of golden yellow marks the transition from summer's many-colored floral palette into the rich reds, golds, and oranges of autumn's foliar hues.

In the late 1800s, there was a movement in this country to name the goldenrod as our national flower. This never came to pass, and it was not until 1986 that a national flower was finally designated: the rose. However, Kentucky and Nebraska did select goldenrod as their state flower; it even appears on the state flag of Kentucky. It is also designated as the state *wild*flower of South Carolina.

Although most species of goldenrod grow in sunny, open habitats, a small number of common, northeastern species are shade-tolerant and, thus, are found commonly in woodlands. One such is the blue-stemmed goldenrod (*Solidago caesia*), which has small clusters of yellow flowers growing in the leaf axils along its arched stem (fig. 369). The stems often have a distinctive lavender-blue color (fig. 370), a feature reflected in both the common and scientific names for the species (*caesia* is derived from the Latin for blue). A second shade-tolerant goldenrod is the zig-zag goldenrod (*Solidago flexicaulis*), with its terminal cluster of flowers atop a stem that appears to zigzag.

Fig. 371. Silver-rod (*Solidago bicolor*) is one of the few goldenrods with white ray flowers. Its scientific epithet, *bicolor*, means two-colored and is descriptive of the flowers: white ray flowers and yellow disk flowers.

An interesting third species in this category is silver-rod, or white goldenrod (*Solidago bicolor*) (fig. 371), the common name a reference to its distinctive white-rayed flowers. The species name, *bicolor*, denotes the combination of yellow disk flowers and surrounding white ray flowers. These goldenrods and a small number of our aster species are among the few native wildflowers found in flower in woodlands in late summer and early autumn.

Solidago ptarmicoides (fig. 372) is a less common species with both showy white rays and disk flowers, and narrow, stiff leaves; it is different enough in

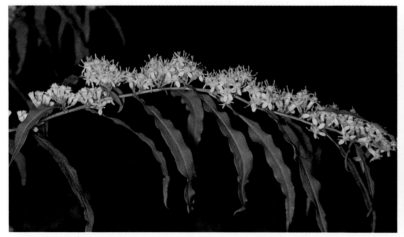

Fig. 369. (Left) A stem of blue-stemmed goldenrod (*Solidago caesia*) showing how the flowers are arranged in tufts in the leaf axils. **Fig. 370.** (Right) A section of the stem of a blue-stemmed goldenrod plant exhibiting the blue color from which the plant receives both its common and its scientific names.

Goldenrods

Fig. 372. The white-rayed flowers of *Solidago ptarmicoides*, a species that inhabits sunny, dry prairies as well as sandy and limy areas; it is more common in the Midwestern section of the range of this book.

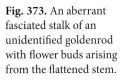

Fig. 373. An aberrant fasciated stalk of an unidentified goldenrod with flower buds arising from the flattened stem.

appearance from most goldenrods that it was once considered an aster. In fact, taxonomically it had been known as *Aster ptarmicoides*, *Oligoneuron album*, and *Doellingeria ptarmicoides* (among other names), before molecular studies showed that the species fit best into the genus *Solidago*, grouped in a section of the genus known as the flat-topped goldenrods. Despite earlier claims of hybrids between *S. ptarmicoides* and species of *Aster* in Europe (known as *Solidaster luteus*), the *Solidago* expert, John Semple, says that while hybrids between species of *Solidago* do occur, no intergeneric aster/goldenrod hybrids exist. If you do come across a goldenrod that clearly cannot be identified as one of the named goldenrod species, it might be an aberrant plant such as the one seen in fig. 373 that has become fasciated (an abnormal fusion of plant tissue resulting in flattened or stunted stems or other plant parts) due to a genetic error or other factor.

Goldenrod leaves are of many types: rough and jagged, as in wrinkleleaf goldenrod (*Solidago rugosa*); smooth, as in showy goldenrod (*S. speciosa*); or smooth *and* succulent, as in seaside goldenrod (*S. sempervirens*). Seaside (or evergreen) goldenrod is a favorite of mine for its relatively large, bright flowers (fig. 374) that add color to the sand dunes, sandy beaches, and salt marshes of the east coast (fig. 375).

Fig. 374. (Left) Seaside goldenrod (*Solidago sempervirens*) growing on a sand dune in Cape Cod, Massachusetts. The flowers of seaside goldenrod are exceptionally large and showy. Fig. 375. (Right) Seaside goldenrod growing in the sand dunes on Lucy Vincent Beach, Martha's Vineyard, Massachusetts.

Aster Family

Fig. 376. A leaf of seaside goldenrod (*Solidago sempervirens*) infected with the orange pustules of a fungal rust.

Fig. 377. European goldenrod (*Solidago virgaurea*), one of Europe's few native species of goldenrod, photographed in the French Alps.

Its large, slightly succulent leaves are sometimes prone to attack by a fungal rust that manifests itself as small, bright orange spots (fig. 376). Rusts that attack goldenrods belong to any of three genera: *Puccinia*, *Coleosporium*, or *Uromyces*. *Coleosporium solidaginis* has been identified on seaside goldenrod.

The generic name *Solidago* is derived from the Latin *solidus* meaning "whole," and *-ago*, one meaning of which is "becoming," thus "becoming whole." Linnaeus applied this name to the genus based on goldenrod's reputed medicinal properties. *Solidago*, as with many botanicals, was employed for medicinal purposes—originally in Europe, but now increasingly so in North America as well. Claims for the effectiveness of most such herbal medicines are based on their antioxidant properties; antioxidants are known to combat the free radicals that contribute to a variety of diseases, from the common cold to cancer. Compounds in *Solidago* have been used particularly for their anti-inflammatory properties and in treating urological and gastrointestinal problems. In Europe, where there are few native species of goldenrod, the most common source of these compounds is the European goldenrod (*S. vigaurea*) (fig. 377). It was used for healing wounds as far back as the fifteenth and sixteenth centuries because of its astringent and antiseptic properties. It has also been used as a treatment for fungal diseases such as yeast infection, for thrush, for skin diseases, and as an antihelminthic drug against intestinal worms.

Nonmedicinal uses of goldenrod include brewing a tea from the leaves. Those who enjoy dying yarns from natural materials know the goldenrods well: yellow is obtained from the flowers and leaves, and brown, orange, and mustard-color from the entire plant. *Solidago canadensis*, an early introduction from North America into Europe, is now being widely used for the above purposes by Europeans as well.

Perhaps the most promising potential use of goldenrod is as a source of natural rubber. Despite being remembered best for the invention of the incandescent lightbulb and for his patents in the field of communication, Thomas Edison spent two years in search of a North American plant that could serve as an alternative source of natural rubber. Even today the entire world's supply of rubber is derived from a single tropical plant species (*Hevea brasiliensis*). With his friends Henry Ford and Harvey Firestone, Edison founded the Edison Botanic Research Corporation in Fort Myers, Florida, dedicated to the search for and production of natural rubber. After testing nearly two thousand species, the team selected goldenrod as the best candidate and began cross-breeding experiments, ultimately producing a plant with leaves that yielded 12% rubber. Much of the research was carried out at The New York Botanical Garden in the Bronx. Research continued after Edison's death in 1931, but with the development of synthetic rubber during the

Goldenrods

World War II, the search for a temperate botanical source of natural rubber ceased. Yet, synthetic rubber cannot replicate some of the special properties of rubber, and proteins in the rubber from *H. brasiliensis* are the cause of an allergic reaction in many people. For these reasons and because of the vulnerability of a worldwide industry based on a single plant species, new research is being carried out in the search for an alternative source of nonallergenic natural rubber. The prime candidate is guayule (*Parthenium argentatum*), another member of the aster family.

Canada goldenrod (*Solidago canadensis*) (fig. 378) is native to the United States, ranging from North Dakota east to Maine and south to Florida, Texas, and Arizona; and in Canada from Nova Scotia to Ontario. It was introduced to Europe as an ornamental as early

Fig. 379. Next year's young shoots are already formed on the underground portion of Canada goldenrod by August.

as the mid-eighteenth century and spread from England to most of western and central Europe by the mid-nineteenth century. It has also become naturalized in Asia, Australia, and New Zealand. In some parts of Europe (such as southeastern Norway) it is considered common and invasive. Canada goldenrod produces viable seed only as a result of cross-pollination, but it also reproduces vigorously by spreading rhizomes, forming large clones. Digging up the rhizomes and roots in late summer may reveal the shoots of next year's plants ready to sprout (fig. 379). The clones are long-lived, capable of reaching an age of 100 years. Allopathic compounds released by the plant have been shown to be toxic to some other plant species, hence suppressing germination and growth of potential competitors; they are also responsible for killing some soil organisms. This property makes it easier for Canada goldenrod to take over a field that was formerly diverse in species. Currently, the allopathic compounds are being studied to determine their potential benefit in combating invasive plants that negatively impact crop plants.

Three other North American goldenrods introduced to Europe as ornamentals in the late 1700s are tall goldenrod (*S. altissima*), late goldenrod (*S. gigantea*), and grass-leaved goldenrod (*Euthamia graminifolia*). Both *S. altissima* and *S. gigantea* have since become invasive there.

Because goldenrods flower in late summer and early fall, they are frequently implicated as the cause of seasonal fall allergies. However, the pollen grains of goldenrod flowers are relatively large and sticky—better

Fig. 378. Canada goldenrod (*Solidago canadensis*), a common goldenrod throughout the Northeast, is now invasive in parts of Europe where it was introduced as an ornamental in the eighteenth century.

Aster Family

Fig. 380. The numerous green clusters of male flowers of common ragweed (*Ambrosia artemisiifolia*) are small and draw little attention, but they produce vast amounts of small, light pollen grains that travel easily on the wind. Ragweed pollen is responsible for causing seasonal allergies in many people. (The tiny female flowers are not readily visible in this image.)

Fig. 381. A field of rough-stemmed goldenrod (*Solidago rugosa*) is both colorful and fragrant, attracting many types of insect visitors.

suited for transport from flower to flower by insects than by wind. The true culprit of the fall allergy problem is ragweed (*Ambrosia* spp.) (fig. 380), which has extremely small, light pollen grains that float about easily on fall breezes. Should you tap a ragweed plant while it is in flower, you will see a cloud of yellow pollen drift from the flowers. The small green, inconspicuous flowers of ragweed generally go unnoticed by most people, whereas the highly visible flowers of goldenrod result in its being indicted unfairly as the culprit responsible for "hay fever".

Walking through a field of rough-stemmed goldenrod (*Solidago rugosa*) on a warm, sunny day is a delight not only for the eyes (fig. 381), but for the nose as well; its sweet fragrance permeates the air, attracting insects intent on gathering pollen or nectar. Literally hundreds of species visit goldenrod to feed on the leaves, flowers, stems, or roots; to gather pollen or nectar; to lay their eggs; or to prey upon other insects (fig. 382a–h). In one field alone, a survey of insects during a single summer season found 241 different insect species. Among insects commonly observed on goldenrod are

Fig. 382a–h. Insect visitors to goldenrod: **a.** An Eastern carpenter bee (*Xylocopa virginica*) taking nectar from the flowers of showy goldenrod.

Fig. 382b. A diurnal moth, the delicate cycnia (*Cycnia tenera*), feeds on dogbane (*Apocynum* spp.) as a caterpillar but takes nectar from various flowers, including goldenrod, as an adult.

Goldenrods

Fig. 382c. (Left) Another diurnal moth, the Virginia ctenuhca (*Ctenucha virginica*) nectaring on goldenrod.
Fig. 382d. (Above) The pretty-patterned Ailanthus webworm moth (*Atteva aurea*) is a seasonal migrant from the southern United States that visits and pollinates many flowers; it is seen here on showy goldenrod (*Solidago speciosa*). As a larva, it feeds on the leaves of the tree of heaven (*Ailanthus altissima*), an introduced tree species from China related to the moth's southern host plants.

bees (including carpenter bees [fig. 382a], which serve as principal pollinators of showy goldenrod), butterflies, moths (figs. 382b–e), bugs (see assassin bug in fig. 382e), beetles (figs. 382f–h), syrphid flies, wasps, and other insects that rely on goldenrods as a source of nectar and pollen at the end of the season. Aside from serving as a source of nectar for many beneficial insects, goldenrod's nectar is also utilized as a food source by mosquitoes. Only male mosquitoes are equipped with feathery mouthparts that allow them

Fig. 382e. An unidentified male moth that has been captured by an ambush bug (*Phymata* sp.) on goldenrod. Note: at the end of the moth's abdomen are feathery plumes called hair-pencils that emit a pheromone used in attracting and tranquilizing a mate.

Fig. 382f. The locust-borer beetle (*Megacyllene robiniae*) is a long-horned beetle that feeds on the wood of black locust (*Robinia pseudoacacia*); it is a common visitor to, and pollinator of, goldenrod flowers, feeding on their pollen.

Aster Family

Fig. 382g. (Left) Several small leaf beetles in the genus *Scelolyperus* on showy goldenrod (*S. speciosa*). **Fig. 382h.** (Above) A spotted cucumber beetle (*Diabrotica undecimpunctata*) feeding on pollen from the flowers of seaside goldenrod (*Solidago sempervirens*). The larvae of cucumber beetles feed on roots and tunnel through stems of members of the cucumber family and are considered an agricultural pest.

to dine on the floral nectar; female mosquitoes, with sucking mouthparts, feed on blood in order to procure the protein necessary to produce their eggs.

Of particular interest are the colorful caterpillars of the owlet moths, such as the calico paint (or brown-hooded) owlet (*Cucullia convexipennis*). These caterpillars feed on various members of the Asteraceae, with a preference for goldenrods and asters (see also the chapter on asters). Although colorfully patterned, owlet moth caterpillars can be surprisingly difficult to spot until you suddenly notice a bit of red pressed against the stem or chewing on the leaves or flowers of the plant. The first brood, in mid-July, is content to feed on leaves (fig. 383a), but the later brood, present in August and September, prefers feeding on the flowers (fig. 383b). If you are fortunate enough to have a nearby field of goldenrod you can visit regularly, you may find that the caterpillars you have been watching might one day appear to have small, glistening "pearls" adorning their backs (fig. 383c). Continue to monitor those caterpillars, as this is one of the most interesting, if diabolical, acts of murder in the insect world. The "pearls" are actually the eggs of a wasp that have been deposited on the caterpillar in order to provide the wasp's larvae with a future, ready source of food. As the caterpillar goes about its business (feeding on plants), the eggs hatch into wasp larvae and begin to suck the life out of the caterpillar

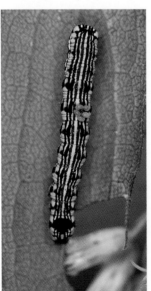

Fig. 383a. (Left) An early instar larval stage of a brown-hooded owlet moth (*Cucullia convexipennis*) feeding on a leaf of goldenrod. Note the cluster of five white eggs on the caterpillar's back. The caterpillar has been parasitized by a wasp.

Fig. 383b. (Right) A mature caterpillar of a brown-hooded owlet moth (*C. convexipennis*) feeding on the flowers of goldenrod.

Goldenrods

Fig. 383c. (Left) Note the three small, pearl-like eggs (and a fourth tiny one just beneath the uppermost one) laid by a wasp on the back of this caterpillar of a brown-hooded owlet moth (*C. convexipennis*).

Fig. 383d. (Right) The three wasp eggs have hatched into larvae that are feeding on the inside tissues of the brown-hooded owlet caterpillar.

(fig. 383d), but not quite enough to kill it—that is, not until they are ready to enter their pupal stage. By that time the tissues of the caterpillar have been depleted, and the wasps form small cocoons right on the caterpillar's carcass (fig. 383e)—this drama all unfolding on the benevolent host of the caterpillar, the goldenrod. There the wasp pupae remain until they are ready to emerge, and as adult wasps begin the cycle again. Certain flies may parasitize caterpillars in much the same way. This is not an uncommon fate for caterpillars, and to ensure the survival of at least some of their progeny, adult moths must lay a large number of eggs.

Goldenrods are also host to a number of insects that spend their larval stage *within* the plant. Cursory inspection of a field of goldenrod will usually result in the finding of odd swellings on the stems of the plants, or an aberrant bunching of the apical, or uppermost, leaves and the adjacent inflorescence. These varied growth structures are called galls and constitute the domiciles and food sources for developing larval forms of gall-making organisms. Up to 53% of plants in a population of goldenrod may be affected. Each gall is caused by one of the three most prominent gall makers that affect goldenrods and can be told apart by shape and location on the plant. If the shape of a swelling on the stem is spherical (fig. 384a), it has

Fig. 383e. (Left) Two wasp larvae have pupated after consuming the inside tissues of the brown-hooded owlet caterpillar. The third fell off. The caterpillar is now dead.

Fig. 384a. The ball gall of a fly larva (*Eurosta solidaginis*) on the stem of Canada goldenrod (*Solidago canadensis*).

Aster Family

Fig. 384b. (Left) A spindle-shaped gall on Canada goldenrod made by the larva of a moth (*Gnorimoschema gallaesolidaginis*). Another species of moth, *Epiblema scudderiana*, makes a similarly shaped gall on goldenrod.

Fig. 384c. Bunch galls (also called flower galls) created by the larvae of a midge (*Rhopalomyia solidaginis*) cause the leaves to bunch together at the apex of a plant of showy goldenrod (*Solidago speciosa*).

been produced by a small fly maggot, *Eurosta solidaginis*; if elliptical, or spindle-shaped (fig. 384b), the occupant is the larva of a moth, *Gnorimoshema gallaesolidaginis*. And if you encounter a goldenrod with clusters of tightly bunched leaves at the apex of the plant (fig. 384c), termed a bunch gall, the gall-maker is a midge (a species of tiny fly) with a name much longer than its size warrants: *Rhopalomyia solidaginis*. Galls develop as a consequence of a proliferation of the plants' tissues in response to a chemical secreted by the larvae as they chew into the plant tissues. An adult female will carefully select the proper host based on sight and smell, lay her egg—or up to 14 eggs in the case of *Rhopalomyia*—on the plant, and depart. Bunch galls (also called flower galls for the resemblance of the clustered leaves to green flowers) and ball galls are both found on tall goldenrod (*S. altissima*). Investigation has shown that there is no competition between the two fly species for this resource. However, if the adult ball gall fly lays more than one egg on a single plant (fig. 384d), the resulting larvae suffer from the effects of competition with their own kind. Ball gall eggs hatch in about 10 days, and the larvae burrow into the stem. *Eurosta solidaginis* overwinters as a

Fig. 384d. (Left) Two ball galls, created by two different larvae of the ball gall fly, *Eurosta solidaginis* on the stem of Canada goldenrod (*S. canadensis*). The larvae will compete with each other for resources and will not do as well as a single larva on a plant.

Fig. 384e. A ball gall cut open to show the larva of the goldenrod ball gall fly (*Eurosta solidaginis*) that would have overwintered in the gall.

Goldenrods

maggot in a state of diapause (fig. 384e), pupates in the spring, and about two weeks later emerges as an adult, a full year after the egg was laid. In contrast, the larvae of the bunch gall fly mature in only 60–70 days, and in warm regions up to four generations may be produced in one year. Ball galls made by different races of *E. solidaginsis* are also found on smooth (also called late) goldenrod (*S. gigantea*) and Canada goldenrod (*S. canadensis*). The different races of gall-makers are host-specific, even if the two species of goldenrod coexist in the same field. Thus, the races of *E. solidaginsis* never come into contact with each other and are therefore reproductively isolated; because of this they may one day develop into separate species. The presence of different types of galls affects the host plants in various ways. Both ball galls and elliptical galls cause an increase in stem material, but plants parasitized by ball gall-makers suffer a decrease in leaf and inflorescence weight (and thus seed production), while those affected by elliptical galls have a negligible decline in leaf material, but a decline in inflorescence material comparable to that of ball gall plants.

The story of goldenrod ball galls is more complicated than what is seen at a glance. Research done by Dr. Warren Abrahamson and his students at Bucknell University has shown that goldenrod gall size plays a role in the success of the larva of the gall-maker. In smaller galls, a parasitic wasp (*Eurytoma gigantean*) is able to insert her ovipositor into the central chamber of the gall where she can lay her egg. When it hatches, the wasp larva eats the fly larva and then feeds on the tissue of the gall itself. However, in the thick-walled larger galls, the wasp's ovipositor cannot reach fully into the central chamber, and, thus, the inhabitant is spared such a fate. Basically, the size of the gall is determined by the genotype of the goldenrod. In areas with abundant numbers of downy woodpeckers and chickadees, both predators of gall larvae, the birds preferentially select the larger galls, knowing they are more likely to be rewarded with a juicy maggot. If both wasps and woodpeckers are abundant in an area, their predation balances out, resulting in a selection for medium-sized galls. As if it is not already perilous enough to be a gall inhabitant, fishermen are aware that they can cut into the galls in winter to find live bait, again with a preference for the larger galls. The identity of the gall fly's home-wrecker is apparent from the hole left in the gall; the hole of a woodpecker is relatively neat, small, and more or less circular; that of a chickadee large and rough; and a fisherman will cut the gall open entirely.

Another wasp in the genus *Eurytoma* parasitizes the fly egg or larva before it has even burrowed into the plant, with the wasp larva then living in the fly larva until fall, when it pupates in the larva and subsequently overwinters in the gall. Add to this scene a species of stem-boring beetle that also preys on *Eurosta*, and it is a wonder that any flies survive to adulthood!

A ball gall larva is freeze-tolerant, which means that it can withstand the conversion of a fraction of its body water into ice, allowing it to survive without energy loss over a period of up to 12 weeks. It does this in late autumn when temperatures drop, by producing glycerol, an antifreeze-like compound that allows the contents of the larval cells to remain liquid while the rest of the animal freezes.

The goldenrod gall moth lays its eggs on the lower leaves of the goldenrod plant in the fall. In spring, the caterpillar hatches, crawls up the new stem of goldenrod and burrows into it, initiating the gall-forming process. It will spend its larval stage feeding on plant tissues within the gall. In late fall, when the gall has turned hard and brown, the caterpillar chews a tunnel through the gall just to the outer surface, plugs it with a mixture of gall material and silk, and then pupates. The hole is beveled, with the widest part at the exterior of the gall, making it difficult for other insects to reach the larva's chamber in the center of the gall. In spring the newly hatched adult moth crawls back through the tunnel and pushes its way out of the plugged entrance,

Fig. 384f. The spindle gall in fig. 384b has been cut open to reveal the pupal case of the moth that emerged from the perfectly round exit hole (seen in fig. 384b) in spring.

Aster Family

leaving a perfectly round hole (see fig. 384b); its pupal skin is left behind (fig. 384f). This happy outcome is possible only if an ichneumon wasp has not used its long ovipositor to lay eggs in the gall. The wasp larva would have then fed on the caterpillar.

The midge-related bunch galls are limited to showy goldenrod (*Solidago speciosa*), Canada goldenrod (*S. canadensis*), and rough-stemmed goldenrod (*S. rugosa*), whereas spherical ball galls may be formed on tall goldenrod (*S. altissima*), smooth goldenrod (*S. gigantea*), rough-stemmed goldenrod (*S. rugosa*), Canada goldenrod (*S. canadensis*), and perhaps other *Solidago* species as well. Elliptical gall-makers deposit their eggs on Canada goldenrod (*S. canadensis*) and tall goldenrod (*S. altissima*). Thus, the presence of certain types of galls can be of some aid in narrowing down the identity of a goldenrod species.

A few species formerly classified in *Solidago* (and still commonly referred to as the grass-leaved, or flat-topped, goldenrods) have been segregated into a separate genus, *Euthamia* (fig. 385). The most easily observed differences are the small, narrow, resin-dotted leaves with parallel venation (if more than one-veined) and the flat-topped groupings of small floral heads. Molecular studies have shown that plants in the genus *Euthamia* are not closely related to those in *Solidago*.

A different type of gall, called a blister-gall, is not as obvious as the ball or bunch galls discussed earlier. It appears merely as a black spot on the leaves of goldenrod plants (fig. 386), particularly those of the grass-leaved goldenrods in the genus *Euthamia*. The blister-gall forms as the result of a tiny cecidomyid midge laying her eggs in the leaf. A fungus carried by the midge is implanted into the leaf along with the egg. The larva then develops between the layers of fungus and, at maturity, exits by making a tiny hole in the gall. Once again, a parasitic wasp may lay its own egg in the gall spot, killing the midge larva and making use of the gall's miniscule resources for its own larva.

As you can see, the story of goldenrod goes far beyond the tale of "the two sisters." The next time you pass a field of goldenrod brightening up a fall day (fig. 387), stop not only to appreciate the beauty of the scene, but also to think about all the life supported by the goldenrod plants in that field.

Fig. 385. A plant of *Euthamia graminifolia* (grass-leaved goldenrod). Note the narrow leaves and the more or less flat-topped inflorescence.

Fig. 386. The characteristic black spots on the leaves of *Euthamia graminifolia* are blister galls caused by an interaction between a midge and a fungus.

Goldenrods

Fig. 387. A field of little bluestem grass (*Schizachyrium scoparium*) laced with showy goldenrod (*Solidago speciosa*), a beautiful sight in late September.

Bittersweet Family

Fig. 388. A group of eastern grass-of-Parnassus plants (*Parnassia glauca*) in bloom adjacent to a small spring in a wet, limy meadow.

Grass-of-Parnassus

Fig. 389. Grass-of-Parnassus has a perfect (bisexual) flower in which the male phase matures before the female phase. In this image the five stamens have already shed their pollen and arched backward between the petals before the stigmas are ready to receive pollen. The ring of yellow-knobbed appendages is composed of staminodes (modified sterile stamens) that attract pollinators.

Grass-of-Parnassus

Parnassia glauca Raf.
Bittersweet Family (Celastraceae)

The home of the mythological Greek Muses was believed to be the slopes of Mt. Parnassus, which were then described as "wet and limy," which would be compatible with the plant's current association with spring-fed, alkaline areas (fig. 388). Although the flowers are small, they deserve close attention so that one can appreciate their delicate green markings and glistening yellow-topped staminodes (sterile stamens) that attract pollinators (fig. 389).

Habitat: Wet, alkaline meadows (fens) near springs and seeps and along streambanks and gravelly shores with water having a high pH.

Range: Most of the northeastern quarter of the United States, with the exception of the Mid-Atlantic states, and west to the eastern Dakotas; adjacent Canada, north to Newfoundland and Manitoba.

Although this is not a grass, nor is it currently growing on the limestone slopes of Greece's Mt. Parnassus, Linnaeus named the genus *Parnassia* based on a reference in *De Materia Medica*, a tome of herbal remedies written more than 2000 years ago by the early Greek physician Dioscorides. Numerous medical uses were ascribed to a plant that Dioscorides called *Agrostis En Parnasso* (translated from the Greek as "creeping in Parnassus"). The plant mentioned is thought to be the circumboreal marsh grass-of-Parnassus (*Parnassia palustris*). Grass-of-Parnassus most likely received its early Greek name, as well as its translated English name, as a result of the ancient application of the word "grass" to all herbaceous plants thought to be consumed by livestock. Today, *Parnassia palustris* inhabits the slopes of mountains in northern Greece, but the calcareous slopes of Mt. Parnassus are now too dry for marsh grass-of-Parnassus to thrive. The lower elevations of the mountain are cloaked in forests of Grecian fir (*Abies cephalonica*), while the upper regions are rocky and barren (fig. 390). Perhaps in antiquity the climate was wetter, and conditions were more hospitable for

Bittersweet Family

Fig. 390. Grass-of-Parnassus is named for Mt. Parnassus in central Greece, said to be a locality for a species of *Parnassia* in ancient times as well as a home to the nine Muses. The clusters of trees are Grecian fir (*Abies cephalonica*).

wetland plants such as *Parnassia*. The 2457 m (8061 ft.) peak of Parnassus is cold enough that a popular ski resort operates near its summit annually through April (fig. 391), but the climate is now warmer than that favored by *P. palustris*.

The most common species of *Parnassia* throughout the Northeast is *P. glauca*. *Parnassia palustris* may be found in the far northern areas, and *P. parviflora* is occasionally encountered in Arctic and alpine regions of the far north. Although *glauca*, the species epithet (the second part of a scientific name), commonly means "with a bluish color," the stalked, leathery leaves of *Parnassia glauca* are a more yellow-green (fig. 392). Leaves are arranged in a cluster at the base of the plant, with just a single smaller leaf located a short distance up on the flower stalk. The upper leaf appears to be sessile, but its petiole (leaf stalk) is actually fused with the flower stalk. In general, each plant has a single flower stalk, topped by a small bud (see fig. 392), that is initially enclosed by five short sepals (fig. 393). As the flower opens, conspicuous green veins on the petals are revealed, with the central veins unbranched, while those toward the edges are forked (see figs. 389, 395, and 396).

Fig. 391. Although Mt. Parnassus is now too dry to provide suitable habitat for grass-of-Parnassus, it may have once occurred there. The summit still has the cold temperatures preferred by marsh grass-of-Parnassus (*Parnassia palustris*), as evidenced here by this thriving ski resort (Parnassus is written in the Roman alphabet as Parnassos in Greece).

Fig. 392. Small flower buds rise above the stalked basal leaves of eastern grass-of-Parnassus. On some plants, a small leaf can be seen on the flower stalk a short distance above the basal leaves (see central bud stalk). Parnassus plants are perennial and evergreen.

Grass-of-Parnassus

Fig. 393. Five short sepals enclose the flower when in bud. The flower seen in this image is about to open, and the forked veins along the margins of the petals can be seen.

The genus *Parnassia* is found widely in north temperate and Arctic latitudes, with the greatest number of species in China. Most species of the 50-plus members of this genus are found growing in similar wet, alkaline conditions. In northeastern North America, there are five species, with one, marsh grass-of-Parnassus (*P. palustris*), found across Canada and the northern United States from Michigan west and south through the western states, plus Alaska. It occurs rarely in northern New York State. This same species grows in similar latitudes and habitats across northern Europe and Asia, extending further south in mountainous areas (fig. 394). In North America at least four additional species are found in the western United States and Canada, making North America the second highest center of *Parnassia* diversity.

You might be surprised (as I was) to see that the Angiosperm Phylogeny Group has tentatively accepted the reclassification of grass-of-Parnassus as a member of the bittersweet family (Celastraceae). The best-known local members of this family are two of our most aggressive alien invaders, Oriental bittersweet (*Celastrus orbiculatus*) and burning bush (*Euonymus alatus*). When I first encountered eastern grass-of-Parnassus, it was considered a member of the saxifrage family, a family in a completely different order from the Celastraceae. *Parnassia* seemed an aberrant member of the Saxifragaceae, and I was pleased when it was later placed into its own family, the Parnassiaceae (along with one other small genus, *Lepuropetalon*). However, the latest molecular evidence shows the members of the former Parnassiaceae family closely allied with plants in the Celastraceae, a worldwide family of trees, shrubs, and woody vines that includes bittersweet and euonymus, and that is where *Parnassia* currently resides. Within the last decade, a small family of herbaceous Australian plants has also been subsumed into the Celastraceae as the subfamily Stackhousioideae. Further investigation is necessary before a final placement can be determined, and several other floras have maintained the classification as the separate family Parnassiaceae.

Fig. 394. Marsh grass-of-Parnassus (*Parnassia palustris*) grows in wet, alkaline areas in temperate and Arctic regions throughout the Northern Hemisphere. Its range extends southward in the mountains. (Photographed in the Dolomites, Italy)

Bittersweet Family

Fig. 395. The sequential "dance" of the stamens has begun: the first stamen has elongated and moved to a position above the ovary to shed its pollen.

Parnassia glauca, like other members of its genus, is a late bloomer, commencing flowering in the latter part of August and extending the blooming period through September, and sometimes into early October. Floral parts are arranged in fives: five sepals, five petals, five stamens, and five staminodes (nonfunctional stamens), which are each divided into three segments (giving the appearance of numerous sterile appendages). The stamens perform a ballet of sorts, each sequentially taking a starring role in which it elongates, bends gracefully over the ovary, opens its four anther sacs to release pollen (fig. 395), and then slowly reflexes back until resting flat between the petals. The anther sacs soon fall off, leaving only the filaments (fig. 396). The dance may begin even before the flower is fully open (fig. 397). I have, however, seen stamens engaged in a botanical *pas de deux*, with two stamens simultaneously arching over the tip of the ovary (see fig. 396). The benefit of sequential ripening of the anthers is that pollen is made available over a longer period of time,

Fig. 396. In this flower, two stamens "vie" for position over the ovary; three others have already completed their pollen release and now lie flat between the petals. One of the stamens has already dropped its anther.

Fig. 397. A stamen has already opened its anther sacs above the ovary before the flower is fully open. The knobs on the staminodes (false stamens) are still green (compare with the mature, yellow staminodal knobs shown in fig. 389 and fig. 399).

Grass-of-Parnassus

Fig. 398. Beetles are said to be among the insects that pollinate marsh grass-of-Parnassus. (Photographed in the Dolomites, Italy).

an important consideration at a time of year when conditions are not always favorable for insect activity.

Despite the self-placement of the pollen directly onto the female portion of the flower, effective pollination is more commonly accomplished by insects. Pollinators include flies and bees, and to some degree, beetles, which are said to feed on the anthers (fig. 398). Flowers are protandrous, meaning that the male parts (stamens) mature before the female parts (stigmas) are ready to receive pollen, a mechanism that favors outcrossing. Insects are lured to the flowers by the staminodes, each of which is divided above its short, fused base into three prongs topped with what appear to be glistening drops of nectar (see fig. 395). The drops turn from green to yellow-green to yellow as the flower matures. In an experiment carried out on the similar marsh grass-of-Parnassus in Norway, scientists found that insect visits to flowers from which they had removed the staminodes declined in number and duration compared with visits to flowers in which the staminodes remained intact. Insects (most often flies) land on the flowers in anticipation of receiving a sugar-filled reward but are deceived, since the apparent nectar drops are dry. The staminodes, however, *do* produce nectar, but the nectariferous tissue is in their undivided bases (fig. 399), so to find the reward,

Fig. 399. The true nectaries are more easily seen in the flowers of marsh grass-of-Parnassus. They are the broad, green areas at the bases of the staminodes, just below the fingerlike projections topped by glistening nectar-imitating globes. The stamens have already relaxed backward, and the tips of the false stamens are now yellow. The staminodes of *Parnassia palustris* have more numerous and more slender filaments than *P. glauca*, and the veins of the petals are rarely forked and not as deeply green.

Bittersweet Family

insects must venture further into the flower. Guided by the green veins of the flower petals, and perhaps by aroma, the insects find the true nectaries, and in doing so, are dusted with loose pollen, which may then be carried on to other flowers. If any of the pollen remaining on the ovary is still viable when the stigmas mature, visiting insects might transfer the flower's own pollen to its stigmas, thereby effecting self-pollination.

This is less common than cross-pollination. Stigmatic surfaces of *Parnassia* are unusual in that they are arranged in four lines directly on the ovary's "seams" (commissures)—junctions which will later split open as the fruit matures; there are no styles. Small butterflies and bees also visit the flowers and, of course, crab spiders are always attracted to flowers where they can lurk until a prey insect arrives (fig. 400). The sepals,

Fig. 400. Crab spiders often lurk on flowers in hopes of snaring a visiting insect.

Fig. 401. (Left) Sepals, staminal filaments, and staminodes persist during the fruiting stage. **Fig. 402.** (Above) The four-part capsules of *Parnassia* will dehisce along their four seams to spill their seeds into the wet marsh. The small upper leaf can be seen on this dried flower stalk.

Grass-of-Parnassus

staminodes, and the filaments of the true stamens turn brown and persist through the fruiting stage (fig. 401). The mature capsular fruit splits along the four sutures, spilling its many irregularly shaped seeds onto the wet ground (fig. 402).

The striking green-veined flowers of our northeastern species of grass-of-Parnassus (*Parnassia glauca*) are worth getting your feet wet to find. Eastern grass-of-Parnassus is always found in wetlands having a high pH, that is, alkaline areas. In fact, eastern grass-of-Parnassus is one of the small number of species considered indicator species for fens, a habitat that is disappearing due to the filling of wetlands. Such areas are given high priority for conservation protection. Once established, grass-of-Parnassus plants increase in number such that when in flower the fen appears to have been strewn with small white stars (an alternate common name for grass-of-Parnassus is bog-star). Such areas, generally rich in sedges, are also habitat for other interesting plants. My closest sites for seeing *Parnassia* are also home to ladies-tresses (*Spiranthes cernua* and *S. lucida*), fringed gentian (*Gentianopsis crinita*), sneezeweed (*Helenium autumnale*), mountain mint (*Pycnanthemum* spp.), bugleweeds (*Lycopus virginicus* and *L. americanus*), and lobelias (*Lobelia siphilitica* and *L. kalmii* [figs. 403–404]).

The unusual flowers of *Parnassia* so impressed early botanists exploring North America that plants

Fig. 403. *Lobelia siphilitica* grows in the same alkaline fens as grass-of-Parnassus.

of a more southern species, *Parnassia asarifolia*, were taken to France in the early 1800s to be planted in the gardens of Empress Josephine.

Fig. 404. The tiny lavender flowers scattered throughout the wet marsh are those of *Lobelia kalmii*, another calciophile (lime-loving plant).

Heath Family

Fig. 405. A clump of pure white Indian pipe almost glows on the dark forest floor.

Indian Pipe

Fig. 406. Indian pipe plants in a Northeastern forest. The flowers have begun to turn upward after being fertilized.

Indian Pipe
Monotropa uniflora L.
Heath Family (Ericaceae)

Indian pipe is a strange-looking forest wildflower that generally grows in clumps (fig. 405). It is capable of living even in dark woods (fig. 406) where most other wildflowers could not survive because it has no chlorophyll and, thus, is not dependent on light in order to manufacture carbohydrates. Rather, it obtains its nutrients via a relationship with soil fungi and nearby trees.

Habitat: The forest floor of coniferous or mixed hardwood/coniferous forest, particularly under hemlocks and beech trees.

Range: Throughout most of the lower United States (except for the Southwest) and in Alaska; Canada (except for the Yukon, the Northwest Territories, and Nunavut); and in Mexico and Central America; eastern Asia.

Despite Indian pipe's eerie-sounding alternate names of corpse plant and ghost flower, it was beloved by Emily Dickinson, who referred to it as "the most preferred flower of life." She described "the sweet glee I felt upon meeting it, that I could confide to none." An illustration of Indian pipe graces the cover of her first book of poetry, titled simply, *Poems* (published posthumously), in which her brief poem, "Indian Pipe—The Most Amazing Flower," appears:

> White as an Indian Pipe
> Red as a Cardinal Flower
> Fabulous as a Moon at Noon
> February Hour

Upon encountering Indian pipe for the first time, some may share Dickinson's appreciation of its striking white beauty gleaming against the dark forest floor (see fig. 405), while others may not even recognize it as a wildflower, mistaking its white, waxy form for some sort of fungal growth. It is only by examining the plant carefully that one can see that it has the "normal" floral components of sepals, petals, stamens, and stigma (fig. 407).

Heath Family

Fig. 407. A view upward into a freshly opened Indian pipe flower showing the floral parts typical of most flowers: sepals, petals, stamens (10), and a broad stigma.

Fig. 408. A bumblebee has packed pollen from Indian pipe flowers into the scopae (special hairs) on its hind legs. The black marks on the flower are evidence of the flower's having received previous visits.

Fig. 409. This bumblebee's head is dusted with loose pollen, which will be carried to the next flower the bee visits where, if it contacts the stigma, it can result in cross-pollination. The pollen ball visible in the pollen basket on the bee's hind leg has been collected for use as food for the larvae in the hive. Held together with nectar, it is moist and unsuitable for pollination.

Fig. 410. A smaller, less hairy bee than a bumblebee does not enter the flower in the same way and is thus probably not an effective pollinator.

Indian Pipe

Fig. 411. The petals of this flower have been removed so the hairiness of the staminal filaments can be seen. Note the paired white nectaries extending from each side of the base of each stamen.

When fresh, all parts of the plant are pure white, or occasionally pale pink (or rarely even red). Picking a flower will result in it soon turning black as will the touch of any visiting insect. Indian pipe's principal visitors (and pollinators) are bumblebees. The sharp hooks on the bees' legs allow them to hang from the downward oriented flowers when collecting pollen or nectar and leave distinctive, small black "bee kisses" on the petals (fig. 408). The bees pack the gathered pollen into the scopae (modified hairs on their hind legs [see fig. 408]). In the process, their heads inadvertently become dusted with loose pollen (fig. 409), which is transported to other flowers, often resulting in cross-pollination. Since Indian pipe generally blooms in mid to late summer, the bees that visit are those of the worker class. The large-bodied queen bees, so important for the pollination of many of our early spring ephemerals, are by then ensconced in their nests and busy with egg-laying duties. Other insects, including smaller bee species (fig. 410), may visit but are probably not as effective as pollinators.

The downward-facing orientation of the flowers may serve to prevent their becoming filled with rainwater, which would dilute their nectar. Nectar is produced by paired 2 mm-long glands called nectaries (the largest in the Ericaceae subfamily Monotropoideae) that surround the base of each stamen (fig. 411). Hairs on the filaments of the stamens and on the interior of the petals (fig. 412) help to prevent evaporation of the nectar as well as deter small visitors (such as ants) that may otherwise pilfer the nectar (fig. 413).

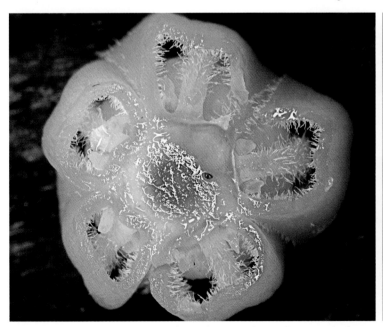 

Fig. 412. (Left) This cross-section of an Indian pipe flower illustrates the hairs present on the interior surface of the petals and on the filaments of the stamens that help prevent evaporation of the nectar and pilfering by small insects, such as ants. **Fig. 413.** (Right) Ants, seen here crawling on the exterior of this Indian pipe flower, are deterred by the hairs from entering the flower to obtain nectar.

Heath Family

Pollen is shaken out of the anthers by visiting bees and is subsequently brushed by them onto the large stigma in the center of the flower. Only the rim of the stigma is receptive; it is covered with a glistening exudate to which the pollen adheres, and which promotes germination of the pollen grains (see fig. 407). The pollen grains grow long tubes down through the style and along the grooved inner walls of the ovary. The sperm nuclei (reproductive cells) travel down through these tubes to fertilize the ovules.

Once fertilized, the flower head begins to turn upward (see fig. 406) until it is 180° from its original position. The exudate of the white stigmatic rim becomes dark (fig. 414), and the petals and sepals fall from the flower. The ovaries enlarge, eventually forming a woody capsule with five vertical slits around the circumference (fig. 415) through which the thousands of minute seeds are shaken out by the wind or other disturbance (figs. 416–417). The seeds measure only 0.6–0.8 mm in length by 0.12–0.15 mm in width and possess a wing at each end that facilitates dispersal by the wind, much like those of orchids (see fig. 156 in the helleborine essay). The embryo within the seed, comprising only two cells, is one of the smallest in the plant world. The seeming superabundance of seeds (fig. 417) is necessary, since, with such scant internal resources, very few seeds survive to produce new seedlings. The only seeds that will survive are those that land in a hospitable forest environment; that is, one where a relationship with a nutrient-providing fungus can be established quickly.

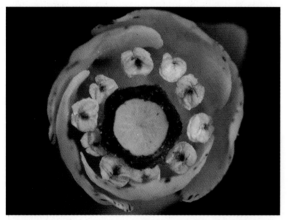

Fig. 414. In an older flower of Indian pipe, the darkened exudate on the rim of the stigma indicates that pollination has already occurred.

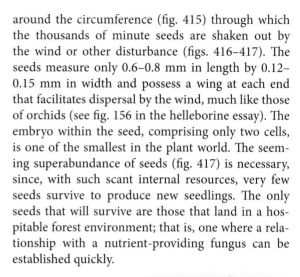

Fig. 415. (Right) Indian pipe fruits are woody capsules that split along five sutures to allow the minute seeds to be released when wind or other mechanical means shake them free from the fruit.

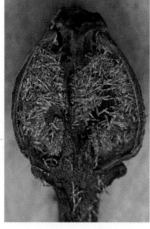

Fig. 416. A longitudinal section through a fruit showing some of the thousands of dustlike seeds inside.

Fig. 417. A capsule from which some of the seeds have been shaken out to show their small size.

Indian Pipe

Indian pipe was given its scientific name, *Monotropa uniflora*, in 1753 by Linnaeus, who based the name and his description of the plant on specimens that had been collected in eastern North America. However, variations in morphology and color of *M. uniflora* plants across its wide and noncontiguous range resulted in it acquiring an assortment of additional scientific names based upon those differences. All those names are currently viewed as synonyms for *M. uniflora* since not enough evidence exists to segregate the various forms and color variants into separate species, or even subspecies; however, that may change as additional material is studied, particularly with the use of molecular techniques. The family placement of *Monotropa* has also changed over the years. It was once placed in its own family, the Monotropaceae, and it had also been included in the Pyrolaceae, along with related forest floor dwellers such as spotted wintergreen (*Chimaphila maculata*) and shinleaf (*Pyrola elliptica*). But most recently it has been recognized as a subfamily, Monotropoideae, of the Ericaceae, as it is presently classified in the Angiosperm Phylogeny Group system.

Molecular studies already show distinct genetic variation among the North American, Central American, and Asian populations, with a probable common ancestral origin in Asia. Two possibilities exist for its broad, but disjunct, range: one is that its tiny seeds were dispersed long distances by the wind; the other suggests that a once single, contiguous population was subjected to geologic and/or climatic events that resulted in geographically isolated populations.

The common name, Indian pipe, and the scientific name, *Monotropa uniflora*, are both based on the appearance of the plant. The common name is easy to relate to the flowers; when in flower, each flower stalk of Indian pipe resembles a white clay peace pipe used by early American Indians (see fig. 405). The generic name, *Monotropa*, is derived from the Greek *mono*, meaning "one" and *tropa*, meaning "turn," a reference to the fact that the flowers of Indian pipe, after being pollinated, make one turn (from down to up) (see fig. 406); the species name *uniflora* means "one flower." Thus, *Monotropa uniflora* is "the single flower that makes one turn."

The only other species of *Monotropa* is *M. hypopitys*, whose species name is from the Greek for "under" (*hypo*) and "pine" (*pitys*), and is commonly called pinesap in English. It is often found growing under pines, both in North America and in Europe, and differs from *M. uniflora* in color—ranging from off-white through yellow to yellow-orange or red-orange—and in having smaller, multiple flowers on each stalk (fig. 418).

Indian pipe lacks chlorophyll, the green pigment that allows most plants to capture the sun's energy and use it to ultimately produce the carbohydrates that plants need to live, grow, and thrive. One might presume that Indian pipe must therefore be a parasite, much like squawroot (*Conopholis americana*) or beechdrops (*Epifagus virginiana*), which have a direct connection to a photosynthetic host plant from which they derive their nutrients. However, there is no such direct parasitism by Indian pipe; rather the story is more complicated. Although Indian pipe does obtain its nutrients from a host plant (or plants)—trees, in this case—it is a subterranean intermediary, a fungus, which serves as a conduit for the nutrients. The fungus absorbs carbohydrates produced by the tree and, in turn, benefits the tree by increasing its ability to take up phosphorus and micronutrients from the soil. Tiny strands of fungus (hyphae) attach to the root hairs of the tree and also to the roots of Indian pipe, enabling the transfer of carbohydrates produced by the tree to the pale, white, nonphotosynthetic plant. Thus, Indian pipe is indirectly parasitic on the tree and provides nothing in return.

Fig. 418. Several multiflowered stalks of a pinesap (*Monotropa hypopitys*), a close relative of Indian pipe.

Heath Family

Previously, botany students were taught that such plants were called saprophytes and that they absorbed their necessary nutrients from decomposing leaves and other dead organic material on the forest floor. Today, that term is reserved for certain fungi, and the more descriptive term "mycoheterotroph" is applied to plants such as Indian pipe. This rather cumbersome name is a reference to the plants' method of obtaining nutrients and is derived from the Greek *myco*, meaning or pertaining to "fungus;" *hetero*, meaning "different;" and *troph*, meaning or pertaining to "food." The result of this attachment of the fungus to Indian pipe is a mass of tangled plant and fungal material at the base of the plant (fig. 419). The association of a fungus with the roots of a plant is termed a mycorrhizal (meaning "fungus" + "root") relationship. Such relationships are more common than once believed and occur even between many photosynthetic plants and fungal intermediaries to aid the plants in supplementing their nutritive absorption.

Fig. 419. (Left) An Indian pipe plant has been excavated to show its ball-shaped subterranean portion comprising mixed plant and fungal material. **Fig. 420.** (Right) A young Indian pipe plant emerges close to several bright red caps of a *Russula* sp. mushroom, its fungal partner.

Fig. 421. (Left) When Indian pipe flowers first emerge from the ground, they have the appearance of eerie white fingers—more fungus-like than flower-like. **Fig. 422.** (Right) An unusual pink-flowered Indian pipe plant.

Indian Pipe

Fig. 423. (Left) These rare red Indian pipe flowers are older and have turned upward after being pollinated. They have lost most of their petals, but their bright red ovaries and stalks are still evident. The white nectar glands can be seen protruding downward at the base of the stamens on the flowers from which the petals have fallen. **Fig. 424.** (Right) Indian pipe fruits remain upright on dark brown "woody" stems throughout winter.

Usually the fungal partner in these associations is unknown, but in the case of Indian pipe, it has been demonstrated to be any of several species of fungi in the genera *Russula* or *Lactarius*—gilled mushrooms in the family Russulaceae that are common in the soils of our forest floor (fig. 420). The aboveground fruiting body of the fungus (the mushroom) is not always evident, but the appearance of Indian pipe is a clear indication that it is present underground.

Indian pipe makes its aboveground appearance any time from late spring to early fall, with mid to late summer being the most usual time to find them in flower. Initially, the still-developing, white forms lie prostate on the ground (fig. 421), appearing most unflower-like. They soon assume their typical, vertical posture and become recognizable as Indian pipe. Only the flower (including its pedicel, or stalk) is visible. Small, white scales arise along the flower stalks (see fig. 405). Although it is a treat to encounter the immaculate fresh white flowers in the dark forest, it is even more special to see the less common pink form (fig. 422), and truly exciting to find a group of red-colored plants (fig. 423). The woody capsules formed by the fertilized flowers often remain throughout the winter (fig. 424) and may persist even into the next summer when the plant's new flowers are present (fig. 425). The fleshy underground mass of Indian pipe and fungal tissue has the ability to remain dormant underground, appearing only in years when environmental conditions are favorable to its growth.

Fig. 425. Sometimes fruits of Indian pipe from the previous year are visible even into the following summer, when the new flowers of the year are in bloom.

Touch-me-not Family

Fig. 426. Orange jewelweed (*Impatiens capensis*) with a flower in the male phase (the white pollen-covered anther cap is visible in the opening to the flower) and an almost mature fruit hanging downward.

Jewelweed

Fig. 427. Yellow jewelweed (*Impatiens pallida*) sometimes intermingles with orange jewelweed in the same wet areas of forest as seen in this image. They are the only two species of the genus that grow natively in the Northeast.

Jewelweed

Impatiens capensis Meerb.
Touch-me-not Family (Balsaminaceae)

The genus *Impatiens* belongs to a small family, the Balsaminaceae, that comprises but two genera, yet consists of more than 1000 species. The species occur in both temperate and tropical regions of Asia, Africa, North America, and Europe. The second genus, *Hydrocera*, consists of only a single species (*H. triflora*), native to Southeast Asia.

Habitat: Moist soils in lightly shaded woodlands, especially along streams, marshes, and other wetlands, provide the most favorable habitat, but the plants can survive in drier or more shaded environs as well if the site is wet enough in spring to facilitate germination of seeds.

Range: Jewelweed is found in most of the lower 48 states excepting the more arid western states of Arizona, New Mexico, Utah, Nevada, Wyoming, Montana, and California. It is most prevalent in the eastern part of the country but reduced to small, isolated areas in Georgia and Florida, the Central Plains states, and the Pacific Northwest (where it may have been intentionally introduced). Jewelweed also inhabits all provinces of Canada, the southern portions of the Northwest Territories, and Yukon, and a small area of Nunavut.

Of the approximately 1000 species of *Impatiens*, only two are native in the northeastern United States and adjacent Canada. The main subject of this essay is the most prevalent of those two, *Impatiens capensis*, often called orange jewelweed (fig. 426) to differentiate it from the second species, *I. pallida*, known as yellow jewelweed (fig. 427).

Jewelweed (this common name is sometimes used with the modifiers orange or spotted) is but one of many common names for *Impatiens capensis*, all of them descriptive. The plant earns the common name jewelweed for any of four reasons: (1) its delicate orange flowers resemble pendant earrings (another name is lady's-earring) (fig. 428); (2) the seeds are the

Touch-me-not Family

color of turquoise; (3) drops of water form silver beads on its leaf blades (fig. 429); and, (4) when held under water, the underside of the leaf takes on the appearance of quicksilver (fig. 430). Other common names: touch-me-not, snap-weed, and quick-in-the-hand, all refer to the ballistic seedpods that shoot their seeds a half meter or more from the parent plant. The scientific name, *Impatiens*, also refers to this characteristic of "impatience," with the second part of the name, *capensis*, (the species epithet) meaning "of the Cape," a reference to Africa's Cape of Good Hope. This error in naming the plant occurred when the author of the species name, the Dutch botanist Nicolaas Meerburgh, mistakenly thought that the collection locality of the plant was the Cape of Good Hope. Thus, it stands as a confusing misnomer.

Fig. 428. (Left) The flowers have an unusual shape and hang from slender pedicels, giving rise to one of their common names, "lady's-earring." **Fig. 429.** (Right) Silvery beads of water form at the tips of the lobes of jewelweed leaves as a result of guttation, a process whereby root pressure forces water out through special structures called hydathodes in order to rid the plant of excess water that results from growing in wet soil. In addition, water droplets "bead up" on the leaf surface.

 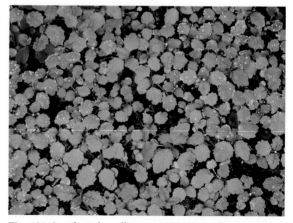

Fig. 430. When a jewelweed leaf is held underwater, the underside shimmers like silver or mercury. This property, and those depicted in the two previous images, may account for the plant's moniker, "jewelweed."

Fig. 431. Jewelweed seedlings germinate synchronously in early spring, often densely covering the forest floor in bare, wet areas.

Jewelweed

Jewelweed is an annual plant having seeds that germinate synchronously in early spring, after having spent the winter months in a state of dormancy (a minimum of four months at cold temperatures is necessary for germination to occur). The seedlings are sometimes so dense that they blanket large areas of bare, wet soil (fig. 431), thereby staking a claim to that territory. To a great degree, their density precludes the later germination of other species. The seedlings develop into a thick stand of plants, thus retaining their local dominance of the habitat; populations may persist in the same area for years, if not decades. The initial set of true leaves, distinguished from the rounded cotyledons (seed leaves) by their more oval shape with small rounded teeth along the margins, appear soon after germination (fig. 432). The seedlings' shallow roots take up moisture from the top layer of the soil.

The stem of jewelweed is smooth, succulent, and hollow. Because the plant doesn't put a lot of resources into building a solid stem, it grows quickly, relying on turgor pressure to remain erect. When the stem is crushed, it exudes a slippery mucilage. In folk medicine, the sap has long been used as a soothing balm for alleviating the itch of poison ivy, and even today a compound from *Impatiens* sap is included in remedies for poison ivy dermatitis and other skin conditions. Recent research, however, has demonstrated that the sap of jewelweed is less effective than washing with soap and water or even plain water in soothing the poison ivy rash. Interestingly, the active ingredient of jewelweed, lawsone, that was examined in the study is also found in henna (*Lawsonia inermis*). Henna has been used for more than 5000 years in North Africa, India, the Middle East, and other nearby areas for dying skin, hair, and fingernails a reddish-brown color, as well as for dying fabrics and leather. The sap of jewelweed can also be used to produce a golden-tan to rusty-orange dye by chopping up the plants and simmering them for about an hour before adding the mordant-treated wool (fig. 433) or fabric to the dye pot.

Jewelweed thrives best in wet, shaded sites but is a weak-stemmed plant prone to wilting, or even expiring, when the soil becomes dry and the sun is strong. The alternately arranged, thin, dull green leaves of jewelweed have rounded teeth, each with a tiny sharp tip, called a mucro (visible on the rightmost leaves in fig. 427). The leaves tend to droop, even on a daily basis, when the temperature heats up in the afternoon. This leaf drooping is an adaptation to conserve moisture in the leaf, as stomata close and less leaf surface is exposed to the sun (fig. 434). The daily fluctuation does not seem to have a negative effect on the plants, which rebound to their turgid condition during the cool evening hours. While jewelweed plants can't tolerate long periods of drought, they are also not able to survive sustained flooding conditions. Such inundation can result in the death of the plants due to a lack of oxygen in the saturated soil. Under such stress, some plants will produce bright red, adventitious roots, arising from the nodes along the stem. These roots are able to

Fig. 432. Soon after the rounded cotyledons have appeared, the plant's first true leaves (more elongated and lobed) arise from the stem.

Fig. 433. The rusty-orange color of this wool sample was dyed using jewelweed with a tin mordant.

Touch-me-not Family

Fig. 434. Jewelweed is a tender, somewhat succulent plant that wilts easily during hot, sunny conditions.

absorb oxygen from the air, much as pneumatophores do for bald cypress or certain tropical mangroves that live with their roots submerged in water. Adventitious roots arising from the lowest nodes of the stem also aid in supporting the stem in the wet soils.

The flowers of jewelweed are most interesting. Their unusual form—somewhat like a mini-cornucopia—is unlike that of any other flower in our flora, and the bright orange color, almost always heavily speckled with rusty-red spots, is also rare in the Northeast. (If you search for orange flowers in any local wildflower field guide arranged by color, you will find few pages). As with the paucity of red flowers in the Northeast, the small number of orange-flowered species may correlate with the lack of diversity of hummingbird species that reside in eastern North America; we have only one, the ruby-throated hummingbird. While hummingbirds, in general, will visit many different flowers that provide a nectar source, they are known to be particularly attracted to red and orange flowers.

The orange-flowered jewelweed appealed to the early English settlers, and early in the nineteenth century, seeds of orange jewelweed were taken back to England, where the species soon became naturalized and was renamed "orange balsam." The species gradually spread along waterways and is now well established in many parts of lowland Britain. By the late twentieth century, jewelweed had been spread by humans to other parts of Europe.

Jewelweed flowers bloom from late summer into autumn, with some continuing to develop until the plants are killed by frost. The flowers hang from slender pedicels in groups of one to three. There are three sepals—two small, translucent yellow-green ones located behind the upper lip, and the third, a spurred saclike structure that appears to be part of the orange corolla (see figs. 428 and 438). The spurred sepal is cone-shaped, with an opening toward the front of the flower that narrows into a thin spur toward the rear; the spur then curves forward under the cone. It is here where the nectar is produced and stored. The corolla has five petals: an upper lip that stands upright, a lower lip comprising two (usually) fused petals that serves as a landing platform for insects, and two small side petals surrounding the opening into the cone-shaped sepal (fig. 435). These lateral petals are fused to the lower lip and are considered by some to be lobes of that lip. The flowers are zygomorphic (bilaterally symmetrical) and, thus, not easily accessible to a wide range of insects. As the sought-after nectar is confined to the small spur portion of the sepal, access to that reward is limited to only those visitors that might be effective pollinators, that is, hummingbirds, long-tongued bees, and butterflies. *Impatiens* flowers are bisexual, with both male and female structures located just under the upper petal. The stamens have their anthers fused into a caplike structure

Fig. 435. When a jewelweed flower first opens, it is in the male phase with its anthers united into a cap just beneath the upper petal. The white pollen is held together by fine threads.

Jewelweed

covered with white pollen grains loosely held together by sticky threads. The anther cap can easily be seen when the flower first opens (see fig. 435); the anthers may even dehisce before the flower opens. Nectar production also begins just before opening. The flowers remain in the male phase for just under 24 daylight hours (usually two days), at which time the stamens and anther cap covering the gynoecium are pushed off the flower by the expanding pistil, thus exposing the narrow green stigma hidden beneath them (fig. 436). The flowers are protandrous (functionally male, then functionally female). The female phase lasts for only about five hours, providing just a short time for pollinators to deposit pollen on the receptive stigma. Thus, although the flowers are perfect in terms of having both male and female organs, the timing of the male and female phases results in little possibility of self-pollination and promotes outcrossing. Shortly after fertilization, the showy flower parts fall and the ovary begins to develop into a fruit, a process that takes from 12 to 35 days, depending on locality and environmental conditions.

Despite the striking orange color and the shape of the flower seemingly being conducive to hummingbird pollination, relatively few hummingbirds are seen to visit jewelweed flowers. They are the only visitors with tongues long enough (ca. 20 mm) to reach the end of the 8.3 mm nectary. Nevertheless, the flowers continuously replenish their nectar supply, refilling the spur, and thus ensuring that nectar remains available to shorter-tongued visitors as well. Jewelweed is considered one of the most prolific nectar producers in the northeastern flora. Its nectar is high in sugar content and amino acids, thereby serving as an important source of energy and nutrition for its insect visitors. Nectar of male-phase flowers has a substantially higher percentage of sugar than that of female-phase flowers, and for this reason the male phase is preferred by insect visitors. The primary pollinators are bees in the genus *Bombus* (bumblebees). Studies of day-long visiting patterns have shown that the shorter-tongued bumblebee species visit earlier in the day than the long-tongued bumblebees. Alien honeybees, other bees, wasps, ants, flies, and butterflies visit the flowers as well. Bumblebees crawl into the opening of the conical sepal (fig. 437) and extend their long tongues into the spur to sip its nectar. In doing so, their hairy bodies get dusted with

Fig. 436. After approximately two days in the male phase, the elongating pistil (the narrow green structure beneath the upper petal) pushes the anther cap from the flower, thereby revealing the virgin stigma at its tip. This female phase of the flower lasts for only about five hours.

Fig. 437. Bumblebees are the most frequent and most effective pollinators of jewelweed. Their bodies are large enough that they are obliged to contact the reproductive parts of the flower as they push their way through the opening of the conical calyx in order to reach the nectar held in its spur.

Touch-me-not Family

pollen from the anthers above when the flower is in the male phase and later, when they visit a female-phase flower, that pollen is positioned correctly to brush off on the stigmas, as the stigmas are no longer covered by the anther caps. Honeybees and certain wasps may also be effective pollinators; however, honeybees, these same species of wasp, and short-tongued bumblebees (as well as large ants) are also able to gain access to the sugary liquid by biting holes directly into the nectar spur. Other smaller bees or flies may then act as opportunists availing themselves of nectar by taking advantage of the access holes made by the primary nectar thieves. Some of these smaller insects can also enter the sepal nectary from the usual frontal opening but are not large enough to contact the reproductive organs when doing so. Honeybees were observed to take pollen from the anthers without performing any pollination service for the species by subsequently visiting a female-phase flower, and a syrphid fly seemed to probe the androecium, perhaps seeking moisture (fig. 438) before drinking nectar from the spur. A study by Richard Rust showed that such nectar-accessing activity, when insects took nectar without also serving as pollinators, did not result in diminishing pollination success or in seed production.

But these showy, charismatic "dangling earring" flowers are not the only flowers produced by *Impatiens capensis*. As with the violets discussed in my earlier book, *Spring Wildflowers of the Northeast*, jewelweed has a mixed breeding system; that is, it has two methods of reproduction by seed. Late in the season—or earlier if conditions are not favorable for the production of showy, pollinator-attracting flowers, or for their pollinators themselves (too much shade, too little water, or poor weather during flowering season)—the plants produce a second type of flower, this one termed cleistogamous (meaning hidden marriage). Cleistogamous flowers are extremely reduced in size (about 1–2 mm in length) and complexity. They grow from leaf axils of stems of the lower part of the plant, never open, and reproduce within the closed budlike flowers without the need of a pollinator. Since the closed system almost assures fertilization, cleistogamous flowers will produce some seeds each year. This is a particularly important guarantee for an annual plant such as *I. capensis*, when adverse conditions that negatively affect the production of the "normal" chasmogamous (open marriage) flowers or the activity of their pollinators would result in little or no seed set for that year.

Both types of flowers have advantages. Since the showy, spotted orange-flowered chasmogamous flowers are attractive to insects and hummingbirds, the resultant cross-pollination yields seed with a new mixture of genes, some of which may prove advantageous to the future plants, perhaps allowing them to thrive in more varied habitats. On the other hand, the production of cleistogamous flowers, which have greatly reduced floral parts, produce no nectar, have only two stamens (compared with four in chasmogamous flowers), and have relatively few pollen grains

Fig. 438. Smaller insects, such as this syrphid fly (*Toxomerus geminatus*), are not bulky enough between their dorsal (upper) and ventral (lower) surfaces to contact the reproductive organs when crawling into the nectar sepal, and therefore are not effective pollinators.

Jewelweed

per anther, requires a lesser expenditure of resources on the part of the parent plant (estimated at 100× less) and ensures that at least some seeds will be produced in a given year. By the end of the season, in both good years and bad, jewelweed plants will have produced many more cleistogamous flowers than chasmogamous, but each resultant fruit of a cleistogamous flower yields fewer seeds than that of an outcrossed chasmogamous flower. One would think that the seeds resulting from the outcrossed flowers would be more vigorous, but in terms of germination and growing into mature, seed-producing plants, Steets and collaborators showed that the opposite is true.

Our other native species of *Impatiens*, *I. pallida*, or yellow jewelweed, is difficult to tell from *I. capensis* until the plants come into flower. Yellow jewelweed is a slightly taller plant, often with slightly larger leaves. When examining the leaves of both plants, it can usually be observed that the upper leaves of *I. pallida* have a shorter petiole (leaf stalk) than those of *I. capensis* (fig. 439). Yellow jewelweed begins flowering a little earlier than orange jewelweed and ceases flowering sooner. It also generally produces fewer chasmogamous than cleistogamous flowers. Once in flower, the two species can be differentiated easily. Aside from the basic color difference, yellow jewelweed flowers are larger with a wider opening into the conical sepal and have a broader landing platform (the lower lip), which is usually bilobed (fig. 440). Only its narrow nectar spur is shorter than that of orange jewelweed (5.9 vs. 8.3 mm) and is either straight (fig. 441) or less curved toward the front of the flower, often forming a 90° angle. In addition, the reddish "speckles" on the flowers of yellow jewelweed are fewer and

Fig. 440. The flowers of *Impatiens pallida* are larger than those of orange jewelweed, with a larger opening to the nectar-holding sepal and a broader lower lip, which serves as a landing platform for insects. Unlike orange jewelweed flowers, the red spots on the corolla are fewer and smaller.

minute (see fig. 440) compared with those of orange jewelweed (see fig. 436). The two species sometimes grow together in the same habitat (see fig. 427), but *I. pallida* is more tolerant of dryer soils than *I. capensis*. Far fewer insects visit the flowers of yellow jewelweed, but both jewelweed species share the same species of bumblebees as their principal pollinators. Because the opening to the spur is larger in yellow jewelweed, bumblebees have an easier time entering the sepal to get nectar than they do in orange jewelweed and can, therefore, visit more flowers in a

Fig. 439. The upper leaves of orange jewelweed (left) and yellow jewelweed (right) may be distinguished by the length of their petioles (leaf stalks); those of orange jewelweed are longer.

Fig. 441. The nectar spur of yellow jewelweed is shorter than that of orange jewelweed (see fig. 428), and either straight or only slightly recurved.

Touch-me-not Family

shorter period of time. In orange jewelweed, the bumblebees must crawl through the narrower opening toward the spur, causing them to squeeze under the reproductive structures, thereby extending their visiting time (see fig. 437). The larger number of species attracted to orange jewelweed may be attributable to its longer spur on which other nectar-seeking "robbers" can perch to avail themselves of nectar through holes made by short-tongued bumblebees or other hole-makers. Ironically, the presence of twice the concentration of amino acids (building blocks for proteins important for insect nutrition) in the nectar of orange jewelweed may result in attracting more visitors, but some of them will be nectar robbers that do not benefit the plant.

Rarely, unusual flower colors are encountered. Some flowers of *I. capensis* are spotless and known as *I. capensis* forma *immaculata*, and others have been encountered with creamy white petals (and nectar sepal) with pink spots (*I. capensis* forma *albiflora*). Both these forms were initially described from Maine but occur occasionally in other parts of the species' range. Likewise, forms of *I. pallida* are also found with cream-colored flowers.

It is, perhaps, the fruits of jewelweed that are best known, usually by the common name of touch-me-not. Much to the delight of children (and a fair number of adults), the fully ripe fruits explode open at the slightest touch. Each fruit (whether resulting from a chasmogamous flowers or a cleistogamous flower) is an elastically dehiscent capsule comprising five valves. As the seeds and the pod mature, mechanical energy is stored in the valves. At maturity, the fruits are translucent and fully turgid, and the ripe, dark brown seeds can be seen through the capsule walls (fig. 442). The seeds are loosely attached to a spongy central material within the ovary, called the columella (the placenta). The valves are highly efficient energy storage structures with the amount of energy stored dependent on the level of moisture in the pod, but the transfer of energy to the seeds has a low level of efficiency (only 0.5% of the energy is transferred), resulting in low dispersal efficiency. Yet, as the pod explodes open, the seeds are flung away from the plant (fig. 443), landing up to 2 meters away. In this process, the valves instantly curl inward, separating from the columella except at the base where they are attached to the fruit stalk (fig. 444).

Erasmus Darwin, a physician and the grandfather of Charles Darwin, was also a poet (though he published his poetry anonymously for fear he would lose credibility as a doctor). His poems often used natural history as a theme, including a poem titled "The Loves of Plants" in which he describes the explosive capsule of jewelweed as a mother who "hurls her infants from her frantic arms." Two generations later, Charles

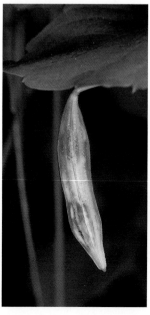

Fig. 442. The fruits of jewelweed are 5-valved capsules that swell and become turgid; at maturity, their ripe, brown seeds are visible through the translucent fruit walls.

Fig. 443. The slightest touch to a fully mature pod will trigger it to "explode," sending the valves and seeds up to two meters from the plant. The coiled valves from a fruit that just exploded when touched were attached to the uppermost pedicle in this image. The valves can be seen flying through the air at the top of the photo.

Jewelweed

Fig. 444. This fruit, triggered to split open while enclosed in my hand, shows the structure of the dehisced fruit with its valves coiled toward the stem end (pedicel) and the central columella (the placenta) to which the five seeds were attached in the center of the fruit. I gently scraped off part of one brown seed coat to show the blue color of the enclosed seed.

Fig. 445. Several seeds of jewelweed demonstrate various stages in their development: The green seed at "eight o'clock" is almost mature, the brown seed in the center is fully mature, and the surrounding robin's-egg blue seeds have had their brown seed coats removed to show their turquoise beauty.

Darwin was one of the first scientists to become intrigued by the efficiency of cleistogamous flowers.

Though most seeds land within a half meter of the parent plant, some are projected as far as 2 meters. The distance is dependent on several factors: the angle and velocity of the launch, the seed mass, and the amount of air resistance. Of course, the seeds launched by the cleistogamous flowers travel shorter distances because of their smaller valves and lower position on the stem. Dehiscence, from the splitting of the pod's valves to complete coiling, takes 4.2 ± 0.4 milliseconds. Jewelweed has a secondary method of dispersal—water. The high lipid content of the seeds makes them buoyant, so that seeds landing on water may travel long distances on the currents of a stream or along the shores of a marsh, pond, or lake.

Jewelweed reveals its last surprise at the end of the plant's life: its seeds. When the thin brown seed coat is scraped off, they are an intense robin's egg blue (fig. 445). I've found the seeds useful in crafting natural ornaments for our Christmas tree—making lovely blue eyes for angels made from acorns, milkweed, pine cones, and other of nature's bounty (fig. 446). The blue color may last for a year or so, but as sometimes happens to infants born with blue eyes, as they age, their eyes become brown. In addition to being beautiful, the seeds of jewelweed are edible, tasting somewhat

Fig. 446. This Christmas ornament angel was made using a pine cone as the body, pods (follicles) of common milkweed as wings, an acorn for the head, and jewelweed seeds for the blue eyes of the angel.

Touch-me-not Family

Fig. 447. The larva of this insect feeds on the jewelweed stem tissue within the gall that has formed around it.

like walnuts—though one would have to pop a lot of jewelweed fruits to obtain enough seeds for a snack. Again, caution should be observed whenever trying a new wild food.

Jewelweed serves as a host plant for caterpillars of certain moths, which bore into their stems (fig. 447) and to gall midges (*Schizomyia impatientis*) that form galls on the midribs of leaves and in the ovaries of flowers when still in bud (fig. 448). A leaf beetle, *Rhabdopterus praetexus*, can do severe damage to the leaves of jewelweed, sometimes causing the death of the plant. But perhaps the most noticeable and damaging

Fig. 448. A cecidomyid midge, *Schizomyia impatientis*, caused this gall to form on a flower bud of *Impatiens capensis*

Fig. 449. The telltale sign of white-tailed deer herbivory on jewelweed is shown here (deer tear the stems, rather than cut them neatly as a rabbit would). The succulent stems of jewelweed are a favorite of deer.

Jewelweed

herbivory is that of white-tailed deer, which feed on the succulent plants (fig. 449), and in so doing may prevent them from reproducing, which in an annual plant is akin to death. It has been documented that jewelweed growing where long-term heavy browsing by deer has occurred compensate for the loss of their apical meristems by increased branching at the lower nodes of their stems.

The impatiens plants used for mass plantings by landscapers and gardeners (e.g., 'Elfin' impatiens from southeastern Africa, and New Guinea impatiens) rarely escape from cultivation, but *Impatiens glandulifera*, from the Himalayas (fig. 450), has naturalized in the United States and has already reached invasive status in some states.

Impatiens species, overall, are relatively similar vegetatively, but their flowers differ greatly and can excite the imagination—leading to such evocative names as Congo cockatoo and parrot plant—for a species from tropical Africa (*Impatiens niamniamensis*) that bears flowers resembling the bright red and yellow beaks of such birds (fig. 451). Congo cockatoo is grown as a tropical ornamental in warmer climates and in display glasshouses. In the wild, the long-spurred flowers are pollinated by sunbirds, Africa's equivalent to the New World hummingbirds.

Fig. 450. A garden escape, *Impatiens glandulifera*, is originally from the Himalayas. It has become an aggressive invasive species in parts of the Northeast and should not be planted in gardens. The flower shown here is being visited by another syrphid fly, *Rhingia nasica*, which appears to be taking pollen from the anthers. These long-tongued flies also feed on nectar.

Fig. 451. *Impatiens niamniamensis*, a spectacular species from tropical Africa, has flowers that resemble the beaks of large birds, such as parrots—the inspiration for the common names of this plant: parrot flower and Congo cockatoo. Note that the basic floral pattern is similar to that of our native species in that the reproductive organs are located under the upper petal and the enlarged sepal's nectar spur is curled under the flower. However, the flowers are sturdier, which protects them from being damaged by their sunbird pollinators when they probe for nectar.

Nightshade Family

Fig. 452. Jimsonweed pollinators are large-bodied moths, in this case the pink-spotted sphinx moth. They do not begin visiting the flowers until the sky is truly dark, more than a half hour after sunset. The moths are attracted by a strong, fragrant aroma that lures them to the flowers, where they drink from the copious nectar produced after the sun sets. In doing so, the moths often carry pollen from one flower to another. Note the cream-colored pollen on the lower two-thirds of this moth's proboscis.

Jimsonweed

Fig. 453. The flowers of jimsonweed begin opening more than two hours before sunset and by sunset are fully open and fragrant. They do not begin to produce nectar until after sunset.

Jimsonweed
Datura stramonium L.
Nightshade Family (Solanaceae)

The sex life of jimsonweed begins after dark, when it wafts its seductive perfume into the night sky (fig. 452). Its flowers, tightly furled by day, begin to open in late afternoon and are fully open before the sun sets (fig. 453). Although the flowers are attractive, the plant itself is toxic, a property that played a role in an interesting episode in American colonial history.

Habitat: Farmlands, barnyards, and other areas with disturbed soil.

Range: Nearly worldwide, in temperate and subtropical regions. There is some disagreement about the origin of jimsonweed; some botanists claim that it has a Central American origin, while others maintain that it originated in Asia. In either case, *Datura stramonium* disseminated quickly and was known from the Americas, Asia, and Africa at an early date; European records indicate that it was already common in Europe by the mid-sixteenth century.

Jimsonweed was known long before Linnaeus scientifically described it in 1753. He named it *Datura*, based on an ancient Hindu word for another species of *Datura* known as "*dhatura*," and *stramonium*, from the Greek words *strychnos* and *maniakos*, meaning "nightshade" and "mad," respectively, a testament to its known toxic and mind-altering properties. It has many common names, none of them flattering; among them are thorn apple, devil's weed, hell's bells, stinkweed, devil's trumpet, locoweed, and prickly burr. The most commonly used name in the United States is jimsonweed—a corruption of Jamestown-weed, which is sometimes written as two words. The origin of this name is derived from an event that occurred in the village of Jamestown in what was then the British colony of Virginia.

Nightshade Family

In the spring of 1676, British soldiers were ordered to Jamestown to suppress a rebellion by colonists who resented what they considered too lenient treatment of the local Native Americans by Virginia's Governor Berkeley. The uprising was known as the Bacon Rebellion, named for its leader, Nathaniel Bacon, a young British planter. The rebels sought revenge for Indian attacks and demanded the Indians be driven from land that the British considered theirs by virtue of treaties. While the soldiers were stationed in Jamestown, they gathered the early green leaves of *Datura* to use in preparing a stew, which they then consumed in large quantities. The result was the total loss of their senses lasting over a period of 11 days. This incident was documented in a 1705 publication, *History and Present State of Virginia*, by historian Robert Beverley, in which the effect on the soldiers is described thus:

> They turned natural fools upon it for several days: one would blow up a feather in the air; another would dart straws at it with much fury; and another, stark naked, was sitting up in a corner like a monkey, grinning and making mows (grimaces) at them; a fourth would fondly kiss and paw his companions, and sneer in their faces with a countenance more antic than any in a Dutch droll. In this frantic condition they were confined, lest they should, in their folly, destroy themselves.

Fig. 454. Jimsonweed seeds remain viable in the soil for many years and may sprout when the soil has been freshly disturbed, as shown here at an archeological dig at the Jamestown historic site in Virginia.

Although Berkeley was routed and Jamestown torched, the rebellion was short-lived and did not result in the expulsion of the Indians. Governor Berkeley again took control but was eventually recalled to England in order to avoid further conflict among the settlers.

Having long known of jimsonweed's historic connection with Jamestown, while visiting friends in southern Virginia, I made time to go to the Jamestown National Historic Site. Aside from learning more about colonial history, I hoped to learn if jimsonweed still grew at the old fort site. My husband and I spent the better part of an enjoyable day touring the historic site but saw nary a plant of jimsonweed—that is, until we were about to leave at the end of the day. There, growing on a pile of soil that had been excavated from a nearby archeological site, were half a dozen large, healthy jimsonweed plants, all having both flower buds and wilted flowers (fig. 454). Park maintenance staff told us that the only time they ever saw jimsonweed, was when it appeared following a fresh excavation; the plants sprouted from freshly disturbed soil, which must have contained long dormant seeds. I later learned from a review in *Weed Science* that a long-term study of seed viability of 41 species of different families, carried out by J. T. Duvel, showed that 91% of jimsonweed seeds remained viable in the soil at the conclusion of his 39-year study.

A robust, rank-smelling weed with furled buds and wilted flowers by day (fig. 455), jimsonweed begins a Cinderella-like transformation as the sun sinks low in the sky. Knowing that the flowers of jimsonweed would not open until almost sunset, I asked a ranger if we could stay late enough to see them—and perhaps their pollinators. We were given permission to stay until sunset: 7:48 p.m. at that time of year. I settled in for the few hours wait while my husband got some exercise by walking the perimeter of the site, unintentionally gathering chiggers along the way. At 5:30, the flowers began to unfurl, just as a parasol would, and by 6:30 the lovely purple-throated, white flowers were fully open (see fig. 453). The seductive gardenia-like fragrance began 15 minutes later—pleasing to the human nose and reportedly irresistible to large moths—but no moths came. I *had* to see the moths, and so our nighttime vigil began. I checked the flowers periodically for nectar, but there was none present at 7:30 pm. However, at 7:35 there was one fleeting visit by a hummingbird, perhaps searching for nectar.

Jimsonweed

Fig. 455. Flower buds beginning to open late in the afternoon. Yesterday's still-attached flowers hang spent and drooping. The historic site of Jamestown is seen in the background with the British flag waving.

Fig. 456. (Below) This pink-spotted sphinx moth inserts its proboscis deep into a jimsonweed flower to sip the nectar.

As expected, the sun sank below the horizon at 7:48—but still no moths. Fortunately, there were also no rangers coming to check that all visitors had left the site. The waning moon was less than one-quarter full and had not yet risen above the horizon; clouds covered most of the sky. Not until 8:25, when the sky was truly dark, did the first moth appear (see fig. 452). By then the nectary at the base of the flower was secreting nectar.

The flowers of jimsonweed are typical of many moth-pollinated flowers; they open at night, are light in color, emit a pleasantly fragrant aroma, and produce copious amounts of nectar (but only at night). Although the funnel-shaped flowers permit easy access to flying visitors, the nectar is held deep within the base of the flower in a chamber accessible only to visitors having a long narrow proboscis (or tongue) (fig. 456). Only such a tongue is capable of slipping into the five narrow channels that lead into the nectar chamber. Certain large moths, known as hawk, or sphinx, moths, fit this description perfectly. The flowers generally open for only one night, but if the following morning is overcast, they may remain open throughout much of that day, attracting bees and other insects that scavenge any pollen remaining in the anthers. Paradoxically, the same species of hawk moths that benefit the species by pollinating the flowers can also have a detrimental effect on *Datura*. The adult female lays her eggs on the jimsonweed plant, and the larvae feed on its leaves.

The moths came in small waves of two to four within a few minutes and spent only a second or two at a flower. Since it was so dark, it was impossible for me to judge through the limited field of my viewfinder which flower of the 80 open that night that a moth was about to visit and still have time to position myself and focus the camera to capture the visit. My husband and I worked out a plan whereby I would remain focused on a few selected flowers at the upper perimeter of the plant. When a moth approached those flowers, my husband would say, "Now," and I would press the shutter

Nightshade Family

Fig. 457. A freshly opened jimsonweed flower with a short style (not visible). Note that the anthers had already released pollen before the flower opened.

Fig. 458. A jimsonweed flower with a style equal in length to the stamens, and its stigma receptive, allowing it to easily receive pollen from the anthers of the same flower.

Fig. 459. A long-styled jimsonweed flower showing the style with its stigma protruding slightly beyond the anthers, making it more difficult for self-pollination to occur in the same flower. Note: no pollen is on the stigma.

button. Fortunately, our system worked! We stayed until almost 10:30 observing visits by two species of sphinx moths: the five-spotted hawkmoth (*Manduca quinquemaculata*), shown in fig. 452, and named for the five spots on either side of its abdomen (also known as the tomato hornworm in its caterpillar stage) and a similar-sized moth, the pink-spotted hawkmoth (*Agrius cingulata*), called such because of its pink and black striped abdomen (see fig. 456). It is also known as the sweet potato hornworm for one of its preferred larval foods; its larvae also feed on jimsonweed.

The anthers of jimsonweed open even before the flower buds unfurl, allowing self-pollination to occur within the flower before the corolla has opened (termed autogamy). However, as with most plants, outcrossing with pollen from another plant is more advantageous for the species, promoting both genetic diversity and increased fitness. Nectar-imbibing hawk moths inadvertently pick up pollen on the distal two-thirds of the proboscis (see fig. 452). The pollen is then transferred to the stigmas of subsequently visited flowers. Variation in the length of the stigma within jimsonweed flowers leads to what is termed a mixed mating system. Despite the seemingly well-adapted moth pollination system, visitation by moths accounts for only 5% of pollination, the remaining 95% occurring as a result of self-pollination (when the loose pollen released by the anthers comes in contact with the stigma—the receptive female structure—of the same flower). It seems unusual that a plant that produces such showy, aromatic flowers, so well designed to attract pollinators, achieves only 5% of its pollination through the services of a pollinator. The high degree of self-pollination can be attributed to a combination of the shedding of pollen prior to the opening of the flowers and variability in style length. Those flowers having styles (the female structures that bear the stigma at their tip) that are shorter than the level of the anthers (the male structures) (fig. 457), or slightly overlapping with them, are almost exclusively self-pollinated. Those with styles that extend only partially beyond the anthers, though having some degree of outcrossing (fig. 458), are still primarily self-pollinating. And only those flowers with styles that project entirely beyond the anthers (fig. 459) are more likely to be outcrossed as a result of moth visitation. Thus, flowers with styles of any length can have some degree of self-pollination. The low rate of outcrossing was verified in a study by

Jimsonweed

Fig. 460. A flower of the lavender-flowered form of *Datura stramonium* with a style that clearly extends beyond the anthers.

Fig. 461. (Above) The open four-parted capsules of jimsonweed remain on the plant into winter and release their seeds when the plant is shaken by the wind.
Fig. 462. (Right) A spiny, developing ovary of jimsonweed is surrounded by a "collar" that is the remnant of the basal portion of the calyx. The long, now-black style persists.

Motten and Stone, in which they manually crossed flowers of different colors (either lavender or white) (fig. 460) to develop man-made hybrids for the study. The parentage of the seedlings that resulted from the crosses could be easily determined by the color of the lower part of the seedling—purple seedlings from the lavender-flowered plants and green from the white-flowered plants—and then correlated with the relative stamen/style lengths of the parent plants to determine which combination of reproductive organs resulted in greater pollination and fertilization success.

Experiments in which the amount of nectar in jimsonweed flowers was controlled showed that moths preferentially visited flowers with greater amounts of nectar and that those flowers produced fruits with more seeds; however, the moths were also more likely to lay their eggs on those plants, resulting in greater herbivory by their larvae.

Datura fruits are spiny capsules that occur in the forks of the branches and open along four sutures (fig. 461). The base of the calyx remains attached to the fruit, forming an encircling "collar" (fig. 462). Small, dark seeds pockmarked with tiny crater-like depressions (fig. 463) are shaken out by the wind. Each capsule may contain as many as 700 seeds (fig. 464). Most seeds fall to the ground, where they produce the following year's crop of plants or lie dormant in the soil (sometimes for many years) until future disturbance produces the right conditions for germination. If the seeds land in water, they may also be dispersed in that way since they are capable of floating intact for up to 10 days.

Nightshade Family

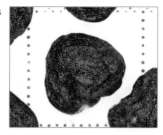

Fig. 463. The pitted seeds of jimsonweed displayed on a 10 mm grid as seen through a dissecting microscope.

Pat Willmer, in her book *Pollination and Floral Design*, states that in those locations where nectar-feeding bats occur, as in the southwestern United States, bats may also visit jimsonweed flowers at dusk.

Jimsonweed is a fast-growing annual with jaggedly lobed leaves. Though some gardeners plant it (or its close relative, the moonflower [*Datura metel*]) as an ornamental because of its attractive and fragrant nocturnal flowers, farmers consider it a weed. It can be a serious pest in cultivated fields, especially where crops belonging to the same family are being grown, for example, tomatoes, tobacco, sweet and chili peppers, and other members of the Solanaceae. Herbicides effective against jimsonweed also kill the closely related crop plants. Furthermore, jimsonweed can serve as an alternate host for certain insects and pathogens of solanaceous crops. Because of its rapid growth and large size, it quickly deprives the crop plants of light and nutrient resources. In addition, the tough stems of jimsonweed can interfere with the mechanical harvesting of crops such as wheat and soybeans, and jimsonweed seeds contaminate the harvested crops. The cultivation and/or use of jimsonweed is banned in some states because of its toxic and narcotic effects.

The foul-smelling and bitter leaves of jimsonweed tend to deter herbivores. Cattle and other livestock rarely eat jimsonweed and will do so only if other vegetation is scarce (fig. 465). When they do eat jimsonweed, they become ill. The principal toxic compounds of jimsonweed are found in all parts of the plant, but especially in the seeds. They include the alkaloids atropine, scopolamine, and hyoscyamine. Young adults, in particular, have experimented with the easily obtainable *Datura* seeds because of their mind-altering properties, and have often suffered severe adverse effects, including tachycardia, coma, and even death. Rather than causing hallucinations, ingestion of *Datura* causes delirium and amnesia, an effect found unpleasant by most users. Others have been inadvertently affected when they have consumed products made from wheat or soybeans contaminated with the seeds of jimsonweed, and farmers reportedly have even suffered severe nausea and marked pupil dilation just from exposure to the dust produced while harvesting fields with crops invaded by *Datura*. Historically,

Fig. 464. (Left) An open fruit of jimsonweed with its ripe seeds exposed. The lower valve of the capsule has been pulled down to allow the seeds to be seen more clearly. **Fig. 465.** (Above) A cow grazing in a field invaded by jimsonweed ignores the spiny, toxic *Datura* plants.

Jimsonweed

the ability of scopolamine to dilate pupils led to the use of a diluted, liquid form of the compound (usually derived from belladonna, a jimsonweed relative) by Italian ladies who thought that large pupils enhanced their beauty—thus "bella donna," meaning "beautiful lady" in Italian.

The same alkaloids are responsible for the effectiveness of *Datura* as a medicinal plant. In small amounts scopolamine is useful as a preventive for seasickness when administered via a long-acting skin patch. It has been used as well as an intravenous drug in the treatment of depression and is currently being tested as a treatment for bipolar disorder. Scopolamine was once a component, along with morphine, used to induce "twilight sleep," a popular method of producing a "painless," or more correctly, a "pain-memory-free" childbirth in the middle decades of the twentieth century. But since this drug mixture was subsequently shown to have negative effects on the infant, this use was discontinued.

In addition to its use in medicine, jimsonweed has other positive value. Its seeds serve as a food for birds, and the plant can be used as a soil remediator. *Datura* has the ability to detoxify trinitrotoluene (TNT), an explosive residue found in munitions waste sites. Studies have demonstrated almost total removal of TNT within 12 hours after it has been added to *Datura* cell culture suspensions. Further research is necessary to determine whether the whole plant would be as effective in the remediation of contaminated soil in such places as military bases.

There are eight other species of *Datura*, all with similar flowers and chemistry (fig. 466). South American tree species once classified in the genus *Datura* are now put into the genus *Brugmansia*, differentiated from *Datura* by having a woody stem and large, pendulous flowers. They are often called angel's trumpets (fig. 467).

Even those who have no interest in the flowers of weedy plants may be familiar with jimsonweed, for it served as a favorite subject of the twentieth-century artist Georgia O'Keeffe. O'Keeffe often painted flowers in a large format to emphasize their beauty and form; several of her best-known paintings are of jimsonweed. She wanted people to look closely at the flowers in order to better appreciate their beauty. When her paintings were to be exhibited in New York, she said, "I will make even busy New Yorkers take time to see what I see of flowers."

Fig. 466. (Above) *Datura wrightii*, a relative of jimsonweed. A native of southwestern United States, it is also pollinated by sphinx moths.

Fig. 467. *Brugmansia arborea*, a small tree native to Colombia. It is related to *Datura* and was once included in the same genus. *Brugmansia* flowers are pendulous rather than erect.

Lily Family

Fig. 468. A wood lily blooming among a field of hay-scented fern (*Dennstaedtia punctilobula*). Wood lilies are the shortest of the three species discussed in this essay. Note that the leaves of the lily plant occur in whorls around the stem.

Lilies

Fig. 469. The dangling bell-like flowers of Canada lily flare out at their tips but do not recurve strongly backward as do the tepals of Turk's-cap lily.

Canada Lily, Wood Lily, and Turk's-cap Lily

Lilium canadense L., *Lilium philadelphicum* L., & *Lilium superbum* L.
Lily Family (Liliaceae)

We often use lilies to exemplify the typical monocot: they have linear leaves with parallel venation, floral parts in three (or multiples of three), and a single cotyledon (seed leaf). Lilies are perennial, reproducing most frequently by bulb offshoots but also by seed.

Habitat: Canada lily (*L. canadense*): wet meadows, woodland edges, streamsides, and margins of bogs, swamps, marshes, and wet roadsides. Wood lily (*L. philadelphicum*): often in power line rights-of-way in the East where open lands are disappearing; in the western part of its range, it is found in high meadows in the mountains and in tall grass prairies in the Great Plains. Turk's-cap lily (*L. superbum*): openings in rich woods, swamp edges, streamsides, and moist meadows and thickets.

Range: Canada lily: New England, New York, New Jersey, Pennsylvania, the Appalachians (becoming less common in the southern part of the range), Arkansas, Nebraska, and Kansas; in the Canadian provinces of Quebec, Ontario, and New Brunswick and Nova Scotia within the Maritime Provinces. Wood lily: Most of the United States excepting Washington, Oregon, California, Idaho, Nevada, Utah, Kansas, Oklahoma, Louisiana, Mississippi, South Carolina, and Florida, but restricted to small areas in the southernmost and westernmost states within that range; in the Canadian provinces of Quebec, Ontario, Manitoba, Saskatchewan, Alberta, and British Columbia. Turk's-cap lily: In the eastern states of the U.S. south of Wisconsin, Michigan, and Iowa, and east of Kansas, Oklahoma, and Louisiana, excepting Maine and Vermont; the species' range was only recently extended to small areas of Missouri, Louisiana, and South Carolina. It is not present in Canada.

Lily Family

Some of our most beautiful wildflowers belong to the genus *Lilium*, the true lilies. The lily family (Liliaceae) is one of those plant families in which many of its former species have been segregated out of the family based on differences discerned using new genetic methods of analysis. The species excluded from Liliaceae have since been assigned to 13 other monocot families. Examples include lily-of-the-valley (*Convallaria*, now in the Asparagaceae; hosta (*Hosta*, now in the Hostaceae); trillium (*Trillium*, now in the Melanthiaceae); bellwort (*Uvularia*, now in the Colchicaceae); and daylily (*Hemerocallis*, now in the Hemerocallidaceae). Fifteen genera remain in the lily family; among those common in the Northeast are true lilies (*Lilium* spp.); trout-lily (*Erythronium* spp.); the monotypic Indian cucumber-root (*Medeola virginiana*), and twisted stalk (*Streptopus* spp.). Tulips (*Tulipa* spp.), which are not native to North America, also remain in the lily family.

To emphasize the importance of using scientific names, there are many familiar plants that have "lily" in their common names but, in fact, do not belong to the genus *Lilium* e.g., lily-of-the-valley (*Convallaria majalis*), trout-lily (*Erythronium* spp.), blackberry-lily (*Iris domestica*, formerly *Belamcanda chinensis*), water-lily (*Nymphaea* spp.), daylily (*Hemerocallis* spp.), calla lily (*Zantedeschia aethiopica*), wild calla lily (*Calla palustris*), and cobra lily (*Darlingtonia californica*). Of these, only the genus *Erythronium* (trout-lily) is even a member of the lily family.

There are about 100 species of *Lilium* recognized worldwide, 21 of them native to North America. The genus is primarily one of northern temperate regions, especially in North America and East Asia, but with a few species occurring in the mountains of the Philippines and India. It is thought that the North American species are part of what is termed the East Asian–Eastern North American disjunct distribution pattern in which the same, or closely related, plants occur in both regions of the world.

Lilies are characterized by their colorful, funnelform flowers with corollas comprising six parts: three petals and three sepals. When there is little apparent difference between the sepals and petals, the flower parts are referred to by botanists as tepals. True lilies have leaves along their stems, either alternately arranged or in whorls. And as with skunk cabbage and some other plants of wetland habitats, many lilies have contractile roots that function to pull the bulbs or corms deeper into the wet soil, thus preventing them from becoming dislodged during times of flooding. Lilies are a favorite horticultural plant, and thousands of hybrids have been produced. Perhaps the best-known lily species is the large white Easter lily (*Lilium longiflorum*), a native of Taiwan and the Ryukyu Islands in Japan, but now widely grown in northern California and southern Oregon for commercial purposes. The genus name, *Lilium*, is derived from the Latinization of the classical Greek name for lily, *lirion*, meaning white lily—perhaps referring to the native Greek lily, *Lilium candidum* (commonly called the Madonna lily for its association with the Virgin Mary). The Greeks used the name *krinon* for lilies of other colors.

The Easter lily as well as other lilies and daylilies are extremely toxic to cats and should not be brought into a home where cats reside—even licking the fallen pollen can result in poisoning causing potentially fatal renal failure. For that reason, and because oily pollen that falls from the anthers may stain both the white lily tepals, as well as table linens or carpets, the anthers are often removed before the lilies are sold—diminishing the flower's beauty in my opinion.

Although strongly toxic to cats, many lily parts (flowers, buds, shoots, and bulbs) are or were, consumed by people in both Asia and North America; however, some lilies are toxic to humans as well, so caution is advised.

Our eastern lilies are of two types; two species having upright flowers: wood lily (*Lilium philadelphicum*) and pine lily (*L. catesbaei*), and the others having nodding flowers: Canada lily (*L. canadense*) and Turk's-cap lily (*L. superbum*).

Fig. 470. The tepals of Turk's-cap lily recurve backward such that their tips meet above and behind the base of the flower. Note the six spreading stamens with long, thin, versatile anthers (anthers attached at their midpoints to the staminal filaments). Just a hint of the green star at the inner base of the tepals can be seen in this view.

Lilies

In northeastern North America, we are fortunate to have three of the most beautiful lilies in our native flora: the intense red-orange wood lily (fig. 468), its flowers facing upward to the sky; the usually yellow or orange Canada lily with dangling bell-like flowers (fig. 469); and the tall, orange Turk's-cap lily with its strongly recurved petals (fig. 470). All three of these lilies have their leaves in whorls along the stems.

In John Burroughs' *Riverby*, he writes of flowers "that one makes special excursions for." The lilies discussed in this chapter are among those that, for me, fall into that category. Burroughs' description of his sighting of one such lily captures the feeling of finding a sought-after flower in the perfect place:

> What glimpses we get, as we steal along, into the heart of the rank, dense, silent woods! I carry in my eye yet the vision I had, on one occasion, of a solitary meadow lily hanging like a fairy bell there at the end of a chance opening, where a ray of sunlight fell full upon it, and brought out its brilliant orange against the dark green background.

Although just a solitary bell-like lily (a Canada lily) enchanted Burroughs, this species frequently has a greater number of flowers in bloom (up to 20) on a single plant (fig. 471). The flowers of Canada lily dangle like delicate yellow-orange trumpets with their tepals curved slightly upward at their tips. The tepals of the outer whorl (the sepals) are not ridged, unlike those of Turk's-cap lily.

Unfortunately, we find fewer of these and other lilies today. Reasons include shading by trees as forests have replaced meadows, herbivory by an ever-expanding deer population (with a seeming penchant for some of our most beautiful flowers), and competition from introduced invasive plants that aggressively crowd out the native species that originally populated the habitats of lilies. Dr. Richard Primack of Boston University used the natural history journals of Henry David Thoreau written more than 165 years ago in Concord, Massachusetts as the benchmark to compare today's flora of that area with the flora of Thoreau's time in the mid-1800s. When Primack revisited sites where Canada lily once occurred in abundance, he found few plants of that species.

In addition to transformation of habitat, changes in climate may also play a role in the loss of species

Fig. 471. Canada lily plants may have many flowers in bloom simultaneously, as seen here on a plant found growing in a roadside ditch in the Adirondacks.

from an area. The decided warming trend over the past decades has resulted in earlier flowering times for many species. Dr. Chuck Davis at Harvard analyzed Primack's data and found that certain plant families—those of orchids, mints, and lilies among them—proved less flexible in flowering time shifts than species of other families that showed a lesser decline. Primack thinks that the time that trees of a forest leaf out—often simultaneous with flower opening in certain lily species and triggered by the plants' ability to track increasing daylight hours and/or warming temperatures—is a more important factor. Plants cued to flower by these triggers gain the advantage of a longer growing season in which to produce carbohydrates; in years when bud development is retarded by cooler than normal temperatures, they are protected from spring frosts by their delayed opening. However, plants experiencing a shift in bloom time may suffer due to a lack of concurrence with the peak flight period of their principal pollinators, and that may lead to a decrease in the plant's reproductive success over time.

A factor contributing to the decline of Canada lily was the overcollection of bulbs by lily enthusiasts in the early twentieth century. The bulbs were commonly dug up and sold to lily lovers in other parts of the world that were conducive to their cultivation. Since lilies grown from seed take several years to attain a size that can support the production of flowers, the bulb collectors were able to shorten the process by a year or more by starting with bulbs. The bulbs of Canada lily are scaly and not covered with a protective

Lily Family

sheath as found in many other bulb species. The outer layer of scales could be removed and used for propagation, forming small bulblets with roots by the end of the first growing season. Canada lily is somewhat rhizomatous and can also increase naturally from the lateral bulblets it produces.

The yellow-orange (to red-orange) tepals of Canada lily are usually marked on their inner surfaces with bright red-orange spots. The staminal filaments remain close to the style, and they and the style are the same color as the tepals (fig. 472). Their dark magenta anthers open to expose rust-colored pollen (fig. 473) either just before or just after the flower opens. Flower opening is the result of cell expansion caused by the breakdown of carbohydrates (starch) stored in the tepals. Flowers can remain open for up to seven days, their tepals slightly curved upward at the tips, but not strongly reflexed. Ruby-throated hummingbirds are the most effective pollinators, but butterflies and large diurnal moths also pollinate the flowers. Fruits are three-parted woody capsules (fig. 474) in which the flattened seeds are arranged in two columns within each chamber.

In the past, taxonomists subdivided *L. canadense* into two varieties: var. *canadense* and var. *editorum*, but in the recent treatment of *L. canadense* in *Flora of North America*, the so-called western wood lily (var. *editorum*) does not merit this distinction because the characters used to distinguish the two varieties overlap such that it is impossible to categorize them as distinct varieties.

Our tallest native lily is the aptly named *Lilium superbum*, a 6-foot-tall plant with large orange flowers that *is* truly superb. Its tepals curve backward to such a degree that their tips touch behind the flower's base (fig. 475). The effect is that of a Turkish turban, thus the common name, Turk's-cap lily. Turk's-cap lily is a distinctive plant within the heart of our northeastern flora, but at the margins of the area covered by this book, similar native lilies exist that must be viewed carefully in order to differentiate between the species. Turk's-cap flowers are visited and pollinated by ruby-throated hummingbirds and large butterflies—among them the spicebush swallowtail, the great spangled fritillary, and the eastern tiger swallowtail.

The native lilies that might be confused with Turk's-cap lily include Michigan lily (*Lilium michiganense*), with a range generally more northerly and midwestern than that of *L. superbum*. The flowers of Michigan lily look much like those of Turk's-cap lily, but the tepals are not as strongly recurved; at most, their tips are just at or near the base of the flower (fig. 476), while those of *L. superbum* extend all the way behind the flower's base (see fig. 475). Most flowers of Turk's-cap lily have a bright green area of nectar-producing tissue at the inner bases of their tepals, which collectively form a star seen most easily when looking upward into the flower (see fig. 470; the angle permits only a glimpse of the green region); in *L. michiganense*, the green area is lacking or greatly reduced. In addition, the anthers of Turk's-cap lily are longer and narrower than those

Fig. 472. (Below) The stamens of Canada lily remain closer to the style for most of their length than do those of Turk's-cap lily (as seen in fig. 470). The tepals are sprinkled with dark red spots on their inner surfaces. The outer tepals do not have ridges.

Fig. 473. (Above left) The anthers of Canada lily dehisce early, exposing their rusty-orange pollen before the style lengthens and its stigma becomes receptive. Cross-pollination is necessary for viable seeds to develop. Pollination is usually accomplished by nectar-seeking ruby-throated hummingbirds or large butterflies.
Fig. 474. (Above right) A three-parted fruit of Canada lily from the previous year still stands erect, but empty of seeds, in a stand of lilies flowering during the following July.

of Michigan lily (compare the anthers in figs. 475 and 476), and its style is pale green (it tends toward red in Michigan lily). Another difference is found in the leaves: unlike the Turk's-cap lily, the leaf margins of Michigan lily (and those of Canada lily) are covered in tiny spicules (spikes), which may also line the veins on the upper side of the leaves; in Turk's-cap lily, the marginal projections are rounded rather than sharp, or sometimes lacking.

A third orange-flowered lily with reflexed petals, Carolina lily (*Lilium michauxii*), is native to the Southeast extending northward into the southern part of our range in Virginia and West Virginia. This species sometimes co-occurs with Turk's-cap lily and resembles a smaller version of it but with broader tepals and somewhat fleshy leaves with wavy margins. Carolina lily prefers well-drained soils, whereas Turk's-cap lily can tolerate wetter soils. When the species do co-occur, Turk's-cap lily often is growing in a moist roadside ditch with Carolina lily on higher ground above it. Carolina lily is the only true lily to serve as a state symbol (of North Carolina). It is also the only fragrant lily east of the Rockies.

Like other lilies, Turk's-cap lily is preyed upon by deer, which nibble the above-ground parts of the plant. If the meristematic bud at the plant's apex is consumed, the possibility of that plant reproducing is eliminated for the year. Repeated browsing on the same plant over time can lead to its demise due to the inhibition of the plant's ability to make and store carbohydrates for future growth. In one study in Virginia, deer were found to consume 28% of the terminal stems and leaves of *L. superbum*; some plants succumbed after just one feeding, while others survived but did not reproduce. A smaller herbivore, the eastern chipmunk, dug up and fed upon the bulbs of the lily, killing 9% of them by doing so. In regions where wild boars have been introduced, such as the Smoky Mountains, the boars dig the bulbs of Turk's-cap lily, destroying the plants and ripping up much of the forest floor. Additional predators of this and the other lilies discussed in this essay include borer moth larvae (that feed within stems and bulbs), aphids, and introduced beetles (discussed below).

Owing to the fondness of gardeners for lilies, it's not uncommon to encounter non-native species growing along roadsides when they escape from gardens. Some of these species superficially resemble Turk's-cap lilies,

Fig. 475. (Top right) Looking at a flower of Turk's-cap lily from above, it is easy to see that the tips of the tepals overlap behind the base of the flower. The anthers are longer and narrower than those of Michigan lily (see fig. 476). The two closely spaced parallel ridges found on the underside of the three sepals can be seen on the rightmost sepal in this view. **Fig. 476.** (Below right) The tepals of the flowers of Michigan lily reflex backward but not to the degree reached by those of Turk's-cap lily; they may touch the base of the flower, but they do not extend beyond it. Note, too, that there is no green area visible at the base of the tepals, and the anthers are shorter and thicker than those of Turk's-cap lily (see figs. 470 and 475).

but if one looks carefully, they are quite easy to tell apart.

Lilium lancifolium (formerly known as *L. tigrinum*) is a native of China and other parts of Asia and is commonly called tiger lily. Tall, like Turk's-cap lily, it has large dark-spotted, orange flowers with strongly recurved tepals (fig. 477) and widely spreading stamens, similar to those of Turk's-cap lily. More than once I have had to turn the car around to check if I have spotted our native *Lilium superbum*. Viewed at less than 55 mph, it becomes readily apparent that the plants differ in many aspects: the leaves of tiger lily are arranged more or less alternately along the stem rather than in whorls and are smooth along their margins. Nestling in the upper leaf axils are small, purple-black bulbils (small aerial bulblets), which, when they detach and fall to the ground, may grow into new plants (see fig. 477). This method of vegetative

Lily Family

Fig. 477. (Left) Plants of the Asian tiger lily (*Lilium lancifolium*) have escaped from gardens and may be encountered growing along roadsides. Their flowers resemble those of Turk's-cap lily in some ways (strongly reflexed, spotted tepals and widely spreading stamens with long, thin anthers), but their tepals are papillose on their inner bases (see the small protrusions silhouetted against the water in the background), and the plants have leaves arranged alternately on the stem (rather than whorled as in Turk's-cap lily), with the upper leaves usually bearing small purple-black bulbils in the leaf axils. **Fig. 478.** (Below left) The European orange lily (*Lilium bulbiferum*) occasionally escapes from North American gardens as well. Like the tiger lily, orange lily has alternate leaves with bulbils in the upper leaf axils (here they are still immature and green), but the flowers differ in being upright at the apex of the stem and in having tepals that narrow into clawed bases. (Photographed in the Dolomites, Italy). **Fig. 479.** (Below right) The bases of the red-orange wood lily flowers are yellow, marked with purple spots thought to serve as nectar guides for visiting insects and hummingbirds. **Fig. 480.** (Bottom right) Each wood lily tepal is narrowed into what is called a claw. In the case of wood lily, the three outer tepals are incurled (their outer surfaces, seen in this image, are lighter in color), to hold nectar, which is accessed by butterflies or hummingbirds that can insert their long tongues into the tubular opening.

reproduction results in plants that are clones (genetically identical) of the mother plant. Since most of these garden plants are sterile triploids, they rarely produce viable seed.

Another orange-flowered lily, this one of European origin, occasionally escapes from gardens. *Lilium bulbiferum* (orange lily) is a few-flowered species with unspotted perianth parts; unlike the previous nodding species, the tepals face upward and are narrowed at their bases (fig. 478). Orange lily usually has bulbils in the axils of its upper leaves as found in *L. lancifolium*.

Lilies are often considered a symbol of purity or love, with orange-flowered lilies denoting happiness and confidence; however, in Japanese flower symbolism (Hanakotoba), "gifts" of *L. bulbiferum* are used as an indication of hatred or revenge. Orange lilies have served as the emblem of the Orange Order, a secret Protestant fraternal order founded as the Orange Society in 1795 Ireland. This order arose during the Protestant-Catholic conflict and still persists today in Great Britain (most notably in Northern Ireland) and in British Commonwealth countries around the world, as well as in the United States and West Africa. Although the orange lily is synonymous with the Orange Order holiday of July 12th, in 2016, to commemorate the 100th anniversary of the beginning of the Battle of Somme on July 1, 1916 (the first World War I battle fought on the Western Front), which lasted 140 days and resulted in a heavy loss of men from Ulster, the Orange Institution imported and planted 180,000 orange lily bulbs throughout Northern Ireland with a hoped-for bloom date close to the July 1st centenary. The planting was successful, but a news article two years later tells of a delivery truck driver convicted of a hate crime for deliberately running over a bed of orange lilies in front of an Orange Hall—on more than one occasion!

Wood lily has the widest distribution of the three native lilies discussed here, ranging across much of Canada and the United States. It is the floral symbol of Saskatchewan and is featured on its flag. Various tribes of Native Americans used the plant in different ways: as a food, for medicinal purposes, and in witchcraft.

Wood lily plants are the shortest of our three northeastern native lilies, but their brilliant, usually red-orange flowers excel in attracting one's attention. There are generally just one to a few flowers per plant. Unlike the previous two species, wood lily is not amenable to cultivation in gardens, so its beautiful red flowers must be sought in the wild. The red to red-orange (rarely yellow) tepals are bright yellow at their bases and dotted with dark purple spots (fig. 479); both features serving to guide pollinators to the nectaries hidden within the narrow, incurled claws of the sepals (fig. 480). The three-lobed stigmas (fig. 481) are a deep purple and usually extend beyond the slightly spread stamens. Anthers are maroon and open to expose rust-colored pollen. The anthers dehisce either just before the flower opens or shortly thereafter. Cross-pollination is necessary for successful development of seeds and fruit in this and other lilies.

The flowers of wood lily differ from those of all our other native lilies except one, the red-flowered pine lily (*Lilium catesbaei*), a plant from the pine savannas of the southeastern United States. Pine lily has the largest flower of any of our native lilies, and, like wood lily, the flowers of pine lily are upright on their stems, rather than nodding. Both species have close relatives in East Asia and probably evolved from a common ancestor that migrated to North America at a time different from that of our nodding lilies. The upright position results in problems for the flowers of these two species, but those problems appear to have been solved through changes in the flowers' morphology over evolutionary time.

The floral arrangement of wood lily's upright flowers (and those of pine lily) has evolved such that the six tepals narrow at their bases into what is called a claw, thus allowing rainwater to flow freely through

Fig. 481. The three-lobed stigma of this wood lily is evident in this image. Compare it with that of the daylily stigma in fig. 490.

Lily Family

Fig. 482. The clawed tepal bases of the upright wood lily flowers allow rainwater to flow through the flowers, thereby both protecting the pollen from being washed away and ensuring that the flowers will not fill with water and topple over from its weight. As further protection, the anthers can close when it rains and reopen when the rain stops.

Fig. 483. The bright red elytra (wing covers) of the Eurasian lily leaf beetle are distinctive (and beautiful), but the introduced beetles are wreaking havoc on lilies in gardens as well as on those in the wild. Here a beetle is feeding on a native Canada lily.

Fig. 484. This lily leaf beetle has one of its red wing covers out of position. The small black head; black, threadlike legs; and black body are all but invisible when, in response to a perceived threat, the beetle defensively drops to the ground and remains motionless on its back.

the flower (fig. 482). If the tepals were tightly adjacent, forming a cup, accumulating rainwater would dilute the flower's nectar, wash away its pollen, and prevent pollinators from visiting. In fact, if a heavy rain were to fill a flower not modified in this way, it might cause it to topple over. Still, exposure to rain could cause some or all of the pollen to wash away because, unlike many of our spring ephemerals, wood lily flowers cannot reclose on cloudy or rainy days (or at night) to prevent pollen loss. But wood lily flowers have an unusual modification to the anthers that further protects the pollen when rain begins to fall. The anthers, which open longitudinally, can actually reclose—something highly unusual for this type of anther. The mechanism for this movement involves the swelling of certain cells lining the anther sacs due to rehydration when it rains. Once the rain has ceased and the anthers have dried out, they reopen to expose the pollen. This is an important adaptation since wood lily flowers remain open for several days—up to 11—with the pollen retaining its viability throughout that time. Pollinator visits are infrequent and often don't begin until the flower has been open for a few days. Thus, it is important that the pollen remains protected and available throughout the blooming period of the flower. As the pollen ages, it may fade in color from rust to orange, to yellow, to almost white. The principal pollinators of wood lilies are large butterflies, for example, swallowtails (especially tiger swallowtails) and monarchs, which, while perching on the platform formed by the incurved portion of the tepals, repeatedly open and close their wings, thereby brushing them against the anthers and stigma as they move from tepal to tepal to imbibe nectar held in the deep nectar cups. The anthers of this species and of other lilies are affixed at their midpoints to the apices of the staminal filaments (see fig. 476) allowing the anthers to move freely when touched. Ruby-throated hummingbirds also visit the flowers, but they are less likely to contact the reproductive parts as they reach their long tongues down into the nectaries. This is true of smaller butterflies and diurnal moths as well.

Lilies

Today, a new menace threatens our native lilies—as well as other members of the lily family: the red lily leaf beetle (*Lilioceris lillii*). This beautiful scarlet beetle (fig. 483) was first documented in North America when it was collected in Montreal, Canada in 1943. However, it is thought to have been accidentally introduced more than 100 years earlier based on the description of a specimen collected in Canada in 1826 but later lost. The lily leaf beetle is native to most of Europe, and to parts of Russia and China, as well as North Africa. From Montreal the beetle spread slowly across Canada and into adjacent parts of northern New York and New England. *Lilioceris lillii* was officially documented in the United States in Cambridge, Massachusetts in 1992. It is thought to have been introduced there via the transport of infected plants imported through Boston. The beetle dispersed throughout New England and into New York, appearing in my garden by 2008. I waged battle with them for three years, inspecting my lily plants daily and squishing all eggs, larvae, and adults that I found. In year four, I threw in the towel and removed all the lilies from the garden. An occasional lily will still sprout where the originals grew, but the beetles manage to find it quickly. Their bright red elytra are easy to spot, but when alarmed, they quickly drop to the ground, flipping immediately onto their backs so that only their black undersides and tiny black heads and legs are upright (fig. 484), which causes them to seemingly disappear into the dark soil.

Gardeners are dismayed by the damage being done to their beautiful Asiatic lilies, but of greater concern to me is the potential loss of our native lilies. When walking on woodland trails, I have encountered the lily leaf beetle feeding on the leaves of Canada lily (see fig. 483) as well as those of twisted stalk (*Streptopus lanceolatus*) (fig. 485), another member of the lily family. Left uncontrolled, the beetle will continue to spread and has the potential to decimate many of our most beautiful lily-related native wildflowers. Unfortunately for us, the beetle has a long and fecund life span, as demonstrated in an experiment that kept one beetle alive by feeding it lily leaves until autumn, when it died because lily leaves were no longer available. The beetle lived for 133 days from the time it was captured. Lily leaf beetles produce prodigious numbers of eggs that they lay in a line on the undersides of leaves (fig. 486). The larvae are juicy grubs, pale yellow in

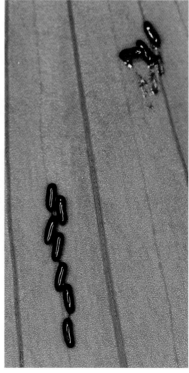

Fig. 485. (Above) These two lily leaf beetles are taking a break during a feeding session. Although the genera *Lilium* and *Fritillaria* are the preferred hosts of lily leaf beetles, here they have been feeding on twisted stalk (*Streptopus lanceolatus*), a plant native to the Northeast and a member of the lily family.
Fig. 486. (Right) The eggs of lily leaf beetles are laid in a connected line, like tiny sausages, on the undersides of leaves of their host plants. They are initially reddish but darken with age.

Lily Family

Fig. 487. The yellow-orange larva of the lily leaf beetle collects its dark fecal pellets and covers its body with them, thus deterring other insects or birds from eating it.

color, but cleverly disguised by what is politely called a fecal shield (fig. 487).

The beetles have long coexisted with lilies in Europe. There, the beetles' natural controls (primarily parasitic wasps) have kept the beetles in check. Investigations into the use of these biocontrol agents in North America have been ongoing in New England since 2003. They appear effective in substantially reducing the pest beetles without harming other beetle species, so perhaps we will not forever be deprived of lilies after all.

Many people don't realize that daylilies are no longer members of the lily family. The daylily genus, *Hemerocallis*, was placed in its own family, Hemerocallidaceae, when the lily family was disassembled based on molecular studies of their DNA. Although the flowers of daylilies resemble those of true lilies, they have other characteristics that differ, the most noticeable of which is that the leaves of daylilies arise from the base of the stem (fig. 488), rather than along the stem as found in true lilies (fig. 489). In addition, the stigma of daylilies is simple (fig. 490), as opposed to three-lobed in lilies. Cultivated daylilies are commonly grown in gardens and used to add color to roadside plantings and supermarket parking lots. A few of these cultivars escape locally, but you are most likely to encounter only one member of this Eurasian genus growing in an uncultivated situation—the common orange daylily, *Hemerocallis fulva* (the etymology of the genus is derived from the Greek for "a day," *hermera*, and *callos*, meaning beauty; the epithet, *fulva*, means orange). Indeed, each of these beautiful flowers lasts for only a day. Flowering from late May into July, depending on the latitude, the bright orange flowers of common daylily line the roadsides in the Northeast and much of the rest of the country (fig. 491), as well as eastern Canada. Fortunately, the lily leaf beetle does not eat the leaves or flowers of daylilies, and the plants do not seem to be as attractive to deer as the true lilies.

As with true lilies in the genus *Lilium*, the flowers of common daylily have six similar flower parts, collectively called tepals, but unlike lilies, the flowers are oriented more or less horizontally rather than strictly upright or nodding, and the three inner tepals (the petals) have wavy margins and obtuse tips. The trigger

Fig. 488. (Left) The leaves of daylilies (*Hemerocallis* spp.) grow from the base of the plant.
Fig. 489. (Right) Leaves of true lilies (*Lilium* spp.) grow along the stem, either alternately or in whorls, as seen here in wood lily.

Lilies

for flower opening is not starch degradation, but instead the degradation of fructan, a polymer of fructose sugar. Daylily anthers are versatile and open along the length of the anther as do those of lilies, but unlike lilies, the anthers of daylilies can reclose only partially if it rains. Daylilies are most commonly pollinated by large butterflies, especially the tiger swallowtail (fig. 492). Seeds of common daylily also differ from those of the lilies discussed in this chapter in that they are hard, round, and black as opposed to the true lily's flat, brown, and sometimes winged seeds; however, seed production in our naturalized orange daylilies is rare as they are of hybrid origin. The plants reproduce primarily by the branching of their underground rhizomes. Most patches of orange daylilies, even if growing "naturally" along roadsides, have as their origin a deliberate planting by humans—perhaps in the long ago past.

Fig. 490. (Above) Daylily flowers are oriented in a more or less horizontal position rather than upright as in wood lily or nodding as in Canada lily. They have a small, simple stigma at the tip of their style rather than a thickened, three-lobed one as found in lilies (see fig. 481). **Fig. 491.** (Left) The common daylily, *Hemerocallis fulva*, has naturalized along roadsides in much of the Northeast.

Fig. 492. (Below) Daylilies are pollinated most commonly by large swallowtail butterflies such as this eastern tiger swallowtail.

Daylily flowers may be eaten fresh in salads, or dried and used to flavor a clear broth. Dried daylily flowers are sold in Chinese markets and online. Almost all parts of the plant have been used as a food source. Flower buds can be stir-fried or steamed like green beans, young shoots can be collected in early spring and cooked in the same way (when harvested early, the plant will have time to produce new shoots), and the tubers, if harvested before the plants flower, can be cooked like potatoes (however, this obviously eliminates the plant). Particularly when harvesting young shoots, it is critical to have the plant properly identified first, and it is always wise to try only small amounts of any new food to ascertain that it does not cause unpleasant symptoms.

Bedstraw Family

Fig. 493. A partridge-berry clone growing on a steep slope along the Hudson River in New York.

Partridge-berry

Fig. 494. Some open flowers retain the pink coloration of the buds on their exterior surfaces.

Partridge-berry
Mitchella repens L.
Bedstraw Family (Rubiaceae)

This low-lying, creeping plant with glossy green leaves is my favorite of the relatively few members of the Rubiaceae family native to the Northeast. It is all the more appreciated when, during an otherwise monochromatic brown or white winter, one encounters its rich evergreen leaves with their strikingly light greenish-white midrib and some persisting red berries.

Habitat: This groundcover plant occupies sparsely vegetated understory sites in deciduous and mixed coniferous-deciduous forests, and occasionally more open areas along forest margins.

Range: Southern Newfoundland and Nova Scotia, west to southern Ontario and eastern Minnesota, south to Florida and eastern Texas. Populations also exist in disjunct deciduous forests in mountainous areas of Mexico and Guatemala.

The family of partridge-berry is the fifth largest plant family in the world, having in excess of 10,000 species. The great majority of Rubiaceae species grow in tropical latitudes; in fact, in some New World rainforest regions, the family comprises the largest number of species in the forest and includes several economically important plants, among them coffee, quinine, gardenia, and other tropical ornamental plants. Most tropical species are trees and shrubs, in contrast to our northeastern United States and Canada species, which are almost exclusively herbaceous, the only exception being buttonbush (*Cephalanthus occidentalis*), a shrub of wetlands in eastern North America and parts of Texas, Arizona, and California.

In the Northeast, the Rubiaceae family is commonly referred to as the bedstraw family because of the prevalence of species in the genus *Galium*, known as bedstraws. However, in Europe, Rubiaceae is known as the madder family because an important red dye (madder) is derived from the roots of one species, *Rubia tinctorum*. Madder was once used to color the bright red jackets of the British Redcoats and is still used as a dye source for the red color in many Turkish rugs.

Bedstraw Family

Although this worldwide family is large and includes plants of many different habits and lifestyles—trees, shrubs, lianas, vines, and herbs—it is relatively easy to recognize a "rube." Plants have a combination of the following features: opposite leaves, stipules, and tubular flowers usually with four- to five-lobed corollas. In addition, the floral parts arise from the top of the ovary (making the ovary inferior), and the stamens are attached to the interior of the corolla. All these features can be seen in partridge-berry.

Partridge-berry, a trailing evergreen herb, is one of only two species in the genus *Mitchella*; the other, *M. undulata*, is native to South Korea, Taiwan, and Japan. Linnaeus named the genus *Mitchella* to commemorate John Mitchell, a mid-eighteenth-century Virginia botanist and physician, perhaps best remembered for his development of a yellow fever treatment consisting of purging and bloodletting. Although this harsh treatment seemed effective during the epidemics of the mid-1700s, its later employment by Dr. Benjamin Rush in the great Philadelphia epidemic of 1794 was deemed to have contributed greatly to loss of life. The epithet, *repens*, makes reference to the creeping habit of the plant, which reaches only about 2 inches in height.

Partridge-berry spreads by readily branching and rooting at the nodes of the stems, but it is not an aggressive colonizer and rarely forms large patches. If the plants are covered by fallen leaves and other debris, they fail to thrive because of lack of light. This condition rarely occurs, since partridge-berry is found most frequently growing on slopes, where leaves tend to be blown off the plants (fig. 493), or in mixed coniferous forests without many trees such as oaks that shed leaves that are slow to decay.

The flowers and fruits of partridge-berry are the most interesting features of the plant. The tubular white flowers occur in pairs joined at their bases by the fusion of their ovaries. They are often tinged with pink, especially when in bud (fig. 494). If the name twinflower were not already widely used for another small, creeping plant, *Linnaea borealis* (in the honeysuckle family), which has a pair of dangling (but not conjoined) pink, bell-like flowers (fig. 495), it would be most appropriate for partridge-berry. Partridge-berry would *seem* an apt name since red fruits are known to be attractive to birds, and the fruits grow close to the ground within easy reach of our native ruffed grouse (sometimes called partridge); however, few fruits are produced each year, and despite the attractive color, the fruits often remain undisturbed on the plant throughout the winter (fig. 496)—and even into the following summer when the plant is again in bloom (fig. 497). Why then aren't these seemingly attractive berries eaten in winter, when food is difficult to find? I know from personal experience why humans don't consume them; they are mealy (fig. 498) and utterly tasteless. But one would think that a ground-dwelling bird, like a grouse, or a small mammal, might welcome such a meal when other resources are scarce. In fact, although both birds and forest-dwelling mammals are *said* to eat them, the berries are obviously not a favorite food.

The pollinators of partridge-berry flowers, as with so many of our native species, are native bumblebees.

Fig. 495. (Left) A plant (ramet) of twinflower (*Linnaea borealis* in the Caprifoliaceae) showing the paired, but completely separate flowers, unlike those of partridge-berry that have conjoined ovaries. **Fig. 496.** (Below) The bright red fruits of partridge-berry are long-lasting; partridge-berry is seen here growing with a small, fernlike moss, *Thuidium delicatulum*.

Partridge-berry

The bumblebees spend about two minutes in a typical patch of partridge-berry, moving quickly from flower to flower, before flying on to another patch. The bees are attracted by the light, sweet fragrance of the flowers. Because the flowers are produced on the tips of shorter, more erect shoots of the plant, bees can more easily access them. The interior of the partridge-berry flower is lined with a dense covering of white hairs (fig. 499), which help to discourage creeping insects (such as ants) from pilfering the nectar at the base of the tube without serving as pollinators. Bumblebees are robust and have long tongues capable of penetrating the hairy barricade to reach the nectar at the base of the flower. The greatest amount of nectar is found in the unopened mature buds, and bees are known to force their way into such buds to seek the larger reward. As the flower opens and ages, the amount of nectar sharply declines, forcing bees to visit several flowers before finding a reward. Patience is needed to observe pollinators since each patch receives an average of only two to three bee visits per hour.

Charles Darwin's early observations of partridge-berry led him to believe that both flowers of a pair must be pollinated in order for a fruit to develop. More recent experiments by David Hicks et al. demonstrated that fertilization of just a single ovary could produce a one-sided fruit. This condition is rarely seen in nature since the bumblebee pollinators commonly visit both flowers of a pair, resulting in both sides of the conjoined fruit developing.

Fig. 497. (Right) Fruits from the previous year often persist into the blooming period of the following year.

Fig. 498. (Above) Partridgeberries are edible but have a mealy, insipid pulp.
Fig. 499. (Right) Two "thrum" flowers (short style, long stamens) have been opened to reveal the densely hairy interior that serves to prevent pilferage of the nectar by ants.

Bedstraw Family

Fig. 500. (Right) Only four seeds have resulted from the potential of eight in this conjoined fruit. It is common that not all of the four ovules per ovary develop into seeds. **Fig. 501.** (Far right) The remnants of the calyx lobes surrounding the location where the two corollas were attached to their conjoined ovaries might be imagined to be the large eyes of a red-faced, E.T.-like creature.

After a flower has been pollinated, the pollen tubes grow down through the style and release the sperm nuclei into the ovule. Each ovary contains four ovules; thus there is the potential for the production of eight seeds per fused fruit. Not all will be fertilized; on average only four or five develop into seeds (fig. 500). The double fruit bears evidence of its dual origin in the two circular depressions and persistent calyx lobes left where the corollas had been attached—giving the appearance of the face of a tiny, red-faced alien (fig. 501). In Newfoundland, where partridge-berry is rare, it is known as two-eyed berry, and the name partridge-berry is applied to a plant we commonly call mountain cranberry (*Vaccinium vitis-idaea*) (see account of American cranberry in this book).

Although partridge-berry can grow under the shade of conifers and occasionally in semiexposed sites, I find it most commonly in the shade of deciduous trees, frequently growing in a bed of moss (fig. 502). The plants, thus shaded, receive little light during the spring and summer months, but in the leafless months of winter the amount of light striking the forest floor increases markedly. The plants, therefore, receive an excessive amount of light—more than they can use for photosynthesis during the cold winter conditions. As a result, the excess light may result in the production of highly reactive forms of oxygen capable of causing harm to the cells of the plant. Fortunately, the activity of other pigments in the leaf (carotenoids) helps to mitigate this effect by permitting the plant to dissipate some of the light energy by converting it to heat. In winter, in shaded sites (e.g., coniferous forests), partridge-berry also experiences a difference in its photosynthetic rate between mild days and extremely cold days.

Fig. 502. Partridge-berry is frequently found growing among mosses. Note again the steep slope on which the plants are growing; slopes are common sites for partridge-berry since slopes prevent the buildup of leaves that would cover the plants and block any sunlight from reaching the partridge-berry leaves.

Partridge-berry

Patches of partridge-berry are clonal, with all stems within the same patch genetically identical. Although all the flowers within the patch are bisexual, having both stamens and pistils, the reproductive structures exhibit the self-incompatibility breeding system known as distyly. In flowers where the style is long, and exserted from the corolla tube, the stamens are short and hidden within the corolla tube. Darwin referred to these flowers as "pin" flowers (fig. 503). Conversely, flowers with long stamens and anthers exserted beyond the corolla tube, have styles that are short and hidden within the corolla tube—referred to as "thrum" flowers (fig. 504). Both the styles and the stamens are functional, but the stigmatic surface on long styles is receptive only to pollen from long stamens—from thrum flowers of another patch—and therefore not compatible with its own pollen. This promotes cross-pollination (outcrossing) and prevents inbreeding. Within a clone of partridge-berry, *all* flowers would be either "pin" or "thrum" and therefore reproductively incompatible. (See buckbean account in this book for a discussion of the breeding system of another plant with distylous flowers.)

Larger clones of partridge-berry may live for 15 years or more and may comprise up to 800 ramets (a distinct plant that is part of a group of genetically identical individuals). It is a treat to encounter such a clone on a cold winter's day; the combination of glossy, dark green leaves and colorful red berries adds a bright spot to the day, particularly if the ground is dusted with a light coating of snow. However, the attractive Christmas-colored beauty of this plant has also had a negative consequence for the plant: partridge-berry has long been harvested for use in terrariums, a popular holiday gift. Such overharvesting could lead to the demise of this favorite evergreen plant, much as has been the case with the spring-blooming trailing arbutus (*Epigaea repens*): a fragrant, early-flowering plant that was once avidly picked for loved ones as a token of spring. Trailing arbutus is now absent or severely depleted from many of its former haunts. So please do not collect partridge-berry from the wild. Rather, look for it at trusted native plant nurseries or ask friends for cuttings from plants on their property; the plant is propagated easily from cuttings and will provide an evergreen highlight for a shady garden that can be enjoyed even in winter.

Fig. 503. The prominent, exserted styles of this pair of "pin" flowers indicate they are functionally female. The stamens are hidden within the corolla tube.

Fig. 504. Functionally male ("thrum") flowers, such as these, have four long stamens and, hidden within the corolla tube, a short style.

Passion-flower Family

Fig. 505. Yellow passion-flower (*Passiflora lutea*) vines climb when their tendrils contact other objects and wrap around them. The plant's leaves are broad and shallowly three-lobed.

Passion-flowers

Fig. 506. Maypops (*Passiflora incarnata*) has fragrant lavender and white flowers and leaves that have three sharp lobes and serrate margins.

Passion-flowers
Yellow Passion-flower and Maypops (Apricot-vine);
Passiflora lutea L. and *Passiflora incarnata* L.
Passion-flower Family (Passifloraceae)

For the most part, passion-flower plants are vines, either herbaceous or woody (then called lianas) that inhabit mild temperate, subtropical, and tropical regions. The genus *Passiflora* comprises about 550 species, with only 20 of those found growing naturally outside the New World (i.e., the Americas), those all in Australasia. Because of the beauty of their flowers, passion-flower vines are frequently planted in other climatically hospitable regions.

Habitat: Yellow passion-flower (*Passiflora lutea*): Open woodlands and forest margins in moderately moist soil.

Maypops (*Passiflora incarnata*): Open woodlands, prairies, and disturbed areas in varied soil types; able to tolerate dry and sandy soils.

Range: Yellow passion-flower (*Passiflora lutea*): Most of the Southeast, extending northward into Pennsylvania, the southern portions of Ohio, Indiana, Illinois, and Missouri, and westward into the southeast corner of Kansas, eastern Oklahoma, and thinly scattered in eastern Texas and sections of the Gulf Coast. *Passiflora lutea* is especially prominent in southern Missouri, Arkansas, and Louisiana.

Maypops (*Passiflora incarnata*) has a similar distribution to yellow passion-flower, but is more prevalent in the Carolinas and Florida, and less so in Texas; it even reaches as far north as southern New Jersey.

We are fortunate to have growing in the southernmost areas of the greater Northeast, two native species of *Passiflora*, a genus known for its exotic tropical flowers and for its use as a popular tropical punch component. Passion-flowers bedazzle the eye with their elaborately adorned flowers and intrigue botanists and ecologists alike, who study their intimate relationships with insect, avian—and even mammalian—visitors, particularly in the tropics. Species diversity of both plants and animals is greater in tropical forests, giving rise to many more complex relationships between plants and herbivores—and their predators—than exist in temperate climes. Countless plant-animal interactions, including those between passion-flowers and the butterflies dependent on them as larval host plants, serve as classic examples in the field of tropical ecology.

Passion-flower Family

Fig. 507. The leaves of yellow passion-flower show some variation in the length of their lobes; the middle lobe is sometimes as long as (see leaf on left), and sometimes shorter than, the two lateral lobes (see leaf on right).

Fig. 508. The flowers of yellow passion-flower show many of the characteristics of the genus: disklike perianths comprising five sepals and five petals (united at their base to form a floral tube, the hypanthium); a crownlike corona having one series of long, white filaments and another of short, stiff magenta filaments that cover the opening to the nectar chamber; their reproductive structures (five stamens, an ovary, and three stigmas) are all borne above the perianth on a stalk (collectively called an androgynophore).

Although our two species whose ranges extend into the southern portion of the Northeast, *Passiflora lutea* (fig. 505) and *P. incarnata* (fig. 506), do not exhibit the full array of these fascinating traits, they are interesting in their own right. The critical feature allowing these two species to survive through winter's subfreezing temperatures in the northern part of their range is their large, subterranean root system. The aboveground portions of the plants die back to the ground each autumn, but the underground portion can remain viable to ambient temperatures as low as −20°C (−4°F). For this reason, these species, especially maypops, are planted in gardens even north of its native range, where the plants may survive for years.

Passiflora lutea grows as a small vine, climbing by tendrils to a length of 3 meters. It is capable of growing in shaded conditions but thrives in full sun. Its leaves are most often shallowly three-lobed, with the middle lobe sometimes equal to or shorter than the others (fig. 507). It is thought that variation in leaf shape within a passion-flower species (sometimes even on the same plant) is a means of deceiving herbivores that rely on shape recognition to locate their hosts. The leaves turn bright yellow in autumn.

The flowers of yellow passion-flower are diminutive, measuring about one cm in diameter, yet they are elegant in form and exhibit the typical morphology of most passion-flowers (fig. 508). The overall appearance of the flower is rotate (circular), comprising five spreading greenish sepals alternating with five similar, but smaller, yellow-green petals. At the junction where the sepals and petals are fused to form a tube is a circular rim from which arises a ring of slender white filaments, each about as long as the sepals and resting on the perianth. An additional inner row of much shorter filaments (white, often colored purple at the base) is oriented upward, forming a crownlike structure that surrounds the opening to the flower's shallow nectar chamber (the floral tube or hypanthium). These filaments are referred to as the corona. Both male and female reproductive structures are held above the flower's perianth on a stalk called an androgynophore (from the Greek for male + female + stalk) that arises from the center of the base of the floral tube. Five stamens arch outward just below the ovary at the summit of the stalk; the three styles, each tipped by a thickened stigma, are atop the ovary (see fig. 508).

This unusual arrangement of floral reproductive parts, as one might imagine, plays a role in the flower's pollination. The great majority of passion-flowers, including our two species, open for only one day and are self-incompatible; that is, a flower cannot be fertilized by pollen from a flower on the same plant. When self-pollen is deposited on a stigma of a flower on the same plant, the process of fertilization is stopped at some point so that seeds never develop. Pollen from one flower must reach the receptive female organ (the stigma) of a flower on a different plant in order for it to produce seed. As with many plants with attractive flowers, the pollination vector is biotic; an insect in this case must play the role of pollen transporter. Two pollination studies of *P. lutea* have been carried out in different parts of its range. Both showed bees or wasps to be capable of transferring pollen, but the studies' conclusions differed in opinion regarding the effectiveness of the pollinators visiting the flowers. In a careful study by J. B. Holland and J. Lanza in Arkansas, the authors observed only two insect species to actually contact the reproductive organs in such a way that pollen could be carried from the anthers of one flower to the stigmas of another; one was a leafcutter bee, *Megachile concinna*, which was observed on only one plant; the other was a small, ground-nesting andrenid bee, *Anthemurgus passiflorae*, a frequent visitor to the flowers on many plants in the study site, and with a geographical range that overlaps that of yellow passion-flower. Both bees visited the flowers to imbibe nectar and to harvest pollen. The other study, in Texas, observed *Anthemurgus* to contact the stigmas only rarely as it gathered pollen from the flowers; it was the researchers' opinion that the bee was of insufficient size to be an effective pollinator and that larger bees, including bumblebees and carpenter bees, were the most likely pollinators. However, *Anthemurgus passiflorae* is so closely linked with *P. lutea* that it was given its species epithet, *passiflorae*, to indicate this strong relationship. Indeed, *A. passiflorae* is said to visit only the flowers of yellow passion-flower and no others.

For pollination to be successful requires the movement of the flower's reproductive organs in space and time. Before the flowers open, the styles are vertically oriented upward, and the stamens vertically oriented downward on the androgynophore. Upon opening, the anthers' filaments move upward to a horizontal position while the anthers, each attached at its midpoint to the tip of its filament, remain in a vertical position. Such anthers are termed versatile and are able to move when contacted by a visitor, thus increasing the likelihood of pollen deposition on the insect. As morning progresses, the styles begin to bend downward until they are at or below the level of the anthers, while concurrently, the anthers curve upward until they are parallel with the perianth of the flower. These positions are retained until late in the afternoon when the flowers begin to close.

Insects begin to visit the flowers soon after they open. A variety of insects visit, but *Anthemurgus passiflorae* is, by far, the most frequent, with its visits peaking in midmorning. This peak coincides with the realignment of the styles and stamens, which are then positioned in close proximity to each other. As the bees forage for nectar and pollen, their bodies get dusted with pollen, which is shed downward and will later brush off onto the stigmas of subsequently visited flowers. The Arkansas study showed that high fruit-set resulted from those flowers visited by the specialized *Anthemurgus* bees. It also demonstrated that, contrary to the belief that *P. lutea* is self-incompatible, a very low rate of fruit-set occurred even when the flowers of yellow passion-flower were enclosed in net bags to exclude visitors. Yet, experimental hand-pollination of flowers with self-pollen yielded few fruits, whereas hand-pollination with pollen from other flowers produced a more usual rate of fruit-set, reinforcing the presumption that *P. lutea* relies almost exclusively on outcrossing.

Passion-flower Family

The ripe fruits of yellow passion-flower are dark purplish-black globes about 1–1.5 cm in diameter (fig. 509). They are full of flattened, dark brown-black seeds, each encased in a fleshy white aril (fig. 510). The fruits are considered pepos, a type of berry. When birds eat these small fruits, the seeds (fig. 511) are dispersed in their droppings. *Passiflora lutea* is one of only a few in its genus to have hypogeal seed germination (a germination process in which the cotyledons remain underground).

Passion-flower vines climb by producing tendrils that grow from the leaf axils (fig. 512) and coil around any object they touch. The energy requirements a plant would need if it had to produce woody tissue to support itself are thereby reduced. But by not being self-supporting, the vine *must* come into contact with a nearby support in order to climb toward sunlight. Making contact is not a completely random event; the tendril makes circular movements in space (circumnutates) until it encounters something solid. When the tip of a tendril contacts an object, whether another plant or an inanimate object, it is stimulated by touch to bend toward the side that has made contact and rather quickly grows to encircle the object (fig. 513). The tendril, now attached at both ends, forms two helical coils, each coiling around its axis in opposite directions at the two ends of the tendril. The coils are joined by what Charles Darwin referred to as a "perversion" (an uncoiled section that allows for rotation and/or additional turns of the helices) (fig. 514).

When Darwin studied the phenomenon of tendril coiling, the sophisticated technology required to

Fig. 509. (Left) Yellow passion-flower fruits are deep blue-black and only about 1.2–1.5 cm in diameter. They develop from the ovary atop the stalk of the androgynophore; the stalk has a thickened region where it is attached to the fruit's pedicle.

Fig. 510. (Left) This immature fruit has been cut open to show the black seeds embedded in fleshy, white arils. Birds eat the fruits, digest the arils, and then excrete the seeds.
Fig. 511. (Right) Like many *Passiflora* seeds, those of *P. lutea* are flattened and have a grooved or pitted surface.

Passion-flowers

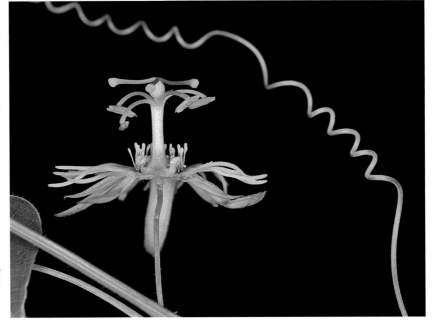

Fig. 512. (Above left) A tendril used for climbing arises from a leaf axil. **Fig. 513.** (Above right) *Passiflora* tendrils bend around anything they touch (in this case, a chain-link fence); the contact causes the tip to grow quickly in order to wrap around the support, thus pulling the plant closer to the support and higher toward the sun. Note the young fruits developing from the raised stalks above the other old flower parts. **Fig. 514.** (Right) This side view of a flower of *Passiflora lutea* shows the arrangement of the floral parts and a tendril coiling in opposite directions toward each attached end with a straighter section between (called a perversion).

Passion-flower Family

determine the underlying mechanism responsible for the coiling did not exist; nevertheless, he recognized that the coil and its perversion had to be determined by the constraints of twisting and stretching at both ends of the tendril. It was only in 2012 that Harvard scientist Sharon Gerbode and her colleagues published a paper based on a discovery by earlier researchers that gelatinous fiber ribbons occur within tendrils. Gerbode elucidated the mechanical behavior that underlies coiling in the cucumber tendril (similar to that of passion-flower). Within the tendril this gelatinous fiber ribbon, two cell layers thick, is flat and supple before the tendril has made contact with an object. Once the tendril has come in contact with an object, the inner layer of cells (closest to the support object) begin to produce lignin, causing the cells to become stiffer and woodier. This rigidity causes the inner side of the ribbon to shrink and thus curve inward, or coil. The tendril then assumes the shape of its inner gel-ribbon. Strangely, if tension is applied to the end of a mature tendril by pulling it lengthwise, rather than causing the tendril to straighten out (as would happen with an old-fashioned coiled telephone cord), the cucurbit coil becomes even more tightly coiled (overwinds). Preliminary studies of *Passiflora* tendrils have shown them to have the same properties as those found in the cucumber family with one exception: young cucumber tendrils *can* be fully stretched out after coiling, but older, drier tendrils will overwind when subjected to tension; in *Passiflora*, both young and old tendrils will overwind.

Coiling serves to shorten the tendril, thereby pulling the plant closer to the support it has grasped and lifting it higher, where it benefits from increased access to sunlight. In addition to tendrils that coil after contact, some tendrils that never contact a support object may form coils on their own. Little is known regarding the stimulus for unsupported coiling.

Maypops (*Passiflora incarnata*) has a flower whose beauty rivals that of any found in the rainforests of South America. The fragrant, white or purple flowers (it is sometimes called purple passion-flower) are circular in form, approximately 5 cm in diameter and have the typical 10 perianth parts (five sepals, each with a projection near the apex of the green outer surface, and five petals). The rim of the hypanthium (floral tube) is embellished with a corona of two series, the outer consisting of long, white-banded, lavender fringe and the inner a series of much shorter maroon-colored hairlike filaments that crowd close to the stalk of the androgynophore and form a lid over the nectar chamber, thus blocking access to the nectar (see fig. 506). Only a strong insect can push its way through the covering of filaments to reach the nectar.

Maypops is visited and effectively pollinated by various species of large carpenter bees (*Xylocopa* spp.) (fig. 515); this is true for many species of *Passiflora* in other areas within the ranges of carpenter bee species (fig. 516). Attracted by both the scent and the lavender and white color, the bees' hardened mouthparts allow them to probe through the compressed filaments and extend their proboscises into the nectar. The bees visit primarily in morning and early afternoon, both before and after the styles of the flower have descended to a position close to the anthers. As the backs of the bees brush against the downward-facing anthers, they become covered with bright yellow pollen. The bee leaves the flower in this state (fig. 517) and, even if the styles of a subsequently visited flower have not yet descended to the lower position, the bee may inadvertently brush against them as it lands on the flower, depositing pollen as it does so. Once the styles *are* at the level of the anthers (see fig. 506), it is almost impossible for the nectar-foraging bee not to come into contact with both organs. As with most passion-flowers, *P. incarnata* is not self-compatible, so pollen

Fig. 515. The flowers of maypops (*Passiflora incarnata*) are efficiently pollinated by large carpenter bees (*Xylocopa virginica* in this case), which are strong enough to probe through the inner whorl of corona filaments to obtain nectar. Note how the bee's back brushes against the pollen-bearing surface of the anthers as it forages.

Passion-flowers

from one flower must be brought to the stigmas of another flower in order for fertilization to take place. The resultant fruit is a yellow to yellow-orange, nearly round, fleshy fruit (accounting for its other common name, apricot-vine) that measures about 5 cm in diameter. The fruit is eaten by birds and mammals and the seeds thus dispersed.

Passiflora incarnata is a more robust plant than *P. lutea*, its vines growing to more than 2 meters in length and its leaves larger and more deeply three-lobed; the leaves have serrate margins (see fig. 506). Like many other members of the genus *Passiflora* (but not *P. lutea*), maypops has paired, continuously secreting nectar glands at the apex of its leaf petioles (fig. 518) and on its floral bracts. Ants are attracted to this sweet resource and, in return for this reward, defend not only the nectar resource, but also the adjacent leaves from herbivorous insects that may otherwise devour them. Some species of ant merely make forays onto leaves being attacked by herbivores scaring them away, but other species actually attack the herbivores.

The two native species extending their range into the Northeast inhabit areas that are generally north of the range of the primary herbivores of passion-flower, the heliconian (long-winged) butterflies (fig. 519). These butterflies are common in tropical America and even in our most southern states (occasionally straying northward into our range). The most common species

Fig. 516. Throughout their range, many passion-flowers are similar in size and shape to maypops and are also pollinated by carpenter bees. In this image, a flower of *Passiflora foetida* is visited by a female of the only species of bee on the Galápagos Islands, *Xylocopa darwinii*. (Photographed in Galápagos)

found in the United States is probably the Gulf fritillary (not a true fritillary, but a heliconian butterfly misnamed because of its silvery spots). The spiny larvae of long-winged butterflies (fig. 520) are obligate feeders on passion-flower vines. As in the monarch/milkweed association, the caterpillars feed on the leaves, ingesting toxins—including cyanogenic glycosides—which

Fig. 517. As the carpenter bee flies from the flower of maypops, it carries a load of yellow pollen on its back that may brush off on the stigma of a subsequently visited flower and result in pollination.

Fig. 518. Two dark green glands are just barely visible on the petiole (leaf stalk) of the leftmost leaf, just below where the petiole meets the leaf blade. A small ant can be seen taking nectar from a gland at the base of the leaf on the right.

Passion-flower Family

Fig. 519. In the tropics (and in the most southerly regions of the United States) the caterpillars of long-winged butterflies are obligate feeders on passion-flower vines. This butterfly is one of many heliconian butterflies in South America bearing similar colors and patterns. (Photographed in French Guiana)

Fig. 520. The spiny caterpillars of heliconian butterflies feed exclusively on passion-flower vines. These are caterpillars of the Juno long-wing (*Dione juno*), which often feed communally. (Photographed in Costa Rica)

Fig. 521. Heliconian butterflies are able to "eat" pollen, and they especially prefer that from members of the cucumber family such as this orange-flowered vine of *Gurania spinosa*. The amino acids in the pollen enable the butterflies to live longer lives and lay more eggs. (Photographed in French Guiana)

provide the caterpillars with protection from vertebrate predators such as birds. Some species are able to sequester these chemicals into adulthood; other butterflies can manufacture them after they become adults, and other species are chemically unprotected but derive some protection through their aposematic coloration, that is, colors that mimic those of toxic species. Accordingly, both caterpillars and adult butterflies usually display contrasting bright colors (such as orange and black) that warn predators of their unpalatability.

Long-winged butterflies are said to be the most intelligent group of lepidoptera. They have large eyes to aid in seeking out nectar sources and host plants and a keen memory that allows some species to trapline, that is, repeatedly follow a specific route between flowering plants, even on sequential days. The butterflies are long-lived, surviving for up to six months, most likely because of their ability to feed on pollen (a source of protein), which they obtain primarily from the flowers of members of the cucumber family. In order to "eat" pollen with its strawlike proboscis, the butterfly secretes enzymes through the proboscis to dissolve the amino acids on the exterior of the pollen grains, and then imbibes the nutrient-rich fluid. I have watched heliconian butterflies battle with hummingbirds over access to this important resource from the flowers of *Gurania* and *Psiguria* vines (Cucurbitaceae) in French Guiana (fig. 521). Thus, as with much in nature, many diverse organisms are interdependent, in this case passion-flowers, long-winged butterflies, and cucurbits.

In addition to its arsenal of toxic compounds, *Passiflora* defenses against herbivory include variation in leaf shape, which confuses butterflies that rely on a particular search image for a known leaf form, extrafloral nectaries that attract protective ants that deter herbivores, and egg mimics. This last-mentioned defense evolved as a means of deceiving adult heliconian butterflies into believing that another butterfly has already oviposited on that plant. The plant produces small, yellow protuberances on its petioles, floral bracts, leaves, or tendrils to mimic butterfly eggs (fig. 522). Because the larvae of heliconians are cannibalistic (as is also true of monarch caterpillars), a female butterfly will carefully inspect plants for previously laid eggs before laying an egg herself. In many cases the eggs are laid singly on the most tender, young tissues of the plant: new leaf stalks, stipules, tendrils, etc. (fig. 523). Other heliconian butterflies, including the Juno long-wing (*Dione juno*), deposit

Passion-flowers

Fig. 522. The undersides of the leaves of *Passiflora biflora* have small yellow spots that mimic the eggs of heliconian butterflies, which will not lay their eggs near any earlier deposited eggs because the older larvae would cannibalize larvae that hatch later.

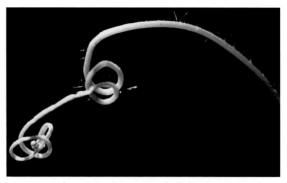

Fig. 523. (Above) A solitary egg of a heliconian butterfly on the tender tip of a *Passiflora* tendril. Fig. 524. (Below) The Juno long-wing butterfly lays her eggs in large rafts on the undersides of older, tougher leaves.

their eggs in a "raft" on the undersurfaces of older leaves (fig. 524), bringing into play survival of their offspring by safety in numbers.

Passion-flower species have been introduced to many other tropical or subtropical regions of the world for both their beauty and their utility. Several species have edible fruits, the pulp of which is eaten fresh, made into juice, or used as an ingredient in desserts. In Brazil, passion-fruit (*maracujá* in Portuguese) is used to flavor ice cream, cakes, and a traditional—and delicious—*maracujá* mousse. The species most often used for culinary purposes is *Passiflora edulis* (meaning edible), a close relative of *P. incarnata*, which is native to southern Brazil, northern Argentina, and Paraguay (where it is the national flower). *Passiflora edulis* has two forms: one with 5 cm yellow fruits (forma *flavicarpa*) (fig. 525), similar to those of maypops; the other with smaller, purple fruits (forma *edulis*) (fig. 526). The yellow-fruited form is self-sterile and the purple self-compatible. The purple fruits are said to have better flavor. When grown as a crop, the vines are grown on trellises so the fruits are easier to harvest (fig. 527).

Commonly, introducing plants from one part of the world to another can have negative consequences.

Fig. 525. Yellow *Passiflora edulis* forma *flavicarpa* fruits for sale in a market in Venezuela. Fruits are ripe when wrinkled like these.

Passion-flower Family

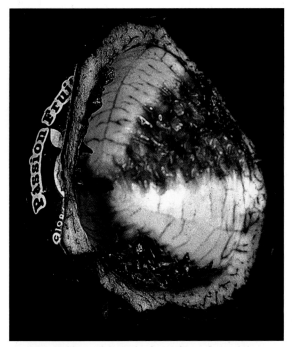

Fig. 526. Half a rind from the fruit of *Passiflora edulis* forma *edulis* showing where the seeds were attached to the wall of the fruit (parietal placentation).

This is the case of a long-tubed, pink-to-red-flowered passion-flower species complex known variously as *Passiflora mollissima*, *P. tarminiana*, and *Passiflora tripartita* var. *mollissima* (fig. 528), native to higher altitudes in the Andes. The fruits of these vines are consumed fresh and made into various juices and desserts. Members of this species complex are grown as a crop in Colombia, and all have been introduced to other parts of South America, as well as to distant lands, among them, New Zealand, Australia, South Africa, the Philippines, China, and the islands of Hawaii. In Hawaii, the species originally called *P. mollissima*, but now designated as *P. tarminiana*, runs rampant, blanketing and shading out Hawaii's imperiled native flora (fig. 529). This is particularly true on the Big Island and Kaua'i. The seeds of the yellow, short-banana-shaped fruits of the plant (known in Hawaii as "banana poka") are widely dispersed by birds and introduced feral pigs that digest the fleshy orange arils (fig. 530) and excrete the seeds. Banana poka has the federal designation of a noxious weed in Hawaii.

In their native range, the passion-flowers that are members of this long-tubular, pink-flowered complex are pollinated by hummingbirds. However, when this

Fig. 527. (Above) Growing *Passiflora edulis* on trellises allows farmers to more easily harvest the fruit hanging beneath the trellises. (Photographed in Amazonian Brazil) **Fig. 528.** (Right) The long tubular corollas of *Passiflora tripartita* var. *mollissima* and similar species are pollinated by hummingbirds. Some other South American species are pollinated nocturnally by bats.

Passion-flowers

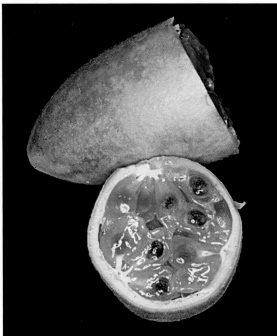

Fig. 529. (Left) Banana poka (*Passiflora tarminiana*) was introduced to Hawaii in the 1940s and has had negative effects on the native forests and other flora there.
Fig. 530. (Below) The orange flesh surrounding the seeds in fruits of banana poka is edible and is also attractive to birds and mammals that eat the fruit and then disperse the seeds. (Photographed in Peru by Scott Mori)

complex is introduced to regions without hummingbirds, honeybees and bumblebees assume the role of pollinator, but far less effectively. Passion-flowers are generally introduced for their beauty or their use as a food, but they are also used medicinally in some cultures—primarily as a sedative.

You might wonder if the common name passion-flower (a translation of the scientific name, *Passiflora*) comes from some property of the plant that could induce any form of passion. This is not the case; the name was given to the plant because it was used by early Spanish missionaries as a symbol of the Passion of Christ. When missionaries arrived in the Americas in the fifteenth and sixteenth centuries, they were intent on converting the native peoples to Christianity. They seized upon the numerous species of passion-flowers as teaching tools to instruct the natives in the story of Christ's last days and crucifixion. They found symbolism in all parts of the plant; the symbolic meanings vary according to the person telling the story, but basically the 10 perianth parts (sepals + petals) were used to represent the 10 apostles (excluding Judas, the betrayer, and Peter, the denier), the coronal filaments stood for the crown of thorns, the three stigmas symbolized the three nails used to nail Christ to the cross, and the five anthers represented the five wounds suffered by Christ (four from the nails, one of which pierced both feet, and one from the spear used to confirm his death). In addition, the plant's tendrils were said to represent the whips used to flagellate Christ as he carried the cross to Calvary, and the often three-lobed leaves represented the hands that wielded the whips.

Thus, there is much to admire about passion-flowers beyond their beauty. They have interesting relationships with their pollinators and their herbivores, they provide a delicious fruit used in the making of many flavorful desserts, and they have a long history of religious significance.

Heath Family

Fig. 531. Pipsissewa (*Chimaphila umbellata*) is an evergreen plant with shiny, unmarked leaves. Here the woody, upright fruits produced from the previous year's flowers have remained into June, when the plant is flowering again. The fruits of all the species discussed here are capsules and each contains about 1000 dustlike seeds.

Pipsissewa and Related Species

Fig. 532. The leaves of spotted wintergreen (*Chimaphila maculata*) are more striped than "spotted." Throughout the winter, spotted wintergreen's evergreen leaves provide a welcome bit of green among the brown leaves and snow.

Pipsissewa, Spotted Wintergreen, and their evergreen relatives

Chimaphila umbellata (L.) W. P. C. Barton,
C. maculata (L.) Pursh, and other species
Heath Family (Ericaceae)

The genus name, *Chimaphila*—from the Greek *cheima* (winter) and *phileo* (to love)—is a reference to the evergreen state of the members of this genus in winter. *Chimaphila*, along with the closely related genera *Pyrola*, *Orthilia*, and *Moneses*, were formerly all included in the genus *Pyrola* and placed into a small family known as the Shinleaf Family (Pyrolaceae). That family has now been subsumed into the Heath Family (Ericaceae).

Habitat: Pipsissewa (*Chimaphila umbellata* (fig. 531) grows in dry, acidic, often sandy, soils in semishaded forest of coniferous trees (usually pines); spotted wintergreen (*Chimaphila maculata*) (fig. 532) is found in coniferous, deciduous, or mixed forest, again usually in dry, sandy soils; the other species generally grow in either dry or moist woods and/or bogs (or in the case of *Orthilia secunda* [sidebells], in swampy forests, heaths, and tundra).

Range: Pipsissewa is a widespread, circumboreal species, occurring from Canada throughout the eastern and southwestern United States and extending in the mountains into Mexico, as well as on the Caribbean island of Hispaniola. It is also native in northern Eurasia. The range of spotted wintergreen is more restricted, extending from southernmost Quebec and Ontario in Canada through the eastern United States (except for the Deep South); it also occurs in the mountains southward through Mexico and into Panama and has been introduced into Europe. Although spotted wintergreen is widespread and considered common in the United States, it is imperiled in certain states and legally protected in others. In Canada it is listed as federally endangered and as an "at risk species" in the province of Ontario. A program to protect and increase the current population in Ontario is underway. Many pyrola species are circumboreal, but a few species are limited to the northern parts of North America. Single delight (*Moneses uniflora*) is a circumboreal species, as is sidebells, which is native throughout Canada, Greenland, Alaska, and most of the lower 48 United States other than the Southeast and the southern Midwest.

Heath Family

To the botanist, scientific names bring precision, but common names can also be straightforward and useful. However, in this case the common names of the two species of *Chimaphila* are often interchanged, causing confusion when people are trying to discuss the same plant but using different names for it. Pipsissewa (an Abenaki Indian word meaning "flower-of-the-forest") is more commonly applied to *Chimaphila umbellata*, while spotted wintergreen refers to *C. maculata*. The adjective "spotted" is used even though the white markings are more striped than spotted (the Latin translation of the species epithet "*maculata*" can mean spotted or stained in some way, as in "*not* immaculate") (see fig. 532). "Prince's pine" and "wintergreen" are other names applied to both plants, leading to further confusion.

The name wintergreen is sometimes applied to *Moneses uniflora*, which is also commonly called one-flowered wintergreen, or (one-flowered) shinleaf, or pyrola. More charming names for this one-flowered plant are single delight (from the Greek *monos* meaning "alone" and *hesis*, meaning "delight"), or wood nymph (for its habitat), and shy maiden (for its downward-facing "bashful" flower) (fig. 533). The species epithet "uniflora," of course, is derived from the Latin for "one flower." *Orthilia secunda*, too, is known by some as one-sided wintergreen; I prefer the common name sidebells, which is more descriptive and less misleading. While all these plants remain green in winter, none have the aromatic wintergreen oils characteristic of another species in the heath family, the *true* wintergreen, *Gaultheria procumbens* (fig. 534). In addition to the multiple and confusing common names, the pyrola group has been a taxonomically perplexing one, but it has been well sorted out by morphological distinctions and differences in chromosome number, something I will not discuss in this essay.

Members of this group resemble each other in habit: all are small, forest-dwelling plants, some of which may be subshrubs having a slightly woody base; some are rhizomatous, and others are considered herbaceous. The flowers of all have five petals, but differences in the flower forms of each genus are obvious. The flowers of most species are symmetrical, with widely open petals that in pipsissewa and spotted wintergreen are rounded

Fig. 533. The reason that *Moneses uniflora* was given the common name "shy maiden" (along with "single delight" and other names) is evident in this image of the solitary, downward-facing, "bashful" flower.

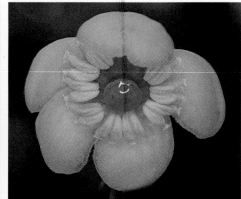

Fig. 534. (Below left) Unlike other plants that bear the name wintergreen, the true wintergreen plant has a pleasing, aromatic wintergreen scent in its leaves. It is in another genus in the heath family and known by the scientific name *Gaultheria procumbens*. Fig. 535. (Below right) Flowers of *Chimaphila* species (*C. maculata* seen here) are characterized by five rounded, but erose (slightly jagged), petals that surround a globose ovary topped by a broad stigma. The anthers of the 10 stamens open by two pores, but rather than pollen, it is nectar secreted by a nectar-producing disk at the base of the ovary that is the reward that insects seek when visiting the flowers. The wet stigma of this flower indicates that it is in a receptive phase, and pollen deposited on it will germinate.

Pipsissewa and Related Species

Fig. 536. (Above left) Flowers of pipsissewa are similar to those of spotted wintergreen, but they are commonly suffused with pink. The pink disk surrounding the base of the ovary secretes nectar (the disk is white in spotted wintergreen and more difficult to see in fig. 535).
Fig. 537. (Above right) *Pyrola* flowers also have five petals with rounded petal tips, but they are more asymmetric, with their stamens clustered at the top of the flower rather than in the center. The stamens surround a long style that protrudes from the ovary; in most species, the style is curved. The flower produces no nectar; pollen serves as the reward for pollinators.
Fig. 538. (Right) Flowers of single delight (*Moneses uniflora*) differ from those of *Chimaphila* and *Pyrola* in that their petals are pointed at their tips and ruffled along their margins. The stigma is five-pronged, and the flowers are always solitary on the stalk. Like flowers of *Pyrola*, there are no nectaries. This circumboreal species was photographed in Italy.

at their apices (fig. 535), and shallowly fringed or jagged (erose) so that the margins appear to be finely torn, rather than neatly cut. Both of our *Chimaphila* species have white to pink flowers, with those of pipsissewa suffused with pink or red (fig. 536), sometimes strongly so. Anthers in both species are prominent, with their filaments widely expanded at the base; those of pipsissewa have small hairs along their sides, whereas those of spotted wintergreen are densely hairy. The styles are short and broad. Flowers of the two species commonly encountered in the Northeast have, at the base of their ovaries, a disk that secretes nectar rich in sugars, a reward for visiting insects (see fig. 536). The bees that visit the flowers collect *only* nectar and not pollen. The broad stigmas are wet when receptive and provide a large expanse of tissue for pollen deposition. While visitors to the flowers of *Chimaphila* are not specifically collecting pollen, they incidentally pick up and transfer some pollen while probing for nectar.

The flowers of most species of *Pyrola* are nodding, somewhat bell-shaped, and white (but greenish in *P. chlorantha*, [green-flowered wintergreen] and pink in *P. asarifolia* [pink pyrola]), with petals generally having rounded tips with smooth margins. Their anthers are clustered toward the top of the flower with a long,

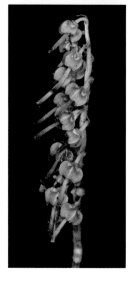

Fig. 539. Sidebells (*Orthilia secunda*) was transferred out of *Pyrola* because its inflorescence differs from those of other members of *Pyrola* by being one-sided. It also differs from *Pyrola* spp. in its wood anatomy and in the differences in its pollinator reward system (it produces both fine pollen *and* nectar for its insect visitors).

usually curved, style projecting downward among them (fig. 537). The single species of *Moneses* (single delight) has widely open flowers with petals that taper to a pointed apex; the margins of the petals are ruffled. In the center of the flower is a conspicuous ovary with a prominent style topped by a five-pronged stigma (fig. 538).

Flowers of sidebells (*Orthilia*—probably from the Greek *orthos*, meaning "straight," and *ilium*, meaning "side") are small, green, nodding, and arranged along only one side of the inflorescence, hence the specific epithet *secunda*—one-sided (fig. 539).

The principal pollinators of all these flowers are bees, especially bumblebees. Despite many observations

Heath Family

in the field, only once have I seen the flowers of any of these species being visited by insects—pink pyrola/shinleaf (*Pyrola asarifolia*) hosting a midsized bee (fig. 540). On examination, the flowers of pipsissewa and spotted wintergreen are both found to have prominent anthers that open by small pores at their tips (poricidal) rather than dehiscing by longitudinal slits as is more common in many other species. I assumed they were likely buzz-pollinated (further described in the following paragraph), as are the flowers of many other plants in the heath family. Indeed, plant species in nearly 70 different angiosperm families have been observed to have anthers that dehisce by pores.

Buzz-pollination (also called sonication) is a pollination mechanism whereby bees vibrate fine pollen from poricidal anthers. There is a noticeable difference in the buzzing sound during buzz-pollination compared with the sound made by a bee when in flight—it is higher-pitched, more rapid, and "angry" sounding. To accomplish this change of pitch, bees of many species first curl the bottom side of their abdomens around the anthers (see fig. 540) and grasp the base of an anther with their mandibles (mouthparts). Then, once in position, the bee is able to vibrate the indirect flight muscles in its thorax, resulting in the pollen being ejected in "salt-shaker" fashion from the anther's pores onto the body of the bee. An electrostatic charge that builds up between the pollen and the hairs on the bee's body may augment the effectiveness of this procedure. Bees then groom most of the pollen from their bodies, forming pollen sacs on their hind legs. As they visit other flowers, any remaining loose pollen may brush off onto that flower's stigma(s), resulting in pollination. Most buzz-pollination occurs in the early part of the day, and in a study of this phenomenon in the pyrola group, Knudsen and Tollsten hypothesize that this might be so because the higher temperatures generated in the bee's body during buzz-pollination may more easily be dissipated during the cooler morning hours. Interestingly, honeybees are incapable of buzz-pollination; they are, however, able to obtain some pollen from poricidal anthers by drumming the anthers with their forelegs. One species of hover fly has been recorded as using buzz-pollination as well.

Because of the poricidal anthers of pipsissewa and spotted wintergreen, buzz-pollination appeared to be a likely means of pollination, but as mentioned earlier, bees have been observed collecting only nectar from these flowers, not pollen. In addition, the flowers' pollen is produced in large clumps that are not fine enough to be vibrated out of the small pores in the anthers.

In contrast, flowers of most species of *Pyrola* and *Moneses* do not produce nectar; rather, they have fine pollen that is easily vibrated out of their poricidal anthers. Thus, bees seen on these species are visiting to collect pollen. The exception is *Orthilia secunda* (sidebells), which has a pollination system that is intermediate between the two other genera, producing both nectar *and* fine pollen; hence both nectar seekers and pollen collectors visit its flowers. Based on this intermediate pollination system, as well as morphological differences and a difference in chromosome number, Haber and Cruise have made a good case for segregating sidebells from the genus *Pyrola*, resulting in a separate genus of one species: *Orthilia secunda*.

It is hypothesized that species of *Chimaphila* are more primitive members of this closely related group of genera. Only pipsissewa, spotted wintergreen, and sidebells produce nectar. All other species of *Pyrola* and *Moneses* are thought to have lost the ability to produce nectar, thus attracting only those specific pollinators able to buzz the flower's anthers to get a reward. This specificity results in a reduction of pilferage of pollen by insects that are not effective pollinators and promotes constancy of visits to flowers of the same species, thus enhancing successful pollination. Perhaps sidebells reflects a transitional stage in the evolution of these flowers from exclusively nectar-rewarding to exclusively pollen-rewarding species.

Aromas produced in the petals and stamens of the flowers of the pyrolas and single delight serve to attract insects. The aromas differ between *Pyrola* and *Moneses*, bolstering the decision to segregate *Moneses* from its former genus placement in *Pyrola*.

Fig. 540. In this lovely pink wintergreen flower (*Pyrola asarifolia*), a midsized bee curves its abdomen around the stamens and vibrates them to shake the pollen from their poricidal anthers, a process called buzz-pollination.

Pipsissewa and Related Species

The flowers of *Chimaphila* spp. produce only a slight scent, and sidebells has no detectable aroma, at least to the human nose. In spite of the above-mentioned rewards, few insects visit any of the species of *Chimaphila*, *Pyrola*, *Orthilia*, or *Moneses*. This is true, in general, of flowering plants growing in dark forests where the flowers can be difficult for insects to locate (fig. 541).

The flowers of all these species are somewhat self-compatible, so some seed is produced via self-pollination, and individual flowers tend to be long-lived if they have not yet been pollinated. Different members of this group are often found growing in close proximity, which could attract a greater number of pollinators. Hybrids sometimes result from this cross visitation. This entire group of species produces capsules (see fig. 531). that typically contain approximately 1000 seeds. The dustlike seeds can easily be transported by the wind because of their tiny size—about the size of orchid seeds, or even smaller—yet most are not transported very far. As with orchid seeds, they lack stored food reserves. In order to germinate and grow, they require the presence of a compatible mycorrhizal fungus in the soil, a system that will then supply the necessary nutrients from nearby host plants via the fungus. Although mycoheterotrophs (plants without chlorophyll) provide nothing in return to the fungus, it is uncertain whether mixotrophic plants (i.e., those with green leaves as discussed in this chapter) "pay back" the fungus once they are mature.

The members of the four genera discussed in this chapter are all thought to be mixotrophic, that is, they obtain only a portion of their carbon requirements through fungal associations with nearby trees; the remainder is provided by photosynthesis. The presence of fungi, both within the underground tissue and surrounding the underground tissue of these plants, has been demonstrated. This system is similar to that of mycoheterotrophic plants, which are totally devoid of chlorophyll and therefore cannot photosynthesize (e.g., Indian pipe, treated elsewhere in this book). However, in a mixotrophic system, the green (chlorophyllous) leaves of the plants *do* have the ability to synthesize carbohydrate, but, because the extremely low level of light in the forest understory does not allow for adequate photosynthesis, the plants benefit from the nutrients provided by the fungus. Plants growing in a more shaded environment are more dependent on the relationship with a mycorrhizal fungus.

Many of these plants reproduce vegetatively. Thus, larger patches of pipsissewa or spotted wintergreen are most likely clonal colonies of genetically identical ramets.

Medicinal uses for all four genera are claimed. Single delight is reported to have been used by aboriginals in Canada to treat coughs and colds, to reduce pain, and to cure cancer and paralysis. Despite this broad range of undocumented claims, a peer-reviewed paper in the *Journal of Natural Products* documented the discovery of a new compound, 8-chloro-2,7-dimethyl-1,4-naphthoquinone (8-chlorochimaphilin for short), isolated from single delight (*Moneses uniflora*), and having activity against both fungi and bacteria. Other similar compounds isolated from this species had already been found in *Pyrola* and *Chimaphila*. One of these, chimaphilin, extracted from *Pyrola*, showed anti-inflammatory and analgesic effects. The same compound, extracted from pipsissewa, was found to have antibacterial and antifungal properties (effective against *Staphylococcus aureus*, *Candida albicans*, and other pathogens), and additionally, antihemorrhagic effects and vitamin K-like activity.

Fresh leaves of spotted wintergreen, pipsissewa, and poultices of shinleaf (*Pyrola*) species have long been recommended as a folk remedy for skin bruises (especially the commonly bruised shins; thus, the common name, shinleaf). However, there are also reports of a contact dermatitis occurring from their use, confirmed in a paper published in *Contact Dermatitis* that reported the compounds in shinleaf have a moderate capacity to produce redness or blisters.

Fig. 541. A typical forest habitat for *Pyrola elliptica*. Most species in this related group of plants inhabit shady forests and rely on mycorrhizal relationships with fungi for a portion of their nutritional needs.

Cactus Family

Fig. 542. Several flowers growing from areoles along the margin of a prickly pear pad.

Prickly Pear

Fig. 543. New pads and flower buds grow from growth pads (areoles) along the edges of the older pads each spring. Note the small, conical leaves on the new pads. The leaves are deciduous and are soon shed.

Prickly Pear
Opuntia humifusa (Raf.) Raf.
(syn. *O. compressa* J. F. Macbr.)
Cactus Family (Cactaceae)

Indeed, there *is* a cactus native to the Northeast. In fact, in addition to most of the Northeast, our local prickly pear (*Opuntia humifusa*), known as eastern prickly pear, is native to most of the eastern United States except for northern Maine, Vermont, and most of New Hampshire.

Habitat: Sunny areas in dry, sandy soils, including dune areas, and in rocky places, such as bald outcrops and the tops of rocky roadcuts.

Range: The most up-to-date flora for all of North America north of Mexico is *Flora of North America*, which shows eastern prickly pear growing in 33 states from New York and Massachusetts west to parts of South Dakota and Nebraska and the very eastern parts of Kansas, Oklahoma, and Texas, and south to Florida and Louisiana. The USDA PLANTS Database, on the other hand, includes Montana, Colorado, New Mexico, and Ontario within the distribution range (although not *native* to New Hampshire, in 2011 prickly pear was found growing on the state's coastal dunes in an area showing signs of disturbance).

Many Northeasterners are surprised to learn that we have cactus growing in our midst, and especially that it is not cultivated, but native. People seldom encounter our only species of cactus growing in the wild because they infrequently visit its preferred habitats, namely the back-dune areas of beaches or exposed rocky outcrops. The plants are prostrate, rarely more than a foot and a half tall, and generally attract notice only when in bloom; then the large, brilliant yellow flowers can't be missed (fig. 542). However, the

Cactus Family

Fig. 544. A young plant of eastern prickly pear growing in the dune area of a beach at Sandy Hook, New Jersey.

plant was noted as early as the mid-1600s by Adriaen van der Donck, an early Dutch immigrant who cataloged and praised the natural history of New Netherland for the purpose of enticing more Dutchmen to settle there.

Cacti are generally associated with arid habitats, and indeed, in the United States, they are most species-rich in our southwestern desert regions. Over time, the Cactaceae have evolved certain features to conserve water: succulent stems that can store water for use in times of drought (fig. 543); leaves reduced to spines so that water is not lost through transpiration; and a special type of photosynthesis that uses crassulacean acid metabolism (CAM), wherein the stomata open only at night to take in carbon dioxide, thus avoiding the hottest part of the day when greater water loss would be detrimental. The carbon dioxide is stored in the plant as an acid until daylight, when it is released for use in photosynthesis. All these adaptations either help to reduce water loss or protect the plant from herbivory.

Although the Northeast has adequate rainfall throughout the year, the specific habitats where prickly pear is found are low in available water because of rapid drainage in sand (fig. 544) or rock (fig. 545).

Fig. 545. A robust colony of eastern prickly pear atop a rock outcrop in New York.

Prickly Pear

Fig. 546. In early spring, the pads of *Opuntia humifusa* still look yellowish and shriveled.

Opuntia humifusa and the very similar midwestern species, *Opuntia macrorhiza*, are two of the few cactus species hardy enough to tolerate cold temperatures. In winter, the fleshy pads undergo a freeze-avoiding process, losing much of their moisture content and appearing wrinkled and yellow-green until the following spring (fig. 546).

There are about 150 species of *Opuntia*, all of New World origin, but introduced widely elsewhere. Even where *Opuntia* is not native, bees are the principal pollinators, the most effective of which are large-bodied bees such as bumblebees and carpenter bees. Eastern prickly pear flowers are visited by large numbers of bumblebees as well as other bee species, for example, small solitary bees, and beetles. The larger bees, with their longer proboscises (tongues), visit the flowers primarily for nectar; the smaller ones to gather pollen (figs. 547a–d).

The large, green ovaries of the flowers, evident beneath the showy tepals (fig. 548), ripen slowly, turning a dull red as they do so. Unlike the plump,

Fig. 547. a. A species of bumblebee (*Bombus* sp.) visits a flower of eastern prickly pear to drink its nectar. b. *Opuntia ficus-indica*, an introduced species in the Mediterranean, is visited and pollinated by bumblebees local to that region. (Photographed in Cinque Terre, Italy) c. In the Galápagos, most pollination of *Opuntia* spp. is carried out by the only species of bee on the islands, a large carpenter bee (*Xylocopa darwinii*). The female bees of this species are large and black; the males smaller and golden tan. (Photographed in Galápagos) d. Other species of bees (such as this male halictid bee) visit the flowers of eastern prickly pear in southeastern New York State to gather pollen.

Cactus Family

Fig. 548. Viewed from the side, the thick, green ovaries of the flowers of eastern prickly pear are evident. If the ovules within are fertilized, the ovaries will develop into red fruits.

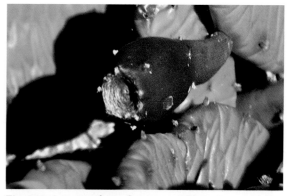

Fig. 549. A mature fruit of eastern prickly pear.

Fig. 550. A prickly pear pad shows evidence of browsing by rabbits. Note how the rabbits carefully eat around the irritating glochid-bearing areoles.

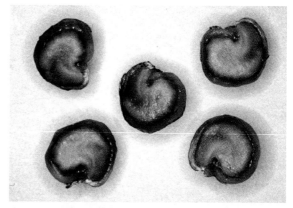

Fig. 551. Each small, kidney-shaped seed of *Opuntia humifusa* is surrounded by a 1 mm wide ring of tissue that derives from the funicle (the stalk that attaches the seed to the placenta of the ovary).

succulent fruits of *Opuntia ficus-indica*, those of eastern prickly pear are more cylindrical in form and do not become juicy and sweet (fig. 549). Eastern prickly pear fruits are occasionally eaten by various animals such as wild turkeys and skunks, which then disperse the undigested seeds. Cottontail rabbits and deer, despite the glochids and spines, are known to eat the pads. Rabbits avoid the prickly spines and glochids by nibbling only the green flesh that surrounds them (fig. 550).

The small, flat, kidney-shaped seeds measure only 3.5–4.5 mm and are surrounded by a narrow band of material derived from the funicle (the stalk that attached the seed to the placenta [fig. 551]). Propagation from seed is very slow; therefore, it is faster to grow new plants from pads, which can be easily rooted by placing the cut end into sandy soil.

In other arid parts of the world, plants of different families (e.g., some members of the Euphorbiaceae [spurge family] in the Old World deserts of

Prickly Pear

Fig. 552. *Euphorbia* species on display in an Old World desert display at the New York Botanical Garden.

Fig. 553. All cacti are thought to have evolved from species having "normal" green leaves, such as those shown in this Brazilian cactus, *Pereskia grandiflora*.

Africa and Asia, and the Didiereaceae [didierea family] in Madagascar) have evolved to have many of the same traits for survival found in cactus plants and thus resemble them in appearance—a phenomenon termed convergent evolution (fig. 552). These plants may easily be distinguished from cactus plants when they flower since their small flowers lack the showy petal-like tepals of members of the cactus family. Moreover, Cactaceae differ from species in *all* other plant families by the presence of areoles on their succulent stems. These small, roundish growth pads may produce spines (thought to be highly modified leaves), glochids (irritating bristly hairs), new stems (pads), flowers, and even small, conical, deciduous leaves that appear in early spring and are then quickly shed (see fig. 543). Indeed, cacti are thought to have evolved from "normal" broad-leaved plants that, over time, adapted to arid conditions by losing their water-wasting leaves. Cactus plants in the genus *Pereskia* still retain this leafy condition and are thought to be more primitive members of the family (fig. 553). The stems of *Opuntia* are jointed, each segment growing from the one below. They are flattened leaflike stems technically termed cladodes.

Most eastern prickly pear plants have few, if any, spines, but when present, they are wickedly long and stout (fig. 554). An abundance of glochids grow from the areoles (fig. 555). Despite their soft, innocuous appearance, the hairs can be a significant source of irritation; if even slightly brushed, glochids dislodge easily and, as a result of the dozens of minute barbs at

Fig. 554. Some eastern prickly pear plants produce long, stiff spines that protrude from their areoles.

269

Cactus Family

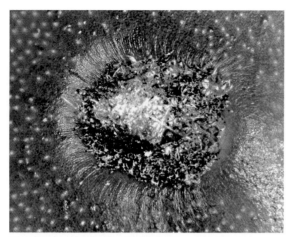

Fig. 555. An areole of eastern prickly pear with delicate, but irritating, glochids. The glochids, with their barbed ends, detach easily to become embedded in the skin of an unwary passerby.

Fig. 556. (Above) Glochids stuck in my finger after I attempted to move an object obstructing my view.
Fig. 557. (Left) Epiphytic species of the genus *Rhipsalis* are the only members of the cactus family native to the Old World, in this case Africa and Madagascar.

their tips, become firmly embedded in the skin (one of the hazards of being a plant photographer [fig. 556]). I have found that glochids gradually work their way out, or at least cease to be an irritant after a few weeks; however, it is recommended that they be carefully removed with tweezers followed by a coating of white household glue, which after drying can be peeled off, helping to remove many of the small glochids. If large numbers of glochids are allowed to remain in the skin, a severe reaction may occur.

The cactus family is almost exclusively a New World family, with the notable exception of *Rhipsalis*, a primarily epiphytic genus sometimes known as mistletoe cactus for the similarity of the small, white fruits of some species (fig. 557) to those of the traditional mistletoe of the Christmas season (*Viscum* spp. in the Viscaceae). Species of *Rhipsalis* grow on rainforest trees from Mexico through South America, with one species in Florida. But the genus is also found in parts of western and central Africa and in Madagascar. Various arguments are made regarding its presence in the Old World, from dispersal by birds blown off course, to rafting on vegetation carried by ocean currents, to early evolution of the genus in Gondwanaland prior to the splitting of that massive landmass into the southern hemispheric continents—all of which, for reasons beyond the scope of this chapter's topic, seem unlikely.

The other cactus outpost that is not geographically part of the Americas is the Galápagos archipelago, 600 miles off the coast of South America, in the equatorial Pacific. Although the islands are territorially part of Ecuador, they are at such great distance from the mainland that the existence of several species of cactus there begs for explanation. The Galápagos, like many other oceanic island chains, were (and still are being) formed by an undersea hotspot. As the Nazca oceanic plate, from which the Galápagos Islands arise, moves slowly southeastward, new islands form to the west as a result of volcanic activity beneath. Thus, the westernmost island, Fernandina, is younger than the other islands and still volcanically active. All the islands were initially devoid of vegetation, being made up solely of bare lava. Over time, erosion resulted in the formation of sandy beaches: black or red from eroded lava, or white from reefs uplifted by seismic activity. Large South American rivers with their origins on the western slopes of the Andes, in particular the Guayas River in Ecuador, empty into the Pacific, carrying in

Prickly Pear

Fig. 558. (Above) *Brachycereus nesioticus* (lava cactus), an endemic genus of only a single species, shown growing on the barren pahoehoe lava fields of Fernandina, a still volcanically active island at the westernmost edge of the Galápagos archipelago. **Fig. 559.** (Right) *Opuntia echios* var. *barringtonensis* has evolved into a treelike form, with succulent pads that are out of reach of giant tortoises. Although giant tortoises are not currently found on the Galápagos island of Santa Fé (where this photo was taken), evidence that they once inhabited the island is found in old whaling ship records, in early reports to scientists on the California Academy of Sciences 1905–06 expedition, and from the DNA of fragments of a tortoise skeleton found on Santa Fé that suggest the tortoise that once lived there was unlike any other now found in Galápagos.

their currents plant debris and even chunks of land carved from the riverbanks during times of peak flow. In this way, mini islands were swept out to sea, some eventually reaching the barren volcanic shores of the Galápagos. Cactus plants, because of their succulence, were particularly well adapted for survival over such long distances without fresh water. When washed up on land, pieces of cactus could establish themselves in lava crevices and on the eroded sands of the islands. Of the three genera of cactus now found on Galápagos, two are endemic, each with only a single species: *Brachycereus nesioticus* (fig. 558) and the tall, columnar *Jasminocereus thouarsii*. The third genus, *Opuntia*,

Cactus Family

Fig. 560. In the Galápagos, finches feeding on *Opuntia* nectar and pollen also incidentally pollinate the flowers. The finch in the above photo is a male large cactus finch feeding on the flowers of *Opuntia helleri* (carpenter bees are absent on the northern islands; thus this activity by birds is especially important for the perpetuation of this cactus species that is endemic to only a few of the northernmost islands in the Galápagos. (Photographed on the island of Genovesa [Tower])

while not itself *endemic* to the Galápagos is represented there by six species, all of which *are* endemic to the islands. *Opuntia echios* is now well established on seven of the islands and has further evolved into five varieties, some treelike and others shrubby. On islands that produce (or once had) populations of giant tortoises, the plants have evolved taller trunks with succulent pads out of reach of the tortoises (fig. 559).

In the Galápagos, pollination of certain species of *Opuntia* is sometimes accomplished by birds, particularly the cactus finch (*Geospiza scandens*) and the large cactus finch (*G. conirostris* [fig. 560]). Mockingbirds and doves have also been observed at the flowers. When a bird (or a bee) comes into contact with the stamens, the stamens move toward the style in response to the stimulus. This rapid response to touch is called thigmotropism (syn. thigmotaxis), and in

prickly pear it helps to ensure that pollen is transferred to the visiting insects. You can witness this response yourself. Select a flower that has not recently been visited by insects (one with widely spreading stamens) and brush the stamens with your finger to observe the result.

Man has introduced several *Opuntia* species to other parts of the world, particularly in the Mediterranean region, South Africa, and Australia. The species most often selected for introduction elsewhere is *Opuntia ficus-indica*, known as the Indian fig for its tasty fruits. Various species of prickly pear were brought to Australia in the late 1700s in an attempt to create a dye industry based on cochineal cultivation (more about this use later). Prickly pear was also grown there as a garden plant and as a living fence. The plants proved to be aggressively invasive, so that by 1919 prickly pear had covered more than 10 million acres of farmland in southeastern Australia and was advancing by 1 million acres per year. Because of the encroaching prickly pear, many farmers had to abandon their farms, and the economy of rural areas suffered. Prickly pear eradication campaigns were launched, first using mechanical means and later chemical attack. Both methods proved either ineffectual or ecologically damaging, and a solution using a biological control was sought. In 1925, a South American moth, *Cactoblastis cactorum*, with larvae known to feed on the pads of *Opuntia*, was selected for introduction throughout the affected region. The insect devastated the acreage invaded by prickly pears, and the prickly pear problem was quickly brought under control. The project still stands as the most successful example of biological control. Currently, this same moth has been accidentally introduced into Florida from the Caribbean and is now dispersing westward. The spread of *Cactoblastis cactorum* poses a threat to native cacti in the Southwest and beyond. The moth could easily spread from our southwestern deserts into Mexico and wreak havoc on *Opuntia* there, resulting in a significant impact on the Mexican economy.

Although our local eastern prickly pear is not an economically important plant, and its fruits, while edible, are not palatable, the prickly pear species from other parts of the New World have many uses. Mexico has the greatest number of species of the genus *Opuntia*, some of which are important locally, and even internationally: for food products; in the production

Prickly Pear

of an alcoholic beverage ("colonche"); as a source of a natural dye; and as a cattle fodder used in times of drought. Prickly pear is so valuable to the Mexican economy that it even has a place on the Mexican coat of arms.

Opuntia ficus-indica fruit (called "tuna" in Mexico [fig. 561]), and its young pads ("nopales") of the same species form an important part of the Mexican diet. The juicy flesh of the fruit is eaten raw—including the small, hard seeds, which are too small and numerous to bother removing. Fruits are also preserved in jams and jellies. The nopales are generally harvested when young and tender, at which point even the outer "skin" can be eaten (after the spines and glochids have been removed, of course). Nopales are eaten raw in salads, brushed with oil and grilled, deep fried, sautéed in omelets, or used in combination with other ingredients in many traditional Mexican recipes. The fruits and pads are served in similar ways wherever this species has been introduced. Both have a tradition of use in folk medicine, and it has been shown that prickly pear fruits are rich in antioxidants, and that the pads might help in lowering blood sugar and LDL cholesterol. In Israel prickly pear fruit is called "sabra," the name also applied to native-born Israelis, who are said to be tough on the outside, but soft and sweet on the inside.

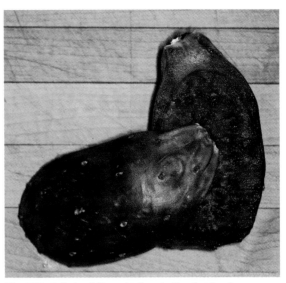

Fig. 561. A fruit of *Opuntia ficus-indica* that has been cut in half to show the juicy red flesh and small seeds.

Earlier mention of using *Opuntia* as a source of dye is somewhat misleading, for it is not the plant itself that provides the bright crimson dye, but the small scale insects (*Dactylopius coccus*) that feed on its juices. The rich red color was highly prized by Europeans in the nineteenth century, and hundreds of tons of cochineal dye were processed in the environs of Oaxaca, Mexico during the 1880s, most of it for export. Even the red jackets of British soldiers were dyed with cochineal, which was traded as an important commodity on the markets of London and Amsterdam. (The European plant, *Rubia tinctoria*, was also used as a source of red dye.)

To yield one pound of powdered dye, large numbers of cochineal insects are needed—more than 70,000 insects! *Opuntia* farming and the raising of cochineal scale insects became the basis of an important economic industry in Mexico. The insects must be handpicked or scraped from the prickly pear pads, a labor-intensive process. They are easy to see because, like many scale insects and aphids, they produce a waxy, white covering that affords them protection from sun and desiccation (fig. 562a). Protected by this covering, the females and the nymphs insert their slender mouthparts into the cactus pads and feed on their juices. As a further deterrent against predators, cochineal insects produce carminic acid, the source of the red color in their body fluid. To obtain the dye from

Fig. 562a. A pad of prickly pear with white wax-covered scale insects from which a red dye is obtained. A sample of the dye color is on the paper to the right of the pad. (Photographed near Teotihuacan, Mexico)

Cactus Family

the insects, they are first killed in hot water and then dried using either sunlight or oven heat. The dried insects are pulverized, traditionally with a stone *mano* and *metate* (as used for grinding corn) (fig. 562b), and the resultant fine powder is placed into a large pot of simmering water. Various ingredients may be added to produce the desired color (fig. 562c).

The advent of synthetic dyes at the beginning of the 1900s caused a decline in the demand for cochineal, but the industry revived when some of the synthetic red dyes used to color cosmetics and food products were shown to be carcinogenic. Cochineal was adopted as a natural, and generally safe, alternative. It appears on ingredient lists as carmine, E120,

Fig. 562b. (Left) A woman in Oaxaca, Mexico uses a stone "mano" on her "metate" to grind cochineal insects into a powder. The powder produces a permanent crimson dye, but by adding various mordants, a range of colors can be produced, from yellow, to orange, to red. **Fig. 562c.** (Right) Lemon juice is added to the cochineal dye bath to brighten the red color of the final product.

Fig. 562d. Periodically, yarn is lifted from the dye pot to check for the depth of color. Wool yarns develop the most intense color.

Fig. 562e. A selection of some of the typical woolen rugs for which Oaxaca, Mexico is famous. The colors are still produced using natural dyes, including the red dye made from cochineal insects.

Prickly Pear

Fig. 563. A giant saguaro cactus (*Carnegiea gigantea*), the popular symbol of our southwestern desert, is silhouetted against a desert sunset (Photographed in the Sonoran Desert in Arizona)

or sometimes as natural red color. Although a natural product, cochineal can produce serious allergic reactions in some people. Since carmine color is produced from an insect, vegans, strict vegetarians, and many Orthodox Jews and Muslims cannot use products that contain carmine color processed from cochineal.

The intense, permanent shades of reds, oranges, and yellows made from cochineal are still used in traditional Mexican weavings, especially those fashioned of wool (fig. 562d) used in the hand-loomed rugs for which the Mexican city of Oaxaca is famous (fig. 562e). Cochineal dye is also produced in Peru, Chile, and the Canary Islands.

A recent finding by researchers at the University of South Florida has demonstrated that the mucilage in the pads of *Opuntia* shows promise as an oil dispersant, which could be used for cleaning up oil spills in the ocean. If proven effective on a large scale, *Opuntia* would provide a safer alternative to the chemical dispersants currently in use for this purpose. *Opuntia* mucilage is already used to remove toxins from drinking water.

Although the Northeast might not have the iconic beauty of a cactus-populated, southwestern desert landscape (fig. 563), it is worthwhile seeking out our lone cactus species in June and July to enjoy its beautiful flowers (fig. 564).

Fig. 564. An open flower of eastern prickly pear.

Pitcher Plant Family

Fig. 565. Pitcher plants growing with another carnivorous plant, the yellow-flowered bladderwort (*Utricularia cornuta*), in a typical bog habitat in the northern Adirondacks.

Purple Pitcher Plant

Fig. 566. The red pitchers (modified leaves) of purple pitcher plants (*Sarracenia purpurea*) are growing here in a mat of red sphagnum moss (*Sphagnum rubellum*).

The family Sarraceniaceae has only three genera, *Sarracenia*, the topic of this essay; *Darlingtonia*, with only a single species restricted to a small area in the Pacific Northwest; and *Heliamphora*, with several species that grow on top of isolated table-top mountains, called tepuis, in northern South America. A current hypothesis proposes that the family originated in South America, migrated to southeastern North America at a time when greater expanses of land were exposed in the Caribbean region between the two continents, and finally radiated to northern and western North America. When the once-contiguous range was disrupted, causing the plants to become geographically isolated, three distinct genera evolved.

The purple pitcher plant (*Sarracenia purpurea*) is the only pitcher plant species that occurs in our range; hence it is sometimes referred to here simply as pitcher plant. The other 10 species of *Sarracenia* are

Purple Pitcher Plant
Sarracenia purpurea L.
Pitcher Plant Family (Sarraceniaceae)

The purple pitcher plant is one of only two North American pitcher plant species without an overtopping hood; thus rainwater is able to collect in its leaves. Insects drown in the water-filled leaves, and their bodies are broken down by a complex web of organisms that release nutrients from the insects, which can then be utilized by the plant.

Habitat: *Sphagnum* bogs, fens, sandy areas in swamps, and other (often acidic) wet places (fig. 565).

Range: Across southern Canada and south through the eastern United States to Georgia.

Pitcher Plant Family

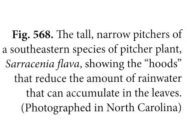

Fig. 567. A leaf of purple pitcher plant with an upright green hood marked with red venation. Note the downward-pointing hairs inside the pitcher that impede insects from crawling out of the pitcher. Pitcher plant leaves have a keel that is evident at the front of the image. The pitcher is full of rainwater.

Fig. 568. The tall, narrow pitchers of a southeastern species of pitcher plant, *Sarracenia flava*, showing the "hoods" that reduce the amount of rainwater that can accumulate in the leaves. (Photographed in North Carolina)

limited to the southeastern United States, a region also known for the greatest diversity of carnivorous plant genera in the world. *Sarracenia purpurea* is the largest and most distinctive of the carnivorous plants in the northeastern United States. It is a perennial with a wide distribution: from the southeastern United States, along the coastal plain in Georgia northward, expanding broadly to Newfoundland, Labrador, Ontario, and Minnesota. It has been naturalized in the Northwest. Despite this extensive range, pitcher plants are not commonly encountered, for they are restricted to open sunny or light shady areas in bogs (see fig. 565), fens, cedar swamps, and similar habitats that few people choose to visit. Nutrient-poor soils and acidic waters typically characterize these habitats, although in some regions, *S. purpurea* may be found in wet areas with neutral, or even alkaline, or serpentine, soils. Such wetlands were long considered worthless and were thus purchased cheaply by developers, who filled them in order to build housing developments and shopping malls. This habitat loss resulted in a marked reduction in the number and variety of specialized plants and animals that have evolved to live in these habitats. There is a fascination with carnivorous plants worldwide, and pitcher plants are now found naturalized in such distant lands as Ireland, Sweden, and Japan.

The pitchers from which these plants derive their common name are modified leaves that have evolved to form cylindrical tanks (fig. 566) in which rainwater accumulates due to the upright position of the hood portion of their leaves (fig. 567). Only the purple pitcher plant and one other (*Sarracenia rosea*) have the ability to collect and retain long-lasting pools of water that are indeed small aquatic ecosystems. All other species of *Sarracenia* have leaves overtopped by a hood that covers the opening and limits such long-term accumulation of water (fig. 568).

People have long been interested in carnivorous plants, not only because of their strange morphology, but also because of their unusual method of obtaining nutrients—a seeming reversal of roles. Amerindians in Canada referred to pitcher plants as grass-toads because they, like toads, "eat" insects. Pitcher plants, as well as other carnivorous plants, do not "eat" insects; they are perfectly capable of producing "food," that is, carbohydrates, through the process of photosynthesis. However, the important elements that most plants derive from the water absorbed by their roots (e.g., nitrogen, phosphorus, potassium, and calcium) are scarce or lacking in the impoverished soils of their bog and swamp habitats. The evolution of carnivory in plants is an adaptation that has taken place in several distantly related plant families. It enables the plants to obtain these necessary elements in mineral-depauperate habitats. The plants might otherwise survive in these areas, but without the ability to obtain nitrogen, phosphorus, and other

Purple Pitcher Plant

nutrients from insects, arachnids, and occasionally small vertebrates, they would not thrive. Insects, the main prey of the larger carnivorous plants, are composed of more than 10% nitrogen as well as smaller amounts of other nutrients and play an important role in carnivorous plant nutrition. In pitcher plants, as in other carnivorous plants, it is the leaves that serve as the trap, performing the function of capturing insects or other small prey to obtain nutrients for the plant.

Variation exists in the morphology of purple pitcher plants, with plants in the northern and southern parts of its range being classified as distinct subspecies of *Sarracenia purpurea*. The northern subspecies is known as *S. purpurea* subsp. *purpurea* and the southern as *S. purpurea* subsp. *venosa*. Dr. Martin Cheek of the Royal Botanic Gardens Kew compares the two subspecies by saying that the "shape of the pitcher is felty and fat in the southern taxon, and glamorous and slender in the northern," though I have certainly observed many northern plants with rather "fat" leaf tanks. The upper margins of the two subspecies differ, with those of the southern subspecies having a more pronounced wavy margin and overlapping lobes. Where the two subspecies coexist, such as in southern New Jersey, intermediates are found. Color and pattern may also vary; some plants of the northern subspecies are clear green with no red markings at all (forma *heterophylla*), while others range from those with varying amounts of red markings (see fig. 567) to individuals that are completely red (see fig. 566). It has been demonstrated that leaves with greater amounts of red coloration are visited by a greater number of prey insects. The flowers of a southern species, *Sarracenia rosea*, formerly considered a variety of *S. purpurea* (*S. purpurea* var. *burkii*), have distinct pink petals and a pale style. Based on both morphologic and molecular evidence, *Sarracenia rosea* has now been shown to be a separate species. For a discussion of the complicated naming of this species, see the sidebar at the end of this species account.

In nature, the pitcher plant itself may have a life span of up to 50 years, with an individual pitcher capable of living for more than a year. New leaves are produced at the center of the plant on an average of one every 20 days. The main function of the immature leaves of pitcher plants, as with most leaves, is photosynthesis. Older leaves are more fully developed into open pitchers and are better able to attract and trap prey. In one study it was observed that the ability of the pitcher to capture prey reaches a peak between 12 and 33 days after opening and then begins to decline (however, its function as a trap may continue even into its second season, after spending the intervening winter in a dormant state). For the first week after opening, the pH of the accumulated water in the leaf measures slightly alkaline (up to 7.0 in some studies), but thereafter it gradually becomes more acidic, stabilizing at about 3.5 after a month.

While the carnivorous leaves of the well-known Venus flytrap are active traps, moving quickly to capture prey that accidentally touches the trigger hairs on the leaf surface, the leaves of pitcher plant traps are passive traps. They attract their prey with measurable amounts of nectar secreted from glands scattered throughout the leaf but especially concentrated near the rolled top edge of the leaf (fig. 569). Nectar secreted from parts of a plant other than the flower is termed extrafloral nectar. Younger, but mature, leaves are more successful in capturing prey—most likely because extrafloral nectar production is greatest then. While gathering nectar from the rim of the leaves, insects may slip down into the water-filled leaves and drown (fig. 570). Escape is difficult; the leaves are

Fig. 569. An ant crawling along the hood of a pitcher plant risks slipping into the water-filled pitcher below.

Pitcher Plant Family

slippery, and the upper portion of the hood is lined with downward pointing hairs that make crawling out of the pitcher nearly impossible (see figs. 567 and 569). Flying insects are prevented from taking off by the narrow mouth of the pitcher. Eventually insects drown and their bodies are broken down either by the action of bacteria resident in the fluid-filled leaves or by digestive enzymes also present (however, only 1% of visiting insects fall into the pitcher and meet this fate). The pitcher plant's digestive enzymes were investigated by researchers at the University of California, who were able to demonstrate that the leaves produce these enzymes when they first open and continue to do so for about a week. If no insects fall into the pitcher during that time, enzyme secretion stops. However, if later on the leaf is successful in capturing prey (in the form of an insect or other small animal), it responds to the presence of animal protein by resuming enzyme production. This response is specific to those leaves of the plant that contain prey. The restriction of enzyme secretion to only those leaves containing prey is an efficient method of conserving plant resources. As more than 50% of pitchers do *not* capture prey, this represents a substantial saving in energy for the plant as a whole. Among the insects that meet this watery death are ants (see fig. 569). If numerous ants are trapped, their decomposition may contribute to the acidity of the water in the pitcher-shaped leaf since formic acid is released from the ants' bodies as they decay.

When the plant comes into flower, nectar production by the leaves ceases, thus reducing the likelihood of potential pollinators being lured away from the flowers. The striking flowers with their deep red petals (fig. 571) are held above the leaves, where they are likely to be more attractive to pollinators than are the leaves. For an insect to fall into the pitcher and accidentally end up as prey would certainly defeat the purpose of attracting the insects to the flowers in the first place. The petals fall quickly (they are often present on plants in a population for only about two weeks), and thus we most often encounter the flowers after the petals have fallen. Some people may not realize they are looking at only the large persistent sepals. Three tiny bracts subtend the sepals (fig. 572). Insects collect pollen from the numerous stamens (fig. 573) and/

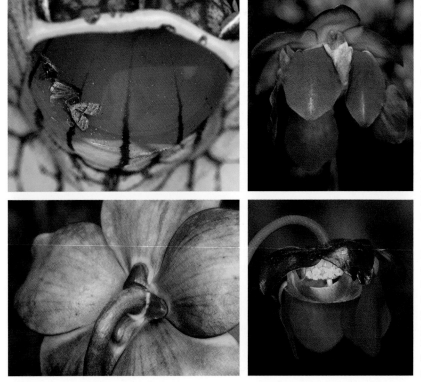

Fig. 570. (Far left) Insects that have drowned in the watery pool trapped within a pitcher plant leaf. The insects may be fed upon by the larvae of a flesh fly at the surface of the water or then sink to the bottom, where they will serve as food for other organisms or be broken down by enzymes secreted by the pitcher and their nutrients absorbed by the plant. **Fig. 571.** (Left) The flower of a pitcher plant has dangling deep red petals that hang between the lobes of the stigma-bearing stylar shield.

Fig. 572. (Far left) A basal view of a pitcher plant flower shows the three small bracts and five sepals. **Fig. 573.** (Left) A purple pitcher plant flower with one petal removed so that the stamens and large, peltate style can be seen. The petals are quickly deciduous.

Purple Pitcher Plant

Fig. 574. The inverted umbrella-like stylar shield and ovary of the flower are visible after the petals and stamens have fallen.

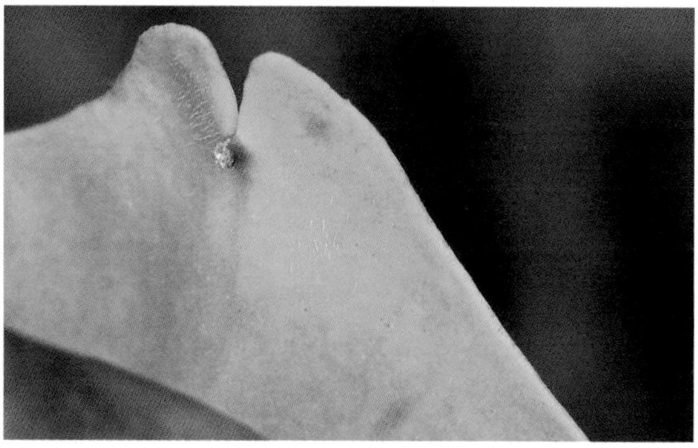

Fig. 575. One of the flower's five small stigmas is visible at an indentation in the shieldlike style. The others would be at similar indentations of the style.

or nectar secreted by both the ovary and the inner surface of the large umbrella-shaped stylar shield (fig. 574). The stigmas themselves are small raised structures located at the indentations of the five lobes of the shield (fig. 575). The petals hang down between the stylar arms (see fig. 571), blocking entrance to the flower and requiring the insect (hopefully covered with pollen from another pitcher plant flower) to enter the flower at the location of one of the stigmas. Videotaped observations have shown that most pollinators come during the day. Far fewer visitors are nocturnal, and those insects that do make night visits do not appear to be taking nectar. It is uncertain as to why they visit the pitcher plants.

Once fertilized, the ovary will develop into a brown capsule that splits open late in the fall, irregularly dehiscing to spill its seeds into the surrounding sphagnum moss where they can germinate and develop into young plants the following year (fig. 576). Viability of seeds is dependent on the amount of carbon produced through the process of photosynthesis, and on the level of nutrients (such as nitrogen) derived from the decomposition of captured prey.

Insects observed visiting the flowers and transporting pollen include bumblebees (*Bombus affinis* and other bumblebee species); small halictid bees, particularly *Augochlorella aurata*; and two species of flesh fly, *Sarcophaga sarraceniae* and *Fletcherimyia fletcherii* (fig. 577). The latter have been observed to use the flowers as overnight roosting places. The relationship of flies with pitcher plants is multifaceted. In addition to roosting and mating in the flowers (and serving as possible pollinators), they utilize the plants as the exclusive hosts for their larvae (fig. 578), which spend almost their entire larval stage in the water-filled pitchers. *Fletcherimyia fletcherii* is one of the very few truly aquatic members of the sarcophagid flies. A viviparous species, the female *F. fletcherii* fly will perch

Fig. 576. (Far left) Young leaves of *Sarracenia purpurea* growing in a mat of *Sphagnum* moss.
Fig. 577. (Left) *Fletcherimyia fletcherii* flies settled on, and flying to, the flower of a pitcher plant where they will roost for the night.

Pitcher Plant Family

Fig. 579. (Left) A hole on the underside of a pitcher plant leaf that was chewed by the larva of a flesh fly ready to leave the leaf to pupate in the surrounding moss.

Fig. 578. (Above) This larva of the sarcophagid fly *Fletcherimyia fletcherii* was found in the water of a pitcher plant leaf. The small, roundish projections protruding from the broader end (near the top of the image) are the spiracles through which it breathes; the narrow, black projections at the lower end are the mouthparts.

Fig. 580. (Above) The mosquito *Wyeomyia smithii* is an obligate inhabitant of pitcher plant leaves during its larval stage.

on the lip of the leaf (usually a new leaf) and deposit a single first-instar larva into the pitcher below. Specialized enzymes in the larva protect it from succumbing to the pitcher's own digestive enzymes. If another larva of the same species is deposited into the same pitcher, the original larva will either destroy the second larva by drowning it or force it to leave the pitcher in order to eliminate competition for resources. The larva develops in the pitcher, remaining at the surface in order to obtain oxygen through its posterior spiracles and feeding upon newly drowned insects and perhaps other organisms at the water's surface until ready to pupate. It then crawls up the side of the pitcher and drops to the ground or chews a hole near the base of the leaf through which it can escape (fig. 579). Pitchers with a hole made in their base will lose all the water above that point and, thus, become less effective as traps. The larva then pupates in the sphagnum moss or soil surrounding the pitcher plant.

If a pitcher plant has been particularly successful in capturing insects, conditions within the leaf may become anaerobic. When the insect inhabitants die, the water develops an unpleasant aroma and turns red due to an abundance of bacteria adapted to anaerobic conditions. The bacteria are photosynthetic and can continue to decompose the prey in the leaves. Other dipterans (members of the fly family) that inhabit the leaves of pitcher plants include larvae of the mosquito *Wyeomyia smithii* (fig. 580), and of the midge *Metriocnemus knabi* (fig. 581). Both these species have a mutualistic relationship with *Sarracenia purpurea*. The larvae of both the mosquito and the midge speed up the decomposition of prey and thus the rate of ammonia production in the leaves. Ammonia and carbon dioxide are taken up by the leaves, which then return oxygen into the water held by the leaves, thereby completing the cycle and helping to maintain an aerobic environment.

Larvae of the three fly family species, all of which are obligates of pitcher plants, can coexist in the

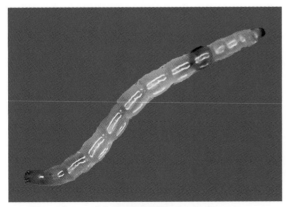

Fig. 581. A larva of the pitcher plant mutualist *Metriocnemus knabi*, a small midge.

Purple Pitcher Plant

same pitchers because, although there is some overlap in the time they are present in the pitcher, each larva feeds on captured prey at different stages of the insect's decomposition and at different levels within the leaf. Thus, the available resources are partitioned. As with the larvae of flesh flies, mosquito eggs are laid in new leaves. The larvae are free-swimming, going to the surface only to breathe. Midge larvae are found in leaves of all ages but are less frequent in newer and older leaves. They inhabit the lowest part of the leaf's tube, feeding on insect remains that have settled on the bottom. They have no need to go to the surface since they absorb their oxygen directly from the water. The midge larvae pupate in a gelatinous mass that adheres to the leaf wall above the water level. Additional pitcher leaf inhabitants include bacteria, rotifers, copepods, mites, and nematodes. Research published in 2000 documented that the presence of *Metriocnemus knabi* midge larvae inhibits the survival of mosquito larvae in species other than *Wyeomyia smithii*. The larval midges were observed to feed on the larvae of *Aedes* and *Anopheles* mosquitoes, but not on those of *W. smithii*. One hypothesis for this discrimination is that the larvae of *W. smithii* have long hairs along their bodies that may make it impossible for the midges to reach the body with their small mouthparts. Mosquito larvae may overwinter in the frozen water of pitcher plant leaves.

Spiders sometimes fall prey to the pitcher plant, but overall their presence is more likely to have a negative effect on pitcher plant success. Some spiders build webs that cover, or partially close, the openings of the pitchers (fig. 582). One study showed that these webs probably reduced the number of functioning trap days by 10%. Crab spiders, sac spiders, and other spider species sit on or near the flowers of *S. purpurea* and capture bees and flies that would otherwise pollinate the plants, thus diminishing the reproductive success of the plant (fig. 583).

Few herbivores feed on pitcher plants, but evidence that some do may be seen by examining the pitchers (fig. 584). The larvae of two species of moth are notable for feeding on pitcher plant leaves. The first, *Exyra fax*, is a relatively common, dull, blotch-marked moth that, during the day, rests within pitcher leaves that are empty of water. Only the males have been observed flying at night; it is thought that the females rarely ever fly. The moths mate within the pitchers, and the female

Fig. 582. A spider web partially closes the leaf of a pitcher plant, unintentionally hindering the leaf from capturing insects.

Fig. 583. The exoskeleton of a spider that was shed on the apex of a leaf.

Pitcher Plant Family

Fig. 584. Damage to a leaf caused by the feeding of an unknown insect.

lays several eggs on the interior wall of the pitcher, generally selecting one of the larger pitchers. The larvae of *E. fax* feed only on the inner layer of the pitcher plant leaf, not chewing through the outer epidermal layer. The distinctive damage pattern, usually on the underside of the leaf, and the presence of light brown frass in the pitcher are evidence of their feeding. Since the female moths are poor flyers, they tend to deposit their eggs in several plants in close proximity, resulting in clusters of damaged plants. The damage does not kill the plant but may result in a decrease in the plant's size in the following year as a result of loss of leaf (photosynthetic) area. The larvae close the mouth of the leaf by spinning a sheet web across the opening or, occasionally, by chewing a circular pathway near the summit of the leaf so that the leaf tissue above, now deprived of water, will wilt and collapse inward, closing the entry. They then feed unseen within the pitchers. Larvae overwinter in the pitcher, and individual larvae, when ready to pupate the following summer, spin individual webs to enclose themselves on the wall of the pitcher plant leaf.

Less is known about the other moth, *Papaipema appassionata*, since it is seldom seen. The moth is considered rare, threatened, or no longer extant in parts of its range. Unlike *E. fax*, it doesn't feed exclusively on *S. purpurea*, but on other species of *Sarracenia* as well. The yellowish-winged adults, seen flying late in the year, lay just a single egg at a time. The ensuing larva, rather than feeding on the leaves, burrows into, and feeds upon, the underground rhizomes. Compared with feeding by leaf-feeders, the effect of these root-feeders is far more severe, causing the leaves to wither and turn black, ultimately resulting in the death of the plant. Evidence of their presence can be seen in the rusty maroon-colored frass that collects in the center of the leaf rosette. When the reddish-brown larvae reach 2 inches in length, they leave the rhizomes and pupate at the base of the plant.

The purple pitcher plant has class II CITES status (the Convention on International Trade in Endangered Species Act), which regulates the traffic of endangered species to protect the species from harmful disturbance. Pitcher plants are threatened not only by habitat loss but also by the harvesting of the plants for the carnivorous plant trade and the harvesting of leaves (primarily of the more southern species of *Sarracenia*) for the "cut flower" trade. Additionally, there has been minor harvesting for medicinal purposes. Native Americans formerly used the rhizomes of pitcher plants in the prevention and treatment of smallpox, a disease introduced by European settlers (recent studies of this remedy proved it to be ineffective). Preparations from pitcher plant leaves were used in easing childbirth and to treat fevers and chills. Early physicians used the rhizomes for their stimulant, laxative, and diuretic properties. The efficacy of these uses has been neither confirmed nor disproven. In the 1940s, a distillate of a suspension of powdered *Sarracenia purpurea* rhizome in an alkaline solution (later to be called Sarapin) used in treating intractable back pain, was analyzed and found to form salts of ammonium chloride and ammonium sulfate. Animal testing showed it to be of help in obliterating pain carried by the C fibers of nerves, which carry input signals to the central nervous system, but not in motor carrying fibers that carry signals away from the central nervous system to the muscles. In a 2001 study, injections of Sarapin in humans showed some positive effect in treating pain of neurologic origin, but further research failed to duplicate the results.

Whether or not compounds from *Sarracenia* are proven to be economically useful to humans, pitcher plants should be admired for their beauty, respected for their evolutionary adaptations for survival in habitats hostile to most other plants, and conserved for all to enjoy.

A puzzle for taxonomists

An early botanist, Carolus Clusius, described the first known collection of *Sarracenia purpurea* in 1601; his accompanying illustration showed only the rosette of leaves. However, Clusius mistakenly put the plant into the same genus as sea lavender, naming it *Limonium peregrinum* (meaning a species of *Limonium* from a foreign land), because its basal rosette of leaves was similar to that of the genus *Limonium*, known from Europe. Sea lavender, however, is a member of the Plumbaginaceae and an inhabitant of salt marshes, with flowers that differ markedly from those of the pitcher plant (fig. 585). The origin of the plant material seen by Clusius is unknown, but it is thought that it might have been collected on one of Jacques Cartier's expeditions to Newfoundland and the Gulf of St. Lawrence in 1534–1541. In the late 1600s, Michel Sarrazin, a French military doctor and botanist working in what is now Canada, sent living specimens to Joseph Pitton de Tournefort, a French botanist, who described the species as constituting a new genus, *Sarracenia* (in honor of Sarrazin). Linnaeus, in his 1753 opus, *Species Plantarum*, a work used as the basis for the names of all plants known at that time, adopted Tournefort's name for the genus, and the plant became known as *Sarracenia purpurea*. The species was later divided into two species by the French-born American botanist Rafinesque (who applied *S. gibbosa* to plants north of Virginia, and *S. venosa* to those in Virginia and south). When Edgar Wherry studied the genus in 1933, he felt that the differences between Rafinesque's two species were not great enough to warrant distinguishing them as separate species, and instead treated them as subspecies of *S. purpurea*: subsp. *gibbosa* and subsp. *venosa*, respectively. Later he modified his views, changing the names to conform to the rules of botanical nomenclature: *S. purpurea* subsp. *purpurea*, applied to the northern plants, and *S. purpurea* subsp. *venosa* applied to the southern plants. Because the specimen upon which Linnaeus based his species name (termed the type specimen) was subsequently lost, a substitute type had to be chosen to represent the species. In 1971 Sidney McDaniel, an American botanist, selected a mid-1700s illustration from Mark Catesby's magnificent work, *The Natural History of Carolina, Florida, and the Bahama Islands*. Mark Catesby was an English naturalist who spent two extended stays in the southern United States and the West Indies. Catesby's description accompanying his illustration of the pitcher plant states, "The Hollow of these Leaves, as well as of the other Kind, always retain some Water, and seem to serve as an Asylum or secure Retreat for numerous Insects from Frogs and other Animals, which feed on them." Linnaeus had seen and mentioned this illustration in his original description of *Sarracenia purpurea*.

Aside from the error in the color of the petals and the shape of the broad style expansion, the use of Catesby's illustration resulted in a taxonomic problem since it portrayed the southern subspecies of *S. purpurea* rather than the northern one originally described by Linnaeus. This goes against the strict code of botanical nomenclature and would have necessitated changing the names of the subspecies, resulting in more confusion. A proposal was made by Martin Cheek of Kew Gardens to conserve the two familiar subspecies names (subsp. *purpurea* and subsp. *venosa*), but to select a different specimen to represent the type of the species. The 1601 plate of Clusius was initially considered since it clearly represented the species as found in the northern part of its range and originally described by Linnaeus. The choice of the Clusius illustration was deemed preferable to the selection of an actual herbarium specimen, such as one collected by Linnaeus's student, Peter Kalm on his North American journeys (and most likely seen by Linnaeus), since it has been documented that Linnaeus had actually seen the Clusius plate, but there was no verification that he had seen a specimen collected by Peter Kalm. If he had, that might have been the specimen missing from the Linnaeus herbarium. Ultimately, since it is preferable to have an actual plant specimen to serve as the type specimen, Cheek selected an herbarium specimen that portrayed the salient features of the species and conserved the subspecies name, *purpurea*, as Linnaeus originally named it. This lengthy, rather convoluted, example demonstrates the difficulties sometimes encountered by taxonomists when trying to determine the correct name for a plant. Much credit must be given to those taxonomists willing to tease out the details of such a complicated history in the interest of accuracy (a 1997 paper on the nomenclatural history of *Sarracenia purpurea* by Cheek et al. is listed in the reference section of this book).

Fig. 585. Sea-lavender (*Limonium carolinianum*), a plant of salt marshes that has a rosette of basal leaves. Because Carolus Clusius perceived a similarity of this basal rosette to the leaves of pitcher plant, he mistakenly placed the pitcher plant in the genus *Limonium*. Sea-lavender is actually in an unrelated family, the Plumbaginaceae.

Carrot Family

Fig. 586. A typical flat-topped inflorescence of Queen Anne's lace (above) and one past bloom (below) that has curled into a bird's-nest form while its fruits mature.

Queen Anne's Lace

Fig. 587. Queen Anne's lace growing along a roadside in Wisconsin.

Queen Anne's Lace; Wild Carrot
Daucus carota L.
Carrot Family (Apiaceae)

The Eurasian plant *Daucus carota* is so ubiquitous along northeastern roadsides that it is assumed by many to be a native plant. Its flowers are admired for their lacelike beauty (fig. 586) and are often included in wildflower bouquets. The seeds of Queen Anne's lace likely were introduced inadvertently to North America and other parts of the world mixed in with the seeds of edible grains—or perhaps they were deliberately brought here for their purported medicinal use. Queen Anne's lace (also known as wild carrot) was recorded as being present in North America as early as the 1600s, but it was not documented in Canadian herbarium collections until the late 1800s. By that time, the plant had already reached pest status in some areas such as Connecticut.

Habitat: Open, sunny meadows and roadsides (fig. 587).

Range: Reported from all states in the United States and all provinces in Canada with the exception of Alberta; it is also not found in Canada's northern territories.

I'll preface this essay with a warning: many of the relatives of Queen Anne's lace—among them poison hemlock (*Conium maculatum*) (fig. 588); fool's parsley (*Aethusa cynapium*) (fig. 589); and the two species of water-hemlock that occur within the range covered by this book, common water-hemlock (*Cicuta maculata*) (fig. 590) and bulbiferous water-hemlock (*Cicuta bulbifera*) (fig. 591)—are highly poisonous and can easily be mistaken for Queen Anne's lace. This is particularly important to remember for those who forage for wild foods since Queen Anne's lace is presumed to be a possible progenitor of the cultivated carrot. While

Carrot Family

I don't recommend eating the roots of wild carrot (those of cultivated carrots are sweeter, more nutritious, and not as fibrous), I feel that it is important to know how to differentiate between *Daucus* and the other genera mentioned here since the look-alikes are deadly. All have compound leaves with sheathing leaf bases and small white flowers arranged in umbels. In technical manuals, members of this family are generally distinguished from each other by the morphology of their fruits, which are not present until season's end and which are not an easy character for the layman to use for identification.

When in flower, Queen Anne's lace and its look-alikes (and most all species in the family) are easily recognized by their distinctive inflorescence (termed an umbel), which comprises several rays arising from a common point, each ray bearing a small umbellet at its apex.

Queen Anne's lace has a large, flat-topped, compound inflorescence with distinctive long, forked bracts beneath the principal inflorescence branches (fig. 592) plus small, usually linear, bractlets beneath each of the smaller umbellets that terminate the umbel's rays. The primary bracts are rough along their margins. However, these identifying features are not present during the plant's first year, when the underground root would more likely be harvested by foragers. When scratched with a fingernail, the narrow,

Fig. 588. (Top) Poison-hemlock resembles Queen Anne's lace enough that it can be mistaken for it—with deadly results if consumed. **Fig. 589.** (Above) Fool's parsley has distinctive downward oriented bractlets on the outer margins of its umbellets.

Fig. 590. (Far left) The leaves of common water-hemlock have toothed margins that distinguish them from the other plants discussed here. **Fig. 591.** (Left) Bulbiferous water-hemlock has very narrow leaves and small bulbils in the axils of the upper leaves. It often has no inflorescences.

Queen Anne's Lace

carrot-shaped white root of Queen Anne's lace smells just like a carrot (fig. 593), but at that point there is only a ground-hugging basal rosette of leaves, making the plant even more difficult to differentiate from the poisonous members of the family. The foliage of Queen Anne's lace is very finely dissected, like that of carrots sold in farmers markets. An alternate common name for Queen Anne's lace, "wild carrot," indicates that it is considered the same species as our garden carrots, but is a different subspecies.

Queen Anne's lace grows in fields and along roadsides especially where the soil has been disturbed. Many plant species have some allelopathic activity resulting from the secretion of chemicals which inhibit the growth of other nearby species; this effect is notably stronger in annual and biennial plants than in perennials. In an experiment that looked at allelopathic effects of 65 different species, both native and non-native, *Daucus carota* was found to produce one of the strongest inhibitory effects. This helps to explain its widespread success in colonizing disturbed soils. It is able to outcompete other nearby species for resources (water and soil nutrients) and thereby develop a large first-year rosette of leaves, which in turn can shade out nearby seedlings.

Among the toxic relatives of Queen Anne's lace is poison-hemlock, which has fewer (ovate to lance-shaped) bracts below its umbels that are *not* forked. Its inflorescence has many rays terminating in smaller clusters of flowers, and if bractlets are present beneath these umbellets, they are smaller versions of those subtending the primary umbel. The leaves are less finely dissected (see fig. 588) than those of Queen Anne's lace. A member of this genus was the "hemlock" responsible for the death of Socrates. Poison-hemlock is an inhabitant of swamps and marshes or wet roadside ditches.

The inflorescence of fool's parsley is distinctive. There are seldom bracts under the primary umbel, but if present, they are linear and not forked. The two to five bractlets below the individual umbellets that make up the inflorescence occur only on the outer side of each umbellet and generally hang downward (see fig. 589). The leaves are similar to those of Queen Anne's lace but shinier. This species is found in waste places.

Common water-hemlock, another denizen of wetlands, has leaves that are compound with individually separate leaflets—with the exception of the uppermost

Fig. 592. The bracts of the primary umbels of Queen Anne's lace are distinguished by their long, forked appearance. This inflorescence is still in bud.

Fig. 593. During its first year, Queen Anne's lace (wild carrot) produces a rosette of feathery leaves and a whitish taproot in which is stored the carbohydrate that the plant will use the following season to produce a tall stem, additional leaves, flowers, and fruits.

Carrot Family

leaves, which may even be entire. The leaflets generally have toothed margins (see fig. 590). Common water-hemlock rarely has bracts under its principal umbel; the many bractlets under the smaller umbellets are narrow and inconspicuous. The sometimes-tuberous roots are generally in clusters. The roots of its relative, bulbiferous water-hemlock, on the other hand, are often not tuberous, and its leaves are very narrow, frequently less than 5 mm wide (see fig. 591). The inflorescences are small or even nonexistent; thus fruits are rarely produced. Bulbous water-hemlock also grows in wetlands.

Queen Anne's lace, fool's parsley, and poison-hemlock are all natives of Eurasia; the two water-hemlocks are the only North American natives among the species mentioned above.

Despite many members of the Apiaceae having poisonous properties, the family also includes numerous edible species. Among the best known are parsnips, parsley, cilantro/coriander (from the same plant, with the citrusy leaves referred to as cilantro and the spicy seeds as coriander), anise, celery, cumin, dill, fennel, and of course, carrots.

The genus name of Queen Anne's lace, *Daucus*, is the name by which it was known in ancient Greece, and in 1753 Linnaeus named the species *Daucus carota*, using for the species epithet an old Latin name applied to carrots. Evidence points to Central Asia as the origin of wild carrots, with the probable wild ancestor from the area now encompassed by Afghanistan and Iran. This area is still the center of diversity for this species. It is likely that a natural variant was selected and bred to increase its palatability and diminish its woodiness. Those efforts resulted in our modern carrot, which may date back to eighth-century Afghanistan. However, prior to that the plant was already widely distributed in other parts of the world. Turkey is recognized as a second center of diversity. Carrot seeds dating back 2000–3000 years have been found in Europe, and the plant is described in a sixth-century copy of Dioscorides' first-century list of herbs and medicines, *De Materia Medica*. It was frequently grown for the medicinal use of its seeds and green leaves before the plant was ever cultivated for food.

Although it is generally assumed that cultivated carrots originated from the fibrous, white-rooted plant now referred to as Queen Anne's lace (or wild carrot),

Fig. 594. Colorful varieties of carrots, often called rainbow carrots, are now easy to find at farmers markets and in many supermarkets. Each color is associated with a different nutritive value.

when modern-day botanists have tried to develop an edible form of carrot from wild *Daucus carota*, they have been unsuccessful. It would seem more likely that there were intermediate forms between the wild and cultivated forms of carrot. Further evidence for this hypothesis is that when cultivated carrots revert to a wild type, they differ in appearance from what we now call wild carrot.

Carrots are an important food source worldwide. Nearly half the total annual production of carrots comes from China, Russia, and the United States. In addition to culinary use, components of carrots are used as a colorant in the cosmetics industry. Carrot root color varies in different parts of the world, ranging from the familiar orange—most likely developed in the Netherlands from Turkish stock in the 1700s—to purple, red, and yellow. Of late, these colorful varieties (fig. 594) are sold at farmers markets and are increasingly available in supermarkets. The variously colored carrots not only make for colorful vegetable dishes and salads but, based on their color, add different nutrients to the diet. Our typical orange carrots have long been known to be an excellent source of beta-carotene, used by the body in the production of vitamin A, important for normal vision. Thanks to the efforts of plant breeders, the amount of beta-carotene in today's carrots is more than double the amount

Queen Anne's Lace

Fig. 595. By looking at the underside of an inflorescence of Queen Anne's lace, it is easy to see the compound umbel structure comprising numerous rays, each supporting an umbellet.

Fig. 596. A typical flat-topped inflorescence of Queen Anne's lace with dark central flowers. Note that the flowers on the edge of the main inflorescence are larger than the flowers toward the center, and the outermost flowers of the umbellets near the edge have larger petals on the outermost side.

present 30 years ago. Carrots of other colors contain pigments credited with different benefits. The anthocyanins in purple carrots provide additional vitamin A and may help prevent heart disease; they also contain beta-carotene. The pigments in red carrots include beta-carotenes and lycopene, a chemical thought to promote a lower risk of some cancers. The xanthophylls and lutein in yellow carrots are linked to cancer prevention and better eye health, and white carrots, while not a good source of health-promoting nutrients, contain considerable fiber, an aid to digestion.

Before the standardization of plant family names requiring that all family names end in -aceae, the Apiaceae family was known as the Umbelliferae, referring to the umbel-shaped structure of the inflorescence seen in most species. To picture an umbel, visualize the underside of an umbrella with its many ribs (called rays in these inflorescences) radiating from a central point (fig. 595). *Daucus*, as with most species in the Apiaceae, has compound umbels with each ray terminating in a smaller umbellet. The stalks of the outer umbellets are longer than those of the innermost, thereby

Fig. 597. Some Queen Anne's lace inflorescences are more dome-shaped than flat.

Fig. 598. The flowers of Queen Anne's lace are sometimes pink in bud but generally become white when they open.

Carrot Family

Fig. 599. Rarely, the flowers of Queen Anne's lace are a deep red in bud. In this photo, most flowers are still in bud, but of those that have opened, some have retained their wine-red color.

producing a flat-topped (or sometimes dome-shaped) inflorescence (figs. 596–597). The tiny flowers in each umbellet are only ca. 1/8 inch in diameter. Occasionally, secondary umbels are produced from lower leaf nodes. The flowers that make up the umbellets are normally white, though in bud, they may sometimes be pink (fig. 598), or even deep red. The flowers within an umbel of Queen Anne's lace are not all alike—those closest to the outside of the inflorescence are larger than the innermost flowers and have distinctly larger petals on the outermost side of the flowers, such that they appear asymmetrical, unlike the radially symmetrical innermost flowers (see fig. 596). On occasion an inflorescence will retain the pink/red color in the fully opened flowers. I happened upon one such individual growing amid several other white-flowered plants not far from where I live. Its flowers, most still in bud, were a striking burgundy color (fig. 599). Of the few flowers that had begun to open, some were white, but others retained their rich wine color. Such deeply colored flowers are rare but have been reported in botanical journals since the late nineteenth century.

A distinction found in many inflorescences of typically white-flowered Queen Anne's lace is the presence of one or more dark flowers in the center of many of the umbels. These flowers contain anthocyanic pigments that cause them to have a darker color—ranging from pink, through red (fig. 600), to a purple so dark that the flowers appear black; these central flowers are sometimes larger than the surrounding white flowers. Such flowers are not found on every inflorescence, but my casual observations inform me that they occur on more than half of them. I have also noted that when plants have been mowed along roadsides and regrow

Fig. 600. Many inflorescences of Queen Anne's lace have a few pink, red, or dark purple (almost black) flowers at their center. Some studies show that these darker flowers attract more insects to the inflorescence.

Queen Anne's Lace

Fig. 601. Queen Anne's lace thrives in open areas, often with other introduced species such as chicory (*Cichorium intybus*) seen in this photo.

to flower again, the new inflorescences lack the darker central florets.

In his poem "Queen Anne's Lace," the Pulitzer Prize winner for poetry, William Carlos Williams (1883–1963), compares a field of Queen Anne's lace to the body of a woman—with each inflorescence comprising "a handspan of her whiteness." The dark central flower is seen as "a purple mole," a blemish caused by "his touch" upon her pure white body.

A summer bouquet gathered from fields or roadsides is bound to include Queen Anne's lace, one of our loveliest wildflowers. It, along with the common ox-eye daisy and chicory (also Eurasian plants), have become widely naturalized and thrive in those open habitats (fig. 601). When Queen Anne's lace is present in large numbers it can give the appearance of a large lace tablecloth spread along the roadside to dry (see fig. 587). Indeed, a simple poem by Mary Leslie Newton, an American educator and writer living in the late nineteenth–early twentieth century, captures this image:

Queen Anne, Queen Anne, has washed her lace
(She chose a summer's day)
And hung it in a grassy place
To whiten, if it may.

Queen Anne, Queen Anne, has left it there,
And slept the dewy night;
Then waked, to find the sunshine fair,
And all the meadows white.

Queen Anne, Queen Anne, is dead and gone
(She died a summer's day),
But left her lace to whiten in
Each weed-entangled way!

Although *Daucus carota* is known most commonly by the name Queen Anne's lace in North America, in its native range of Europe and southwest Asia, it has several other common names including wild carrot, bird's nest, bishop's lace, and devil's plague (a name bestowed by farmers whose fields it had infiltrated).

Carrot Family

Fig. 602. The broad inflorescences of Queen Anne's lace provide a place for insects to feed or to mate. The mating pair of bugs shown here (*Graphosoma semipunctatum*) were photographed on an inflorescence of Queen Anne's lace in Italy, within the plant's native range.

There is some disagreement regarding which Queen Anne is being commemorated in the plant's common name, whether Queen Anne of Denmark (1577–1619) who eventually became Queen consort of King James I of England, or Queen Anne of England (1655–1714). Mary Leslie Nelson ascribed to the second choice, Queen Anne of England, who, in fact, did die on a summer's day in August of 1714, while Anne of Denmark died in March of 1619. In any case, the common presence of a small dark reddish flower or flowers in the center of the white-flowered umbel is said to represent a drop of blood that resulted from Queen Anne's pricking her finger while making lace.

The true significance of these dark, central flowers has been a subject of debate since Darwin's time. Darwin himself felt that the dark flowers had no function. Since that time other researchers have posited various theories: that they help to attract insects (especially flies and small beetles) that view the dark flowers as an indication of either a food resource or a mating opportunity or, conversely, that the dark flowers serve to deter insects, which might perceive them as possible predators. The results of various different studies conflict with each other, with some confirming the attraction hypothesis and others negating it; however, the majority of these investigations did show an increase in visitation to those inflorescences having dark central flowers.

Queen Anne's lace is a generalist when attracting pollinators, providing easily accessible pollen and nectar for many species of insects (among them flies, beetles, ants, bees, and wasps). Its compound umbel functions as a single unit that provides a broad, flat, landing place for insects to graze on pollen or nectar or to find mates (fig. 602). Most visitors eat pollen, but some take the easily accessible nectar. The flowers are generally cross-pollinated by insects but may self-pollinate when pollen from the flowers of one umbel is transported to the flowers of a different umbel on the same plant.

One study done within the native range of Queen Anne's lace showed that small beetles were more numerous on larger inflorescences (which is generally true of larger inflorescences) and on those that had more dark flowers. Inflorescences that had experimentally had their dark flowers removed were subsequently visited by few beetles—but if dead beetles of the same size as the removed flowers were glued to the inflorescences in place of the removed dark flowers, more beetles were attracted. This held true unless a larger species of beetle was substituted for the small ones; in that case the smaller beetles were deterred from visiting, presumably because they perceived the larger beetles as predators or superior competitors for the floral resources. The investigators concluded that the dark flowers served to attract the small beetles, possibly by mimicking them.

Beetles account for a large number of visitors to Queen Anne's lace. In a five-year census conducted in Mississippi, 92 different beetle species were found on the inflorescences of Queen Anne's lace. Examination of the beetles' gut contents revealed that almost all had been consuming the pollen of the flowers, including the two predatory beetles. Flower-loving, long-horned

Queen Anne's Lace

beetles (cerambycids) are especially adapted for feeding on flowers by having elongated heads and hairy mouth parts that aid in retaining pollen as they feed. The long-horned beetles observed on Queen Anne's lace were present only on flowers, not the vegetative parts of the plants, either feeding or mating, sometimes simultaneously (fig. 603).

The presence of so many insects foraging on Queen Anne's lace inflorescences has the effect of attracting predatory insects or spiders, such as ambush bugs and crab spiders. Foraging insects have been shown to spend less time, or avoid completely, Queen Anne's lace inflorescences when an ambush bug is present.

Perhaps the most interesting insect/plant relationship involving *Daucus carota* (as well as many of its familial relatives) is that of the black swallowtail butterfly (*Papilio polyxenes*), the state butterfly of both New Jersey and Oklahoma. Black swallowtails have a range similar to that of Queen Anne's lace. The caterpillars of this yellow-patterned black butterfly feed almost exclusively on members of the Apiaceae, including those grown for food in gardens (e.g., carrots, parsnips, dill, etc.). They eat the leaves of the plant but show a preference for flowers and young fruits. It has been documented that the caterpillars grow larger and are more successful if they feed on Queen Anne's lace and wild parsnip as opposed to poison-hemlock. Much like the monarch butterfly's exclusive dependence on toxic glycoside-bearing milkweeds as a larval food, black swallowtail caterpillars feed only on toxic furocoumarin-containing plants, including the Apiaceae under discussion here. Yet, unlike the monarch larvae, the caterpillars of black swallowtails do not sequester the toxins. Instead, black swallowtails have evolved the ability to metabolize these furocoumarins at a very high rate, utilizing enzymes (cytochrome P450) to detoxify them in the process, and eliminating the residue quickly from their bodies. The mature caterpillars are among our most strikingly patterned: bright green (or creamy white) with transverse rows of black dotted with yellow spots (fig. 604). They differ markedly from the earlier instar larvae, which more closely resemble brown and white bird droppings and are thus usually overlooked by birds. The caterpillars are not toxic if eaten by birds, but they can deter predators by everting a pair of foul-smelling orange horns from behind their heads when threatened. The black and yellow adult butterflies are also not toxic but are usually unmolested by birds because they resemble pipevine swallowtail butterflies, which *are* toxic. Black

Fig. 603. Some species of long-horned beetles are known as anthophilous, meaning that they are flower lovers. They often have specialized feeding parts that allow them to more efficiently feed on pollen. The beetles here (*Typocerus lunulatus*) were feeding and mating at the same time.

Fig. 604. The later instars of black swallowtail caterpillars are colorfully marked. They feed on Queen Anne's lace and other members of the Apiaceae. The one shown here was feeding on the few remaining fruits of *Angelica atropurpurea*.

Carrot Family

swallowtail caterpillars do not congregate in large numbers, and to my mind, it is worth sacrificing some of your garden plants to these caterpillars in order to enjoy the beauty of the adult butterflies that will develop. The introduction of Queen Anne's lace may have benefited black swallowtail butterflies by providing an abundant source of larval food.

The presence of many butterfly-supporting wild members of Apiaceae growing in close proximity to cropland has aroused concern among environmentalists with regard to the use of *Bt* corn, that is, corn that has been genetically modified to contain the bacterium *Bacillus thuringiensis*. *Bacillus thuringiensis*, used to prevent damage by the larvae of European corn borers, is a known killer of lepidopteran larvae. It has been used in fighting outbreaks of gypsy moths and other "pest" caterpillars.

In an experiment designed to test the effect of *Bt* corn pollen on the larvae of black swallowtails, no difference in survivorship of the caterpillars was noted between those that fed on pollen-covered leaves of a particular variety of Monsanto *Bt*-treated corn and those that fed on non-*Bt*-treated corn. Yet, experiments with other Monsanto and Novartis *Bt* products resulted in caterpillars of a smaller size with a much lower survivorship rate. Additional studies are needed to assess more accurately the effects of these products under field conditions and on other larval lepidopteran species.

Daucus carota leaves contain steroids and antibacterial properties, which serve as a parasite repellent and egg-laying inhibitor for blood-sucking mites. The leaves reportedly are used by European starlings in nest building, the steroids and antibacterial compounds helping to reduce the number of nest parasites, which in turn decreases blood loss in the nestlings and results in higher survival rates.

Although I have picked many stems of Queen Anne's lace and have never experienced an adverse reaction, the leaves of the plant, especially when wet, may cause skin irritation in some people when the contacted skin is exposed to sunlight. This is true of some other members of the Apiaceae as well (e.g., wild parsnip, giant hogweed, and cow-parsnip), which can cause an even more painful rash. The reaction is caused by the presence of furocoumarins in the leaves, phytochemicals that enter the nuclei of skin cells and form a bond with the DNA, causing the cells to die. The skin, thus unable to protect itself from sunlight, is subject to a blistering, burning inflammation (equivalent to a second-degree sunburn) of the affected area.

Like many other toxic plants, Queen Anne's lace (known in the herbal trade as QAL) has been used traditionally for medicinal purposes. Since the time of Hippocrates (more than 2000 years ago), the seeds have been used as a means of birth control. Recent research on this property, carried out in China using rats as the study animals, has shown that the seeds actually act as a "morning after" abortifacient, blocking the synthesis of progesterone, a hormone necessary to prepare the lining of the uterus for implantation of the fertilized egg. Much more research is needed before the safety and efficacy of this use can be determined.

While generally considered biennial plants, *Daucus carota* may actually complete their life cycle in anywhere from one to three (or occasionally more) years depending on environmental conditions. Seeds require large amounts of water in order to germinate (which generally occurs within two years); however, the seeds may remain viable in the soil for up to seven years. After germination, a cluster of lacy leaves forms a basal rosette. These leaves use the sun's energy to produce carbohydrates that are then stored in the plant's carrot-shaped, white root. Queen Anne's lace overwinters in this stage and resumes growth the following spring. Its flowering year is dependent on the size of the rosette and the root, and on how quickly the plant is growing. When conditions are right (usually in June–July) the plant begins to elongate (bolt), developing a stem, additional leaves, and flowers. Queen Anne's lace commonly reaches 1 meter tall.

When the flowers open, their stamens are functional immediately. Once the pollen is shed or removed by an insect, the stamens fall. The stigmatic surfaces on the styles then become receptive and the flowers' carpels begin to mature. Such flowers are called protandrous, meaning that their male parts release their pollen before the female parts are receptive. This system aids in preventing self-pollination.

About five weeks after flowering, the umbel, now containing hundreds of developing fruits, begins to curl upward, forming a bird's nestlike structure (the source of another of the species' common names) (fig. 605). The degree of curling is affected by the relative humidity of the air; the greater the moisture absorbed from the air, the more tightly curled the

Queen Anne's Lace

Fig. 605. Once the flowers have fallen and the fruits have begun to mature, the infructescence curls inward, forming a structure resembling a bird's nest.

Fig. 606. The fruits of Queen Anne's lace are spiny along their ribs and can easily catch on the fur of passing animals or the feathers of birds, but many remain on the plant into winter, when the wind blows them across the crusty snow.

inflorescence becomes; a response termed hygroscopic movement. Some inflorescences respond more rapidly than others to changes in relative humidity.

Each fruit is a schizocarp, that is, a fruit which, when mature, splits into separate carpels (termed mericarps). In Queen Anne's lace each of the two mericarps contains one seed. The mericarps are spiny along their ribs and easily catch onto the fur or feathers of passing animals or the clothing of humans (fig. 606). Those umbels that close more tightly in humid conditions may retain mericarps into winter when they can be dispersed by strong winds that blow them long distances over the snow. By winter's end, the old inflorescence is generally bare of fruits (fig. 607). *Daucus carota* is a monocarpic species, that is, after flowering, it dies; thus, all reproduction is by seed.

Queen Anne's lace is present, often in large populations, in most states east of or bordering the Mississippi River and along much of the West Coast. According to the WeedUS database, the species is considered highly invasive in the states of Minnesota, Michigan, Ohio, Washington, and SE Alaska and invasive to a lesser degree in Georgia, Florida, Illinois, Kentucky, Maryland, South Carolina, and Tennessee. In states where efforts have been made to remove or diminish Queen Anne's lace through the use of herbicides, some plants of *Daucus carota* have developed a resistance to the chemical used (2,4-D); a resistance that is variable even within a population of *D. carota*.

If one looks closely at any of our common wildflowers, even those that some consider weeds, a world of interesting relationships demonstrates the interconnectedness of our natural world.

Fig. 607. By winter's end, almost all the fruits have been dispersed.

Orchid Family

Fig. 608. A two-flowered showy lady-slipper plant (*Cypripedium reginae*) growing in a northern fen.

Showy Lady-slipper

Fig. 609. The flowers of showy lady-slippers do not attract many visitors, but the majority of insects that *do* visit are not effective pollinators—including this eastern tiger swallowtail butterfly (*Papilio glaucus*), whose size and activity do not result in contact with the reproductive structures.

Showy Lady-slipper

Cypripedium reginae Walter
Orchid Family (Orchidaceae)

The flowers of showy lady-slipper easily attract the eye and the admiration of the lucky person who encounters them in a fen or bog in the northeastern part of the United States or adjacent Canada. Showy lady-slippers are the tallest of our native orchids and, most agree, the most beautiful.

Habitat: Openings in hardwood and coniferous fens, streamsides, and wet meadows; most commonly underlain by alkaline soils.

Range: Maine to North Dakota and south in the higher, cooler mountains to North Carolina, Tennessee, and Arkansas; in Canada from the Maritimes west through Saskatchewan.

The tallest of our Northeastern lady-slippers, the showy lady-slipper can reach a height of nearly a meter. To my eye, it is also the most beautiful of our slipper orchids. Although the flowers of the pink lady-slipper (*Cypripedium acaule*) are larger, the showy lady-slipper's size and combination of a pink pouch-like lip with white petals and sepals is striking; the common name "showy lady-slipper" is well deserved. Because it is primarily a denizen of wetlands (fens and swamps), showy lady-slipper is not commonly seen; however, when first sighted it never fails to elicit expressions of wonder and delight. Plants have either one or two (fig. 608) (or rarely three) flowers. Flower size has been shown to have no relation to the height of the stem or the number of leaves.

The derivation of its scientific name, *Cypripedium reginae*, is appropriate: *Cypripedium*, from the Greek *cypris*, refers to the island of Cyprus, and *pedilon*, from the Greek for shoe or slipper, to the shape of the flower. This is a reference to the myth that Aphrodite,

Orchid Family

the Greek goddess of love, arose fully grown from sea foam (*aphros* in Greek) off the island of Cyprus—thus, "the slipper of the Cyprian." The species name, *reginae*, means "of the queen," and showy lady-slipper is, without doubt, a flower worthy of a queen.

Like all other members of the genus *Cypripedium*, showy lady-slipper's beauty is deceitful; it attracts insects to its flowers with its flashy colors and a faint aroma of methyl salicylate but offers them no reward for their visit. Many insects learn this after a few disappointing visits, and thus most lady-slippers have few visitors, especially visitors that might be effective pollinators. The paucity of visitors is a contributing cause to the low reproduction rate of this species (and of many other orchid species as well). Some orchid species are able to be pollinated by members of only a single genus of insects, or even by only one gender of a single species, as in some tropical species that rely solely on male bees that collect aromas from the flowers to use in their mating displays.

I have personally witnessed only three species of insects visiting the flowers of showy lady-slipper, and none were the coevolved pollinators. Research papers report that midsized bees are the optimal pollinators because of their size, their hairy bodies, and their activity while within the flower. So, I was surprised when I witnessed repeated visits by eastern tiger swallowtail butterflies (*Papilio glaucus*). The butterflies probed behind the column (fig. 609) or into the lip with their proboscises, as if they were searching for nectar, or more likely imbibing droplets of dew (fig. 610). Sometimes the same butterfly would visit a few different flowers before flying off. Other insects I have observed visiting showy lady-slippers are a bumblebee (*Bombus* sp.), which was too large to exit through the rear opening and thus unable to come into contact with the reproductive structures, and a honeybee (*Apis mellifica*). Honeybees, a species introduced from Eurasia, have not coevolved as pollinators of showy lady-slipper, but they are opportunistic and have become successful pollinators of many of our native flowers. Because they are of the proper size (similar to showy lady-slipper's native bee pollinators, such as leafcutter bees), they should be able to transfer pollen effectively from one showy lady-slipper flower to another. The one honeybee that I observed landed on the outer rim of the pouch and then crawled into it via the large opening on the top of the lip. (Like many orchids, the flowers of showy lady-slipper are resupinate; that is, the flower, while still in bud, twists 180° before it opens [fig. 611]. Thus, the labellum [pouch/lip]—technically the upper petal—becomes positioned at the bottom of the flower, with the opening easily accessible to visiting insects.) The flower functioned as a trap, with the honeybee unable to exit by its entry route because of the inrolled margin of the lip opening. It buzzed frantically about within the pouch-like lip for close to two minutes before following the

Fig. 610. An eastern tiger swallowtail butterfly with its proboscis inserted into the flower of a showy lady-slipper. Butterflies are attracted by the size and brilliant color of the flower.

Fig. 611. Prior to opening, the flower bud of showy lady-slipper twists 180° so that the lip (or pouch) will be on the bottom of the flower with its opening facing upward.

Fig. 612. The magenta lines seen on the interior of the lip may serve to guide insects to the rear exit holes through which they can escape.

Showy Lady-slipper

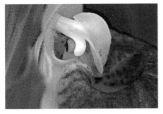

Fig. 613. (Left) A close-up of the side view of a showy lady-slipper flower showing an exit hole and a glossy anther and stigma (white) hidden behind the staminode (the flat white and yellow structure that hangs over the opening of the lip). **Fig. 614.** (Right) A group of showy lady-slippers growing in association with cinnamon fern (*Osmunda cinnamomea*) in an alkaline wetland.

contrasting parallel lines at the bottom of the pouch to the exit holes at the rear of the flower (fig. 612). The bee climbed up the back surface of the pouch and attempted to exit through one of the smaller openings on either side of the column—where the anthers and stigma are located. By following this route, the bee passed through a narrow area where it came in contact with the stigma (fig. 613). Any pollen it had carried into the flower would have been scraped off by the edge of the stigma. A pollinator must squeeze past the anthers as it exits, thereby picking up more pollen on the hairs of its dorsal (back) surface. After three tries, the bee that I watched finally escaped and flew off—with a tale to tell back at the hive. Occasionally a bee will "cheat" and escape by chewing a hole in the lip.

Reported visitors to showy lady-slipper include flies, beetles, and various small bees and butterflies. In a detailed study carried out in Missouri, Dr. Peter Bernhardt and others (see references) found that despite visits by flies, beetles, skipper butterflies, and different-sized bees, the only effective pollinators were those bees that measured about 4.5 mm wide × 3.5 mm high, a size that caused them to come into contact with the pollen masses as they exited the flowers through one of the smaller openings at the back of the flower's lip. Unlike many other orchids, lady-slipper orchids have their pollen in loose masses rather than in coherent masses known as pollinia (see chapters on the fringed orchids and helleborine). The Bernhardt et al. study found several bees to be effective pollinators; included were six bees in five species belonging to four different genera: *Anthophora*, *Megachile*, *Hoplitis*, and *Apis* (honeybees), all just slightly larger than the flower's rear exits, and thus having to squeeze out past the reproductive structures. Bumblebees were too large to fit through the smaller openings and left through the large pouch opening, and other bees were too small to make contact with the anthers if they left through the smaller openings. In a Vermont study, syrphid flies of a certain size were found carrying pollen of *Cypripedium reginae* and may be effective pollinators as well.

Fig. 615. Showy lady-slipper flowers with pale pink lips.

As in other orchids, the fertilized ovaries develop into capsular fruits that split along their seams to release many thousands of dustlike seeds they contain (estimated to be 48,000 per capsule).

Although visitors are few, more nonpollinating insects visit the flowers than do true pollinators. They are attracted to showy lady-slippers by the large, colorful flowers and a slightly sweet aroma. Some insects, such as flower beetles, may feed on the floral parts, whereas others, such as skippers, may come to drink from the lip after rain. Other insects may just land on the flowers by chance.

Cypripedium reginae is most commonly found in fens (calcium-rich wetlands)—often in large numbers (fig. 614); fens may occur either as open meadows or as forested wetlands. Showy lady-slippers also grow in white cedar swamps and other wetlands with calcareous ground water. The saclike lips of the flowers display a range of colors—from pale pink (fig. 615) to a deep magenta (fig. 616). Occasionally pure white forms

Orchid Family

Fig. 616. Two showy lady-slipper plants, one having a flower with a medium pink floral lip and the other with a deeply colored magenta lip.

Fig. 617. A showy lady-slipper with a white flower (a faint blush of pink can be seen near the opening to the lip).

Fig. 618. The glandular hairs seen in this close-up of a flower stalk at the top of the image may cause an irritating rash.

occur (fig. 617); such flowers are known as *C. reginae* forma *albolabium*, or forma *album*. However, the pure white-flowered form is not recognized as distinct from the straight species by *Flora of North America* authors.

Showy lady-slipper's leaves are strongly veined and clasp the stalk with their bases. The glandular hairs that cover all parts of the plant (fig. 618) contain a compound known as cypripedin, which may cause a rash on the skin of those who dare to pick or dig these lovely natives of the fens. The collecting of showy lady-slippers, as well as most other orchids, is a major factor in their decline. Such stolen beauty is usually short-lived, for when planted in a garden setting with soils differing from those of their preferred habitat and, thus, lacking the symbiotic fungus on which the orchids depend for some of their nutrients, they will eventually die. Other factors contributing to the decline of lady-slippers include the draining or pollution of the water of their wetland habitats and their apparent tastiness to our burgeoning population of white-tailed deer. Deer are credited with totally extirpating populations of showy lady-slippers at sites in Michigan. Showy lady-slipper is a rare species throughout much of its range. The only persons permitted to disturb the plants are those having state or federal permits to collect the plants for research purposes. When plants are offered for sale, one should ascertain that they were grown from tissue culture rather than wild-collected. The plants are not only difficult to grow, but also expensive due to the long period of time necessary for the plants to reach flowering size. Thus, harvesting plants from the wild for a quicker reward is a temptation for the unethical seller. Plants of blooming size grown using tissue culture methods sell for a minimum of $40, and more than four times that much for exceptionally large, mature plants that are at least 10 years old.

Showy lady-slipper has long been a favorite wherever it grows. In her 1900 book, *Nature's Garden*, Neltje Blanchan describes the flowers thus: "Wine appears to overflow the large white cup and trickle down its sides. Sometimes unstained, pure white chalices are found." It was called by the early North American botanist Asa Gray, "the most beautiful of the genus," in his *The Manual of The Botany of the Northern United States Including the District East of the Mississippi and North of North Carolina and Tennessee* (1890; revised edition of earlier publications from 1848, 1856, and 1867). And in John Burroughs' 1884 book, *Riverby*, he waxes poetic (if a bit flowery) when he speaks of first sighting showy lady-slippers: "Never had I beheld a prettier site—so gay, so festive, so holiday-looking.

Showy Lady-slipper

Were they so many gay bonnets rising above the foliage? or were they flocks of white doves with purple-stained breasts just lifting up their wings to take flight? or were they little fleets of fairy boats, with sail set, tossing on a mimic sea of wild, weedy growths? Such images . . . only faintly hint its beauty and animation."

Because of its beauty and its commonness in the northern state of Minnesota, showy lady-slipper was designated its state flower in 1902. No other state has named an orchid as its state flower (but in the Canadian province of Prince Edward Island, showy lady-slipper is similarly honored as the provincial flower). The orchids were once so common in Minnesota that the flowers were frequently used to decorate church altars throughout their blooming season. However, in 1925 it was recognized that the orchids were being depleted, and a ban was passed forbidding the digging of the plants or the picking of their flowers. Within the entire range of this beautiful species, it is considered "secure" (i.e., not threatened) only in Ontario.

The conditions in fens where showy lady-slippers are found provide favorable habitat for other orchids as well. One may encounter yellow lady-slipper orchids (*Cypripedium parviflorum*), which have finished blooming by the time the showy species comes into flower (at least in the southern part of their range); and the tall northern white orchid (*Platanthera dilatata* var. *dilatata*) (fig. 619), which flowers concurrently with showy lady-slippers. *Platanthera dilatata* has comparatively tiny, pure white flowers that, despite their small size, emit a sweet fragrance that can be perceived from several yards away; in fact, another common name for this plant is "scent bottle." Fens are always interesting places to explore for both plants and certain animals (such as bog turtles) that share this habitat. Other plant associates may include pitcher plants, marsh marigold, Culver's root, foamflower, water avens, and false hellebore (*Veratrum viride*)—with leaves remarkably similar in appearance to those of showy lady-slipper (fig. 620)—as well as various wetland ferns (fig. 621).

If you are fortunate enough to live close to one of these special places, pull on your boots and visit in early summer—always being cautious not to tread on the special plants that live there. You may be rewarded by an audience with the queen.

Fig. 619. (Left) *Cypripedium reginae* growing with tall northern white orchid (*Platanthera dilitata*) in a rich fen habitat.

Fig. 620. (Right) False hellebore (*Veratrum viride*) (in front) flowering in the same habitat as showy lady-slipper. Note the similarity of the leaves.

Fig. 621. A view of a fen with showy lady-slipper orchids and cinnamon ferns visible. Sensitive fern is also present in this fen.

Mallow Family

Fig. 622. The large pink flowers of *Hibiscus moscheutos* brighten a freshwater marsh in New York. Also visible are several immature fruits enclosed in their accrescent sepals and a day-old wilted flower (lower right).

Swamp Rose-mallow

Fig. 623. The white flowers with crimson "eyes" belong to swamp rose-mallow, seen here along a boardwalk in a brackish marsh in Cape May, New Jersey. One of its companion plants is the pink-flowered seashore-mallow (*Kosteletzkya virginica*), another member of the Malvaceae.

Swamp Rose-mallow
Hibiscus moscheutos L.
Mallow Family (Malvaceae)

The mallow family includes many species with beautiful flowers, the showiest of which, at least in the Northeast, is swamp rose-mallow (fig. 622). As the name suggests, swamp rose-mallow is a native of wetlands, inhabiting the margins of brackish and freshwater marshes (fig. 623). The genus name, *Hibiscus*, is derived from an ancient Greek name for a member of this plant family, and the epithet, *moscheutos*, from its musky smell. The Malvaceae include species with economic value, either as ornamentals (e.g., colorful, tropical *Hibiscus* species and their cultivars), as sources of fiber (e.g., cotton [*Gossypium* spp.] and jute [*Corchorus* spp.]), and as food (e.g., okra [*Abelmoschus esculentus*]). The family has recently been expanded to include many species that formerly had been placed in other families, notably the Sterculiaceae (the chocolate family), the Bombacaceae (the kapok family), and the Tiliaceae (the linden family), according to the current classification in *Flora of North America* (FNA), an ongoing project to classify and describe all species occurring in North America north of Mexico. *Hibiscus moscheutos* is considered in the FNA treatment a single species comprising two former species, *Hibiscus palustris* (now *H. moscheutos* subsp. *moscheutos*) and *H. lasiocarpos* (now *H. moscheutos* subsp. *lasiocarpos*).

Habitat: Brackish and freshwater marshes.

Range: New York, Massachusetts, and southern Ontario west to southern Michigan, southern Wisconsin, and southern Missouri, then south to northern Florida and the states bordering the Gulf Coast. Small populations occur in Kansas, Oklahoma, Texas, and New Mexico, and there are populations of *H. moscheutos* subsp. *lasiocarpos* in California.

The large, colorful flowers of swamp rose-mallow (fig. 624) give it the appearance of a tropical plant, one that gardeners would covet for their gardens. And, indeed, cultivars of this flashy native have been developed to satisfy that yearning. Because the native species inhabits wetlands, those sold for use in gardens

Mallow Family

Fig. 624. (Left) The flowers of *Hibiscus moscheutos* are one of our largest and showiest native wildflowers, and as such are appreciated for their horticultural potential. Three tightly closed buds can be seen at the bottom of the image. Pink-flowered plants of this species are more common in the northern part of its range, and white-flowered plants in the southern part (both may or may not have red centers). **Fig. 625.** (Below) Many cultivars of swamp rose-mallow are sold for use in gardens. They have often been hybridized to make them more tolerant of a range of soil types and to enlarge and reshape the flowers.

Fig. 626. (Above) Other members of the Malvaceae have also found wide horticultural use. Shown here is a yellow-flowered hollyhock (probably a hybrid with *Alcea rosea*), which is of Asian origin but was introduced into Europe as early as the 1500s and brought to North America by early settlers.

Fig. 627. Rose-of-Sharon (*Hibiscus syriacus*) is a commonly grown shrub in eastern North America; it is of Chinese origin.

Swamp Rose-mallow

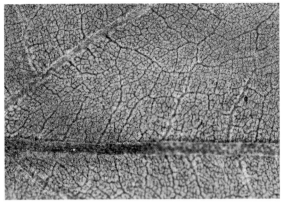

Fig. 628. (Left) Musk-mallow (*Malva moschata*) is one of many pretty, pink-flowered malvas used in gardens. All our mallows have been introduced from Europe/Eurasia and have naturalized in our landscapes. **Fig. 629.** (Above) The lower side of this leaf of swamp rose-mallow shows the hairiness found on the undersurface, particularly near the veins. With a 10× hand lens the starlike tufts of hairs (a character of Malvaceae) are visible.

are often hybrids able to tolerate a range of soil types (though they must be kept well watered). The hybrids have been bred to be even more showy than the native species, with larger, saucer-shaped flowers in a variety of colors (fig. 625) and sometimes purple-colored leaves. Both in the wild and as a garden plant, swamp rose-mallow provides a burst of color in late summer, with a bloom period that peaks from mid-August through early September. Because of its beauty, swamp rose-mallow and its cultivars have been introduced to other regions of the world, such as Europe and Asia.

Many other species in the mallow family are also used in gardens, among them, common hollyhock (*Alcea rosea*) (fig. 626), a species most likely of Chinese origin and imported into Europe as early as the fifteenth century; rose-of-Sharon (*Hibiscus syriacus*) (fig. 627), a commonly planted shrub introduced from Asia; and various mallows (*Malva* spp.) (fig. 628). As seen in figs. 626–628, the floral plan of these relatives is similar to that of swamp rose-mallow, with a style exserted from a united staminal tube surrounded by five petals. As long ago as the Egyptian era (2000 years BCE), the roots of a plant known as marsh-mallow (*Althaea officinalis*) were boiled with honey to yield a mucilaginous substance used in the making of the original marshmallows. Because of the difficulty of making these sweets at that time, marshmallows were a treat reserved for royalty. Today's marshmallows use gelatin and/or egg white rather than mallow roots as the thickening agent and are mass produced, making them available to all. Other members of the family share this mucilage-producing quality—for example, the vegetable okra, which is used to thicken gumbo.

Belying its exotic-looking flowers, our native swamp rose-mallow is cold-tolerant and able to survive northeastern winters in all but the coldest regions. It is also a prolific seed producer with seeds that establish easily. *Hibiscus moscheutos* grows to be 1–2 meters tall, thriving in open, sunny situations with a preference for sandy or silty soil having some organic matter.

Swamp rose-mallow stems are hairy on the upper portion and usually smooth below. The leaves are palmately veined, that is, having their main veins arising from a common point at the base of the leaf. The leaf surfaces are hairless and bright green above but lighter in color and covered with tufts of stellate (starlike) hairs on the undersurface (fig. 629). Although these hairs are difficult to see in the image of the underside of the leaf in fig. 629, they can easily be seen by viewing the lower surface of a leaf with a 10× hand lens. The leaves are broadly lance-shaped or even wider and shallowly lobed with leaf margins that are crenate (having rounded teeth) (fig. 630). Swellings (pulvini) are present at both ends of the petiole, allowing

Mallow Family

Fig. 630. The palmate venation of a rose-mallow leaf is evident in this image, as is the crenate margin of rounded teeth.

Fig. 631. In *Hibiscus* and many other Malvaceae, the male parts of the flowers (the stamens) are fused into a tube of staminal filaments with only their anther-bearing tips diverging from the tube; the tube surrounds the style, the apex of which emerges from the tube and splits into five branches, each bearing a stigma (the receptive female organs).

the leaf to assume different orientations, perhaps to capture more sunlight as the sun moves across the sky during the day.

The 7–12 cm (3–5 in.) diameter flowers are not only attractive but also curious in form. In *Hibiscus* the filaments of the stamens are fused for most of their length into a long white tube with only the apical, anther-bearing portions of their filaments arising free from the tube at multiple points along its length. The style is enclosed within the staminal tube, ultimately emerging from the top where it then splits into five branches, each branch capped by a rounded stigma (fig. 631). The male and female organs are thus separated, reducing the chance of self-pollination in this self-compatible species. This spatial separation of male and female parts is known as herkogamy.

Flowers are borne on a stalk (the pedicel) that has a thickened joint near its midpoint (fig. 632). A leaf may arise from the pedicel as well. The 5-petaled flowers are shaped like a broad funnel and colored pink or white, commonly with a deep red center at the base of the petals (fig. 633). Usually only one or two flowers open on a stem on any given day, but since a plant has many stems arising from its base, there may be several flowers open simultaneously. The flowers remain open for only one day, then wilt and soon fall. Subsequent flowers arise from, or just below, the petioles (stalks) of the leaves. In bud, the flowers are enclosed within a 5-parted calyx, the sepals of which are joined for about half their length. This softly hairy calyx is persistent after the flower falls, remaining attached even after the fruit matures. Fruits are ovoid capsules with a pointed apex; they split apart along the middle of each of their five locules to reveal the numerous kidney-shaped seeds, which are attached to their walls by short stalks called funicles (fig. 634). There may be sparse hairs along these sutures. The seeds themselves are hairless (unlike those of cotton and kapok) but patterned with an intricate, pebbled appearance, usually termed verrucose to describe its tiny wartlike projections (fig. 635). By movement of the plant by wind, water, or other means, the seeds are gradually shaken from the fruits, falling to the ground or into the water beneath. Moving water, especially high tidal currents, can disperse the buoyant seeds for long distances, thus maximizing genetic diversity in newly established sites when seeds from many different stands are carried to the same area. Birds, including ducks, consume the seeds.

Swamp Rose-mallow

Fig. 632. (Above left) The sepals have been pulled down on an immature fruit of *Hibiscus moscheutos* to reveal its more or less ovoid shape with a pointed apex that is a remnant of the style. The rounded bulges on the fruit's surface are caused by the developing seeds inside the capsule. Note the thickened joint on the pedicel. **Fig. 633.** (Above right) Both pink and white flowers of *Hibiscus moscheutos* may have bright red markings at the bases of their petals, forming a "crimson eye." The small bright pink flower to the left of the *Hibiscus* flower is that of seashore-mallow (*Kosteletzkya virginica*).

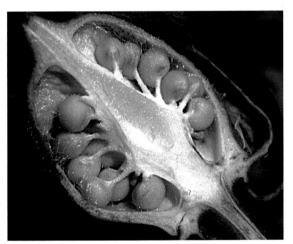

Fig. 634. A longitudinal section of a swamp rose-mallow fruit shows how its many seeds are attached within each chamber (locule) of the ovary. This type of attachment is called axile placentation.

A character that distinguishes many species as belonging to the family Malvaceae is the presence of an epicalyx beneath the normal leafy calyx. The epicalyx comprises a whorl of green bractlets, often 10–14 in this species (fig. 636). The form of the bractlets differs

Fig. 635. Several seeds of *Hibiscus moscheutos*. Small holes of undetermined origin (possibly exit holes of the rose-mallow beetle) can be seen in a few of the seeds. Note the finely sculptured pattern of the seed surface.

Mallow Family

Fig. 636. Here, it is easy to see both the broad, leaflike sepals that remain attached to these young fruits of swamp rose-mallow and the subtending epicalyx, comprising linear bractlets.

Fig. 637. Note the forked tips of the bractlets of the epicalyx that subtend the calyx of *Hibiscus bifurcatus*, an Amazonian species.

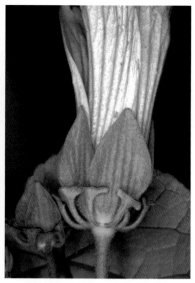

Fig. 638. This similar-appearing Amazonian species, *Hibiscus sororius*, can easily be told from that shown in fig. 637 by its shorter epicalyx, the apices of which terminate in a knobbed T-shape.

from species to species. Those of *Hibiscus moscheutos* are linear and curve upward, surrounding but separate from the calyx. In var. *lasiocarpos* the bractlets have short, fine hairs along their margins. As a further example, illustrated in figures 637 and 638, are two species of *Hibiscus* I have encountered in tropical regions that have similar flowers but can be easily told apart by the shape of their bractlets.

There are about 350 species of *Hibiscus* worldwide, with 21 of them found in North America; all but a handful of those are native. Within the Northeast the only two native species are *H. moscheutos* and *H. laevis*. Species diversity increases in warmer latitudes, and garden escapes of tropical species cultivated in the Southeast are occasionally found in the southern reaches of our greater Northeast region. Examples include *H. radiatus* from southeast Asia, cultivated for fiber; and roselle (aka sorrel) (*H. sabdariffa*), a species which most likely originated in Asia but is widely grown in the American tropics as both a fiber source and for its fleshy red calyx (fig. 639), which is used to add flavor and color to beverages (fig. 640). Roselle is also being investigated for a variety of possible pharmaceutical uses.

If you stop to admire the flowers of *Hibiscus moscheutos*, you are likely to observe insect visitors—especially on the typical warm, sunny days that characterize its blooming period, when large-bodied bees visit the flowers to collect nectar and pollen. In the northern part of the plant's range, the bees that visit to take nectar are usually members of the genus *Bombus* (bumblebees). In the process of inserting their proboscises into the gaps at the base of the petals to reach the nectary below, they get dusted with pollen and carry it on to other flowers (fig. 641). As it departs, or at the next flower, the bee may brush against the receptive stigmas and pollen grains will be left there. Bumblebee visitors differ in species according to geographic locality, but those I have observed in New York's Lower Hudson Valley are generally *Bombus impatiens*. Further to the south, the primary bee visitors belong to the species *Ptilothrix bombiformis* (fig. 642), a solitary bee that resembles a bumblebee to such a degree that its specific epithet, *bombiformis*, means "like a *Bombus*." These bees are called rose-mallow bees because they are obligately bound to swamp rose-mallow, which provides the sole source of food (pollen) for its young (though these bees have

Swamp Rose-mallow

Fig. 639. (Left) Roselle (*Hibiscus sabdariffa*), originally from Asia, is widely grown as a fiber source in the West Indies and in other parts of tropical America, where it is known as sorrel. Its fleshy, red calices serve as an ingredient used to flavor and color beverages.
Fig. 640. (Right) The bottled "Sorrel" beverage is displayed amid the calices of the roselle plant at a market in St. Thomas, U.S. Virgin Islands.

Fig. 641. In the northern part of its range, the principal visitors to (and pollinators of) swamp rose-mallow are bumblebees (*Bombus* spp.). The bees take nectar from the flowers and inadvertently get dusted with pollen that is brushed off on the receptive stigmas of other flowers, often resulting in pollination.

Fig. 642. The rose-mallow bee (*Ptilothrix bombiformis*) is an obligate partner of *Hibiscus moscheutos* in the southern part of its range. It collects nectar and pollen from the flowers and provisions its nests with balls made of the pollen, which serve as the sole food source for its larvae.

Mallow Family

Fig. 643. The pollen grains of swamp rose-mallow are large enough to be seen with the naked eye. The small beetles gleaning pollen from the stigma of this flower are rose-mallow beetles (*Althaeus hibisci*).

Fig. 644. *Hibiscus moscheutos* is a near obligate host for these small bruchid beetles (*Althaeus hibisci*). Their larvae feed on the seeds of the plant, and the adults feast on pollen. There are reports of velvet-leaf/butter-print (*Abutilon theophrasti*), an introduced species of Malvaceae, being consumed by the beetles as well.

occasionally been reported to gather pollen from morning glory flowers). *Ptilothrix* bees dig nests in bare soil near a *Hibiscus* colony and during the nesting season make continuous trips back and forth between flowers and nests to provision the nests with pollen. Although the rose-mallow bee is a solitary nester, it often builds its nest within close proximity of others of its species. This lifestyle differs from that of our native bumblebees and European honeybees, both of which are colonial nesters. Many species of solitary bees are also dietary specialists, feeding on only one, or a few related, species of plants.

Male rose-mallow bees perch on flowers or patrol above them, always on the alert for females coming to feed. They move from flower to flower, aggressively chasing any other males from the area. This aggressive behavior often leads to scuffles within the flowers with the result that the males are more likely to become covered with the large, spiny pollen grains and hence carry greater pollen loads when they leave the flower. They are also more likely to come into contact with the stigmas of the flowers during these battles. Although brushing against the stigmas can also result in some pollen grains being dislodged from their surface, the overall outcome is greater pollen deposition and higher pollination success. The same situation occurs when an attempted mating with an unreceptive female results in a fight between them. The effect of greater pollen movement is even more pronounced if the scuffle takes place later in the day when the petals have begun to close. The resulting narrower opening is more likely to cause the bees to come into contact with the reproductive organs.

The large pollen grains can easily be seen with the naked eye (fig. 643). Viewed through a microscope, the outer surface appears to be covered with sharp spikes; this helps the pollen to adhere to the hairy bodies of the bees. Studies of bee visitation and pollen deposition in swamp rose-mallow have shown that the flowers receive an average of two to four visits by potential pollinators per 15-minute interval. About one-third of the visits result in the bee's contacting a stigma of the plant with a median number of 70 pollen grains deposited in each of these encounters. Since not all pollen grains germinate and fertilize ovules, an excess of pollen grains must land on the stigmas in order to achieve full seed set. This number was determined to be 2.6 grains per ovule, and the stigmas were

Swamp Rose-mallow

found to accumulate this number or more within two to three hours after bee visitation had begun.

Once a female *Ptilothrix* bee has mated, she begins to build a nest. Sometimes the soil in the nesting area is hard packed, requiring the bees to gather water from nearby ponds or standing water in puddles and bring it back to the nesting site to soften the soil before they are able to begin to dig it out with their mandibles. Her water gathering technique, with widely spread legs "skating" over the surface, results in the bees being said to "walk on water." It may take up to 14 trips to bring enough water back to the nest area before digging can commence. As it digs, the female bee pushes the soil up and out of the hole with her legs, then packs it with the end of her abdomen into a turret-like structure surrounding the hole. *Ptilothrix* is a member of the tribe Emphorini, commonly known as chimney bees for the appearance of their nests.

A single female may dig several nests in a season, each a vertical tunnel as much as 85 mm deep (depending on whether one or two cells will be made at the bottom). The cells, urn-shaped and lined with wax, are provisioned with a spherical ball of *Hibiscus* pollen (usually held together with a bit of nectar) carried during several trips back to the nest by the female bee in the hairy scopae on her hind legs. A single egg is embedded in the bottom of the pollen ball. As each cell is completed, the female bee closes it with a spirally patterned cap made of a wax/soil composition. Once the cells are complete and provisioned with food and an egg, the bee caps the opening to the nest with

Fig. 645. Rose-mallow beetles gleaning pollen that has been dislodged from the anthers by visiting bees and fallen onto the corolla.

Fig. 646. Pollen grains are consumed from the stigmas of swamp rose-mallow by rose-mallow beetles, possibly having a negative effect on seed set.

a concave cover of soil, usually made from freshly moistened turret soil.

When the egg hatches, the larva will consume the pollen ball until it is ready to pupate (this may take as little as two weeks). The larva then excretes the contents of its digestive tract, consisting primarily of the outer layer of the pollen grains, and uses this material to line its chamber. The exterior layer, the exine, of pollen grains is the strongest and most resistant plant substance—the reason that pollen grains are often well preserved in ancient fossil records. Thus, the nesting chamber is well protected by this coating. The unique patterning on a pollen grain's external surface allows scientists who study pollen (palynologists) to identify the plants that once inhabited the region. The larva will then spin a brownish cocoon within the chamber and overwinter there as a pupa.

Another insect commonly seen in the flowers of *H. moscheutos* is a tiny bruchid beetle called the rose-mallow beetle (*Althaeus hibisci*) (fig. 644 and also seen in figs. 643 and 645–646). While in their larval stage, they feed almost exclusively upon the seeds of swamp rose-mallow. Adults may be seen crawling about in the flowers and seeking cover when threatened by slipping between the bases of the petals and sepals. As adults they feed on *Hibiscus* pollen, both by scavenging pollen knocked from the anthers by bees (fig. 645) and by purposely removing pollen grains from the flowers' anthers and stigmas (fig. 646). In late summer the beetles lay their minute eggs on the outside of a still green

Mallow Family

seed capsule. When the eggs hatch, each tiny larva will enter the capsule and burrow into a seed, where it will feed on the contents until ready to pupate, still within the seed. The larva consumes the entire contents of the seed, but not the hard seed coat, so the seed appears normal when the fruit opens. Within a few weeks the adult beetles emerge from tiny holes (1.25–1.5 mm) that they make in the seed coat from within (small holes can be seen in a few of the seeds in fig. 635, but it's not known if these were caused by emerging beetles). By this time the capsules have opened, allowing the beetles to leave without having to chew their way out of the fruits. The adults presumably hibernate until the next *Hibiscus* flowering season begins.

A second seed predator, this one a weevil (*Conotrachelus fissinguis*), also specializes on swamp rose-mallow. In this case a single larva may consume several seeds before chewing a 2 mm hole through the capsule and dropping to the ground, where it constructs a pupal case in the soil. Because it pupates in the ground, the weevil is generally found on plants in the nonflooded part of the *Hibiscus* population.

Holes in the flowers or leaves of swamp rose-mallow are usually the work of Japanese beetles (*Popillia japonica*), for which *Hibiscus* is a favorite food (fig. 647). They also selectively chew off the anthers from the staminal tube. Japanese beetles are the most noticeable pests of *Hibiscus* plants grown in gardens. White flies, aphids, sawfly larvae, and caterpillars are among the other insects that attack members of this genus. Among the caterpillars that feed on the leaves and stems of *Hibiscus* are the saddleback caterpillar and the larvae of the Io moth (both of which sting if touched).

Frequently, seashore-mallow (*Kosteletzkya virginica*), a member of another genus of Malvaceae, shares the brackish marsh with swamp rose-mallow (see fig. 623). It can be almost as tall as swamp rose-mallow when the plants are fully mature, but its bright pink flowers are smaller than those of *H. moscheutos* (note size

Fig. 647. The holes in the white corolla and the denuding of the anthers from the staminal tube are the work of the Japanese beetle seen at the base of the flower. Japanese beetles are a major pest of most species of *Hibiscus*.

Fig. 648. Another native member of the Malvaceae, seashore-mallow (*Kosteletzkya virginica*), is frequently found growing in marshes with swamp rose-mallow (see fig. 623). Its flowers are smaller than those of swamp rose-mallow and its staminal tube is bright yellow.

Swamp Rose-mallow

difference in fig. 633). The flowers of the two species are similar in form, with an open funnel-shaped corolla of five pink petals and their reproductive parts consisting of a staminal tube surrounding the pistil, which in seashore-mallow are bright yellow (fig. 648). Its flowers also open for only a day. The leaves of seashore-mallow are more narrow and gray green in color, and the epicalyx is shorter (fig. 649) than that of swamp rose-mallow. The fruits of the two species are markedly different, with the capsules of seashore-mallow comprising a flat ring of five segments (fig. 650), each containing only a single irregularly shaped seed. As in some swamp rose-mallows, the seeds are dispersed by water. Seashore-mallow is a true coastal plain species, growing in brackish or fresh water.

Both species may co-inhabit marshes along with narrow-leaved cattail (*Typha angustifolia*) (as seen in fig. 651), several species of cordgrass (*Spartina* spp.), wild rice (*Zizania aquatica*), saltgrass (*Distichlis spicata*), and other plants of fresh and brackish wetlands. One of my particular favorites in this habitat is the lovely lavender-flowered saltmarsh fleabane (*Pluchea odorata*), a member of the aster family.

A walk in the marsh in late summer provides an opportunity not only to enjoy the beauty of all of these plant species, but also to observe their very interesting insect companions.

Fig. 649. The epicalyx of seashore-mallow is smaller and narrower than that of swamp rose-mallow.

Fig. 650. The fruit of seashore-mallow differs from that of swamp rose-mallow in that it is a flattened capsule containing just one seed in each of its five segments.

Fig. 651. Swamp rose-mallow is seen here growing among narrow-leaved cattails in a freshwater marsh on the Hudson River in New York.

Onion Family

Fig. 652. In April, a dense carpet of wild leek leaves covers the forest floor in a mesic forest in New York.

Wild Leek

Fig. 653. Not until July do wild leek's umbels of small white flowers bloom, by which time all sign of the leaves is gone.

Wild Leek (Ramp)
Allium tricoccum Ait.
Onion Family (Alliaceae)

Wild leeks are an early spring delicacy eagerly anticipated each year by those fond of garlic-flavored foods. One of the earliest fresh greens available, wild leeks may be found in deciduous woodlands throughout the Northeast. They are celebrated as a true sign that winter has released its hold on eastern North America.

Habitat: Rich woodlands.

Range: From Maine to North Dakota and adjacent Canada, south to Missouri, Oklahoma, Alabama and Georgia, with the exception of Nebraska, Kansas, Arkansas, Mississippi, and South Carolina.

It is said that driving southeast on Highway 55, with the windows rolled down, you know when you are nearing Richwood, West Virginia, the self-proclaimed "Ramp Capital of the World." The air is heavy with the pungent, pervasive, almost overwhelming scent of ramps. Each year, Richwood's Feast of the Ramson (an alternative name for ramps or wild leeks) is a spring ritual that takes place in April or May in many similar small towns throughout the Appalachians and beyond. In 2013 there were 52 such festivals listed on the Richwooders.com website (www.richwooders.com)—including one in my home state of New York. Such festivals that celebrate a local food are popular in other parts of the country as well. They include blueberry festivals in Maine, festivals for cranberries in New Jersey, pumpkins in Michigan, watermelons in Georgia, hot sauce in Louisiana, strawberries in Wisconsin, and the famous garlic festival in Gilroy, California, among numerous others.

Onion Family

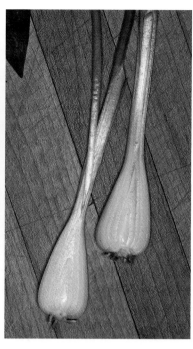

Fig. 654. (Above) By early May the leaves of wild leek have begun to senesce, turning yellow before they decay. The other leaves visible are those of seedlings of *Eurybia divaricata* (white wood aster) that will not flower on the woodland floor until late August. **Fig. 655.** (Right) A wild leek bulb sliced vertically to show that the bulb comprises layered leaf bases. The roots have been cut off in these plants.

Richwood has the most famous ramp fest of them all, now in its 81st year (2019). In addition to reasonably priced ramp suppers served at the local high school, the festival offers traditional music, clogging (a type of dance), an arts and crafts show, and of course, all manner of ramp concoctions—from the traditional potatoes and ramps cooked in bacon fat, to ramp salsa and ramp jelly. More than 1000 attendees enjoy feasting on ramps each year. Local people of the southern Appalachians have long believed that ramps have a spring tonic effect, the claim being that they "cleanse the blood." Actually, this folk wisdom may have some basis in fact in that many scientific papers have been published documenting the health benefits of consuming plants in the genus *Allium*. Both in lab cultures and in animal tests, the chemical compounds (particularly the sulfur-based constituents) found in *Allium* (Latin for "garlic") have been shown to have anticarcinogenic effects; provide positive benefits to the heart; reduce blood glucose and cholesterol levels; and protect against a variety of infections. However, they are strong-flavored, and thus best eaten cooked, since cooking volatilizes some of the pungent sulfur compounds.

New York's festival was first held in Hudson, New York in May of 2011. Appreciation of the combined garlicky/oniony flavor of ramps is shared by the attendees of both festivals, but the ambience was quite different. Held in an art center, the Hudson festival was sponsored by local businesses (wineries, beermakers, etc.). Some of New York's most respected chefs from the Hudson Valley, Manhattan, Brooklyn, and beyond showcased their talents in cooking ramp dishes there. To avoid depletion of this wild resource, the chefs were asked to devise recipes that use only the leaves rather than the entire plant. Unfortunately, the Hudson ramp fest ended after its fifth year, but another upstate New York town took up the challenge—Roscoe, New York hosted its third annual Wild Ramp Fest in May of 2019.

Ramps are available only in spring, the season when their leaves carpet large swaths of forest (fig. 652), particularly in the southernmost reaches of their range. No flowers are present at this time. When I was first learning the local flora, this was a difficult species to key out—all I had to go on were the leaves in the spring—*or* the flowers in summer (fig. 653). This made it difficult to identify either stage of the plant without the presence of the other because most plant keys required using the two features in combination. With today's interactive, multiple-entry keys available on the internet, this botanical mystery could have been more easily solved. Soon, however, I was able to

Wild Leek

Fig. 656. These bulbs of wild leek were exposed by erosion due to heavy rain. Exposure to light has caused the green pigment, chlorophyll, to develop in the normally white bulbs.

Fig. 657. The flower stalks of wild leek begin to develop as the leaves are turning yellow (see fig. 654), but they do not begin to open for another two months. By the time the buds are about to open, the plant's leaves have all but disappeared. The other leaves seen here are those of a sedge, *Carex plantaginea* (seersucker sedge).

Fig. 658. Two flower buds protrude from the translucent, papery sheath that encloses the inflorescence of wild leek, a first step in flower bloom.

put two and two together, and I learned that the flowers of wild leek don't bloom until weeks after the leaves have turned yellow (fig. 654) and disappeared. I also learned that by crumpling a small piece of leaf, I could produce a garlic-like aroma that was a dead giveaway. Although wild leeks are perennial plants that bloom in summer, they take advantage of the increased light available in the forest before the trees leaf out, much as our spring ephemerals do.

Wild leek grows from a bulb. Like other true bulbs (e.g., onions, tulips, and daffodils), the bulb of wild leek is formed underground from the bases of the leaves, wrapped tightly in layers around one another (fig. 655). If bulbs are exposed to light as a result of erosion in the wild leek's habitat, the bulbs become green and are able to carry on photosynthesis (fig. 656). When not disturbed, the plants increase in number by the formation of smaller bulbs adjacent to the original bulb. As the bulbs mature, the plants (ramets) maintain their connection to neighboring plants, forming one large clonal colony of genetically identical individuals. After 10 days to two weeks, when the leaves begin to turn yellow and fade away, flower scapes are produced (fig. 657) but remain as pointed buds for weeks (fig. 658); the flowers do not open until July (fig. 659). Few other forest plants flower at that time, so the small flowers of wild leek provide a welcome

Fig. 659. An inflorescence in partial bloom. Note the remnant of the sheath that formerly enclosed the rounded umbel of flowers.

Onion Family

source of nectar for bees during that period (fig. 660). The fact that they are self-compatible increases the likelihood of seed production. Once pollinated the three-parted capsular fruits begin to form (fig. 661), not maturing to expose their seeds until autumn.

Fig. 660. This bumblebee must hang under the flowers to sip nectar. In doing so, it is dusted with pollen, which it transfers from one flower to another. Other species of bees and flies also visit the flowers.

Fig. 661. Immature, mostly three-parted, capsular fruits beginning to develop on a mature flower stalk.

Occasionally I had eaten "ramps" without knowing it, as in the northern part of our greater Northeast, they are known as wild leeks—a flavorsome wild relative of the garden leek traditionally used in making vichyssoise. The Scottish and English ancestors of many Appalachian settlers brought the word "ramp" with them when they immigrated to the Americas. Ramp is an Old English variant of the previously mentioned alternate name: "ramson," the common name applied to a tasty European relative (*Allium ursinum*). The origin of the Old English name is said to mean "son of ram" and was applied to *Allium* because the plant emerges from the ground during the astrological sign of Aries, the ram.

I've used only small amounts of ramp leaves to season various dishes and boost the flavor of something in need of just a hint of garlic or onion. Traditionally, when ramps are collected in the spring, the entire, shallowly rooted plants are pulled intact from the ground to maximize the amount of flavor obtained relative to the effort entailed in collecting them. The plants look much like scallions but are tinged with magenta at the base of the leaves (fig. 662). The leaves, though, are much broader than those of a scallion and might be mistaken for the leaves of some orchids or blue bead lily (*Clintonia borealis*)—at least until they are broken and sniffed.

The growing popularity of wild foods in cooking has conservationists worried about the sustainability of this commodity if it is harvested by pulling the whole plant. Not only have ramp festivals proliferated throughout the southern Appalachians where ramps have always been popular, but the plants have gained a certain cachet in the rarefied dining rooms of some of New York City's top-rated restaurants, the menus of which now feature a variety of "wild" foods. Ramps are generally wild-collected, as are many of the wild mushrooms and other "locally sourced" foods featured on menus. Local "pickers" have found a way to derive income from the surrounding woodlands with little cost other than their time. I so wish that it were the invasive introduced garlic mustard (*Alliaria petiolata*), a member of the mustard family, that had caught the public's attention and palate.

Ramps still grow profusely in much of the southern Appalachian highlands, but in certain parts of that region they have been overharvested to the extent that a moratorium on collection had to be enacted. Even in Tennessee, formerly a state with many annual ramp

Wild Leek

fests, ramps, recognized as victims of commercial exploitation, are now declared a species of special concern. Some long-time ramp fests have had to be canceled, particularly since 2004, when ramp collecting was banned in Great Smoky Mountain National Park. The French-influenced province of Quebec enjoys the flavor of wild leeks, but because ramps are not as abundant in the northern extent of their range, they have been designated a protected plant since 1995, and no commercial sales are permitted. Unfortunately, poachers take their plunder across the provincial border to Ontario where they may still be sold to restaurants.

Sustainable leek harvesting can be accomplished only with the cooperation of foragers, buyers, sellers, and consumers. Plants seldom bear fruit until they are five to seven years old, so replacement by seedlings is minimal. As with many resources (e.g., commercial fishing), overharvesting is often carried on until the resource is dangerously depleted. With the public interest in locally sourced, organic foods, demand is high, and prices have soared. A bunch of seven leek plants in my local farmers market was being sold for $5.00 in 2014 (see fig. 662). Although farmers are being encouraged to grow ramps as a crop on wooded parts of their land, the tradition and ease of wild collecting is hard to give up.

Those flower scapes that do form seeds shed them in early fall (fig. 663) with the hard, black seeds generally not germinating until after a year (or two) of dormancy. Most seedlings fail to survive their first year. This low reproduction rate is compounded when plants are harvested on a large scale. In the few studies carried out to gauge the effect of wild harvesting, it was determined that colony size declined measurably when as little as 5%–10% of the population was harvested.

Perhaps the interest in wild foods is a passing fad, but with more and more people concerned about the origin and safety of their food, it seems likely that ramps and other wild foods will remain popular. It is important then to consider how to maintain this resource in a sustainable fashion, before it, like ginseng (discussed elsewhere in this book), becomes threatened or extirpated from much of its range.

Fig. 662. (Above) A basket of leeks for sale at a local farmers market in April. Note the magenta leaf bases characteristic of this species. Each bunch of seven leeks cost $5.00 in 2014.

Fig. 663. In autumn, the dry capsular fruits split open to reveal the hard black seeds that often remain on the plant until shaken off by wind or other disturbance.

Bean Family

Fig. 664. Wild lupine (*Lupinus perennis*) growing in the Albany Pine Bush region in upstate New York.

Wild Lupine

Fig. 665. Wild lupine flowering among young oak trees in upstate New York.

Wild lupine
Lupinus perennis L.
Bean Family (Fabaceae)

Wild lupine (also known as sundial lupine or blue lupine) is a beautiful native wildflower (fig. 664), but little known because of the relative scarcity of its dwindling habitat that occurs only in scattered regions of the northeastern and midwestern United States. The organization NatureServe categorizes the species as vulnerable (S3) in New York State, where there are only 20–35 sites and 3000–5000 individual plants statewide. Its global conservation rank is G5, indicating that despite its limited habitat, it is not considered globally threatened or endangered.

Habitat: Sandy dune areas or dry, sandy, open oak-pine woodlands of the Northeast and Great Lakes regions (fig. 665) with nutrient-poor soils, extending somewhat into the dry, sandy prairies of the northern Midwest.

Range: In the broad sense, wild lupine encompasses two subspecies: subsp. *perennis* in the northern part of its range, and subsp. *gracilis* in the southern part. Thus, the overall range of the species (including both subspecies) in the United States extends from New England south along the Atlantic Coastal Plain to northern Florida, and westward along the Gulf Coastal Plain to easternmost Texas. The northeastern subspecies of wild lupine discussed here also grows westward though the states bordering the Great Lakes into eastern Iowa and eastern Minnesota. In Canada, *Lupinus perennis* is found in southern Ontario, the Maritimes, and Newfoundland.

Because of its declining habitat, wild lupine is considered Rare, Threatened, or Of Special Concern throughout most of its range; in the state of Maine it is most likely extirpated. Wild lupine grows inland, inhabiting sandy clearings, open pine/oak woodlands, and oak savannas, on nutrient-poor, acidic to neutral soils. For two to three weeks in late spring such areas are cloaked in an undulating carpet of varying shades of blue (fig. 666), and occasionally pink and white (fig. 667)—or rarely multicolored flowers (fig. 668)—

Bean Family

Fig. 666. Wild lupine at peak bloom in a clearing in the Rome Sand Plains in New York.

Fig. 667. A patch of wild lupine showing the variation in flower color, from lavender/blue to pink.

during wild lupine's brief flowering season. Plants associated with our native lupine include pines (notably pitch pine [*Pinus rigida*] (fig. 669) and jack pine [*P. banksiana*]); oaks (including scrub oak [*Quercus ilicifolia*], red oak [*Q. rubra*], black oak [*Q. velutina*], and white oak [*Q. alba*]); bracken fern (*Pteridium aquilinum*); blackberries and dewberries (*Rubus* spp.); and various members of the heath family (Ericaceae).

Wild lupine is sometimes called "sundial lupine" for the resemblance of its palmately compound leaves to the rays of a sundial (fig. 670). The leaves, comprising 7–15 leaflets, also orient themselves toward the sun. Wild lupine, like clovers, vetches, and sweet peas, is a legume, a member of the pea or bean family (Fabaceae), a large, worldwide family of more than 19,000 species. The genus *Lupinus*, itself, includes about 220 species, most inhabiting mountainous regions of the western United States and South America, the highlands of tropical Africa, and the Mediterranean region. As with many legumes, the roots of wild lupine are endowed with nodules (fig. 671) that form as a result of infection by nitrogen-fixing bacteria (generally referred to as rhizobia). The bacteria inhabiting the nodules are able to convert nitrogen (N_2) from atmospheric nitrogen in the soil into ammonium (NH_4^+) or nitrate (NO_3^-), both of which are nitrogenous nutrients used by plants. Nitrogen fixation permits many legumes to thrive in nutrient-poor

Fig. 668. (Far left) A multicolored inflorescence of wild lupine. Fig. 669. (Left) Wild lupine beginning to form fruit on the lower part of its inflorescences, growing on an open hillside beneath pitch pines (*Pinus rigida*) in the Albany Pine Bush.

Fig. 670. (Above) The arrangement of the leaflets of wild lupine gave rise to another common name, sundial lupine.

Wild Lupine

soils, which would otherwise be inhospitable to the majority of plant species. After a leguminous plant dies, the nodules break down and release the rhizobia and their nitrogenous products into the soil. It is for this reason that legumes (such as alfalfa) are used as cover crops in crop rotation, as they serve to enrich the soil prior to the growing of food crops.

The ability of lupine (and other legumes) to thrive in poor soils was once attributed to their ability to rob or "wolf" the soil of its nutrients. Thus, *lupus*, the Latin word for wolf, was employed in the naming of this genus of plants. Of course, it is now known that the opposite is true: legumes serve to *enrich* the soil. This attribute allows some species of lupine to colonize recently denuded areas, as when Pacific lupine (*Lupinus lepidus*) appeared on the barren slopes of Mount St. Helens shortly after its 1980 eruption.

These soil enrichment properties have led to the planting of relatives of wild lupine, most notably bigleaf lupine (*Lupinus polyphyllus*), as a way to enhance the soil. Most residents of the Northeast, when they think of lupine, picture not our delicate native species, but rather the robust, showy plants of *L. polyphyllus* (or hybrids that include that species as a parent), whose blue, purple, pink, or white flower spikes brighten gardens, roadsides, and coastal landscapes in New England and eastern Canada (fig. 672). Although bigleaf lupine's common, eye-catching plants and its hybrids *appear* to grow as native species along moist roadsides and in open meadows, they are, in fact, native to the *northwestern* United States and Canada. Because of their colorful blooms, these plants have been widely introduced as garden plants (now with many additional color forms), and like many such beloved garden plants, they have subsequently escaped into roadside meadows where they often become invasive.

It was David Douglas, the famous Scottish plant explorer (whose surname is used in the common name of the Douglas fir [*Pseudotsuga menziesii*]), who, in 1826, introduced bigleaf lupine to Great Britain, where it quickly became a garden favorite. More than 100 years later, bigleaf lupine served as one of the parents of hybrid crosses made by George Russell, now known as the Russell hybrids (*Lupinus ×regalis*), spectacular plants with dense stalks of long-lasting flowers in a myriad of colors.

In addition to our northeastern coastal region, bigleaf lupine and its hybrids are also considered invasive

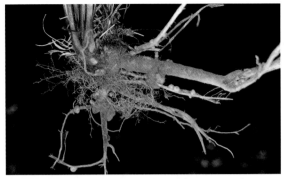

Fig. 671. The small light-colored nodules seen on the roots of *L. polyphyllus* contain nitrogen-fixing bacteria that process nitrogen from the air and convert it into forms of the compounds that allow lupine to grow in otherwise nutrient poor soils. These nodules also occur on the roots of *L. perennis*.

Fig. 672. A hillside on Prince Edward Island, Canada has been taken over by a stand of a northwestern species of lupine, *Lupinus polyphyllus* (bigleaf lupine) and/or its hybrids, which are widely cultivated for their beauty. Unfortunately, such garden introductions often become invasive colonizers, replacing the flora native to the region.

Fig. 673. As in the previous image, this lovely field of bigleaf lupine (and/or its hybrids) is far out of its native range—on the shore of Lake Tekapo on the South Island of New Zealand.

Bean Family

in several other countries, among them: Germany, Finland, Norway, Australia, and New Zealand (fig. 673). Although lupine is typically high in toxic alkaloids, newer varieties low in alkaloids have been developed that can safely serve as forage for livestock. The toxic seeds have also been used as a source of protein since ancient times, when Greeks, Romans, and Egyptians learned to detoxify the seeds by soaking or boiling them in order to leach out the poisonous alkaloids.

Although only half as tall as the introduced lupine from the Northwest, the beautiful blue-purple spikes of our wild lupine (fig. 674) rival those of bigleaf lupine for beauty. Unfortunately, few people ever venture into the native lupine's somewhat barren habitat to see them.

Our native Northeast/Midwest lupine is perhaps best known for its obligate association with one of the butterflies that feeds on its leaves during its caterpillar stage—the federally endangered karner blue butterfly. The life histories of the two species are intricately intertwined, and one cannot talk about one species without discussing its relationship with the other. It is this close association with the karner blue butterfly (fig. 675) that has brought attention to the importance of preserving not only the host plant of the karner blue but also its ecological community.

The karner blue butterfly has undergone a number of taxonomic changes since its discovery. In 1944 the butterfly was officially designated a subspecies of the midwestern melissa blue and named *Lycaeides* (syn. *Plebejus*) *melissa* subsp. *samuelis* by avid lepidopterist and novelist Dr. Vladimir Nabokov (Nabokov is probably better known as the author of *Lolita* than for his taxonomic research in the field of entomology). The type specimen (the specimen on which the description is based and which represents the species) is from the area surrounding Karner, New York, located northwest of Albany, in the township of Colonie. This is the heart of a region known as the Albany Pine Bush, an area of inland sand dunes (some up to 100 feet high) deposited by the huge postglacial Lake Albany about 12,000 years ago. The Albany Pine Bush, dominated by pitch pine and scrub oak, is considered the premier example of the globally rare pine bush ecosystem (only 20 such inland pine bush ecosystems exist). Years later Nabokov came to the conclusion that the karner blue was actually a separate species, rather than a subspecies of the melissa blue, but he did not make that taxonomic change. Disagreement over the taxonomic status of the karner blue

Fig. 674. Although not as large as bigleaf lupine, our native *Lupinus perennis* is every bit as beautiful.

has existed since that time. In 2008, researchers determined to put an end to the taxonomic doubt surrounding the status of the karner blue. They studied the genetic relationships among the karner blue, the melissa blue, and a third closely related species, *Lycaeides idas*, the Anna blue. Their results were published in a paper titled "After 60 Years, an Answer to the Question: What is the Karner Blue Butterfly?" The investigation showed that the levels of gene flow between the karner blue and the melissa blue were similar to those between the melissa blue and the Anna blue, already classified as separate species. Based on that finding the authors determined that the karner blue should be elevated to the species level as *Lycaeides samuelis* Nabokov. This decision aligns with previously observed morphological and ecological differences between the species: the wing pattern and genitalia of the karner blue differ from those of all other *Lycaeides*, and they are obligately dependent on only one host species (*Lupinus perennis*), while *L. melissa* utilizes a number of host plants.

Wild lupine grows most profusely in open, sunny areas located within piney woodlands or oak savannas. Male karner blue butterflies (fig. 676) prefer to fly and feed in such sunny areas, venturing into the shaded parts of the habitat only when in search of mates. But the female karners show a preference for shaded sites. This is especially true of the second brood of the season, which develops from eggs preferentially laid on plants growing in the more shaded sections of the site. The shaded plants tend to flower later and, for reasons not fully understood, provide a

Wild Lupine

Fig. 675. (Far left) The dependence of the federally endangered karner blue butterfly, still listed as *Lycaeides melissa samuelis*, on wild lupine as the sole host plant for its caterpillars has spurred efforts to protect both the butterfly and its host plant by conserving what remains of their habitat. Seen here is a female of the species, identified by the orange crescents and just a wash of blue on its otherwise brownish upper wing surfaces.

Fig. 676. (Left) The male karner blue butterfly can be told from the female by the black-margined deep blue upper wing surface and the lack of orange crescents on the upper surface of the hind wing. Both male and female butterflies are similar in color and pattern on their lower wing surfaces.

higher quality food resource for the caterpillars (perhaps because those plants growing in the sun senesce more quickly). Second brood butterfly larvae that feed on the shaded plants develop up to 22% more rapidly, a trait beneficial to the caterpillars since a shorter time spent as larvae equates with a shorter period of susceptibility to predation.

Two features of the larvae aid in reducing predation. The caterpillars are well camouflaged, being both small (reaching only up to 2 cm when ready to pupate) and the same green as the lupine leaves; thus, not easily seen by birds and other predators (including parasitic wasps and spiders). The larvae feed in such a way that the upper cuticle of the leaf remains in the areas fed upon, giving the leaves the appearance of having translucent windows.

Karner blue caterpillars benefit from the services of protective "guards" as well. Like other members of the genus *Lycaeides*, the karner blue has a relationship with ants of various species. The caterpillars secrete a drop of sweet liquid from a gland on their abdomens. Ants "tend" the caterpillars, imbibing the sweet substance. In return they defend the caterpillars against their insect predators by aggressively attacking the predators. Despite this beneficial relationship, ants of the same species that protect the caterpillars have also been observed to sometimes feed on the eggs, or even the caterpillars.

Once the caterpillar has reached full size it crawls to the ground and forms a chrysalis in the dead leaf litter near the base of the lupine plants. Development from egg to adult may take 25–60 days depending on environmental conditions, with males spending two or three fewer days in the larval stage than females. Throughout much of their range, karner blue butterflies have two broods, one in late May to early June and another in late July to early August. By the time the second brood is ready to lay eggs, lupine leaves have begun to die back, and the eggs are often deposited on the stems and fruits (fig. 677) of the plant, on other species of plants, or even on the ground. Like the fruits of many other legumes, the fruits are pods (in this case, hairy) that contain pealike seeds. The eggs overwinter, protected from frigid temperatures by a blanket of snow. When the eggs hatch the following May, the young caterpillars seek lupine plants to feed on. During their brief adult life stage (a mere 3–5 days), karner blues must find a mate and lay their eggs. It is during

Fig. 677. The fruits of wild lupine develop rapidly after the flowers wither and fall. As with many legumes, the fruits resemble pea pods, but in wild lupine they are hairy.

Bean Family

Fig. 678. Bumblebees are the primary pollinators of wild lupine. Their bodies are heavy enough to cause the reproductive structures to be revealed when the bee lands on the flower, thus allowing for the transfer of pollen to the bee, which transports it to another flower. Bees also collect the pollen as a food source for their larvae, as seen here by the bright orange packet of pollen packed into the pollen basket on the bee's hind leg.

this time that they feed on nectar, not strictly that of wild lupine, but from the flowers of dozens of species.

In the pollination of wild lupine, bumblebees are the primary agents of pollen transfer. When a bee lands on a lupine flower (fig. 678) its robust body is heavy enough to cause the flower's lower keel petals to open (fig. 679), revealing the pistil and stamens within. Wild lupine has no nectar; bees visit only to collect pollen. As a bee probes the flower, it triggers a springlike mechanism that causes the stamens to eject their pollen onto the bee's body. The bee grooms some of the pollen from its body and places it in pollen baskets on its hind legs. It carries this pollen back to the nest to feed its larvae; that not groomed into the pollen baskets remains on the bee and is transported to subsequently visited flowers. Honeybees and some other bee species large enough to trigger the flowers are also potentially effective pollinators (fig. 680).

Unfortunately, the open sandy habitats of wild lupine have declined by nearly 90% in the past century. Concomitantly with the decline of habitat, the presence

Fig. 679. This lupine flower has been artificially opened to show the style and anthers that would be exposed by a bumblebee visit.

of lupine has also plummeted—and with the lupine, the lovely little karner blue butterfly. Since karner blues are obligately linked to *Lupinus perennis*, with their caterpillars preferentially feeding only on the leaves of wild lupine, this dependence on a single plant species places the karner blue in peril when its habitat is negatively affected. If we wish to save this small butterfly from extinction, we must conserve its host plant and, indeed, the entire habitat of both species.

Local scientists and nature lovers have long been aware of the decline of both wild lupine and karner blue butterflies, but initially they had limited success in drawing attention to the problem or in raising money to fund conservation efforts. As far back as 1975, Spider Barbour, a resident of upstate New York's sandy lupine habitat, wrote a song encouraging listeners to protect the declining population of karner blues. The song was titled "Shepherd's Purse." Barbour was a songwriter/musician with the rock band "Chrysalis" and had always had a keen interest in insects and other arthropods (reflected both in his nickname and in the name of his band). He urged people to

> Get out your shepherd's purse.
> It's time to pay your nature dues.

The karner blue butterfly was placed on the List of Endangered, Threatened and Special Concern Fish & Wildlife Species in New York State in 1977, but it was not until 1992 that it was added to the federal list. It was its official designation as Endangered under the Federal Endangered Species Act that finally brought widespread attention to its plight.

Pinelands are often viewed as barren wastelands, but in the eyes of developers, they become prime real estate for development. Hence, the vast regions of lupine habitat that once covered the sandy hills west of Albany, New York have been fragmented by highways, housing developments, shopping centers, and a campus of the New York State University System. This fragmentation takes its toll on the remaining populations of karner blues. During its entire lifetime a karner blue butterfly may venture no more than 700 feet from its place of birth; thus fragmentation of habitat sharply limits the possibility for karner blues to breed with other nearby populations of their species. The resulting inbreeding that occurs within the limited breeding pool of the butterfly population inhabiting

Wild Lupine

the same "patch" of lupine is detrimental to the health of the population; the lupine population suffers similarly when isolated from other lupine populations.

Another threat comes from the suppression of the natural fires that formerly maintained the patchy open habitat needed by lupine. Fire suppression allowed the understory shrub cover to become too dense, and eventually, a tree canopy to develop, thus shading out the sun-loving lupine.

A new peril threatens the relationship of the karner blue butterfly with its lupine host plant—that of climate change. Because the butterfly's development is dependent on the stage of the plant during the larval feeding period, any event that causes one organism to be out of sync with the other (e.g., a long, cold spring that delays the emergence of the lupine during the time the larvae are present, or a lack of snow cover to protect the eggs of the second brood) could result in the loss of a population of both species.

Conservation efforts are underway in upstate New York, New Hampshire (where the karner blue has been adopted as the official state butterfly), and in various midwestern states. Through the efforts of the Albany Pine Bush Commission, land within this ecosystem is being purchased and protected. Where wildfire suppression or natural succession has resulted in the loss of the clearings necessary for the establishment of lupine, trees are being cut and lupine seeds planted. Studies have shown that a mosaic of open and treed areas is necessary to support viable karner blue populations and that successful pollination is dependent more on the density of the plants than on the size of the colony. Such mediation efforts in the Albany Pine Bush region have yielded positive results, with the number of karner blue butterflies increasing by sevenfold between 2000 and 2014. The Albany Pine Bush Preserve provides numerous educational programs for school groups and the public in its new "green building" Discovery Center. Museum exhibits and outdoor displays explain the crucial relationship between lupine, the endangered butterflies, and the pine bush habitat, as well as detailing the cultural history of the region. Hiking trails in pineland habitat are open to the public and admission is free.

The larvae of three other butterfly species (and those of several moths) feed on wild lupine as well, most notably the frosted elfin (*Callophrys irus*) (fig. 681). This species has only one flight period per season, which ordinarily occurs earlier than that of the karner blue.

Fig. 680. (Right) The flowers of wild lupine are also visited by honeybees that are capable of effecting pollination as well. **Fig. 681.** (Below) The larvae of other butterflies, including that of this frosted elfin (*Callophrys irus*), also feed on wild lupine and therefore receive protection from any measures taken to conserve the karner blue butterfly's habitat.

The frosted elfin is not entirely dependent on lupine but will also utilize wild indigo (*Baptisia tinctoria*) another legume with alkaloids similar to those found in wild lupine, as a host plant. Frosted elfin caterpillars feed on lupine flowers and developing seedpods, but since they are active early in the season, and most lupine plants produce their flowers after the elfin's feeding period, they have minimal impact on lupine seed production. While not federally protected itself, the frosted elfin is rare where it does occur and has been extirpated from Maine and Canada. As a co-resident of the habitat of *Lupinus perennis*, the vulnerable frosted elfin stands to benefit from efforts to conserve lupine and its habitat for the benefit of the karner blue. The two other butterfly species are skippers; they likewise have little adverse effect on wild lupine.

To conserve the diminishing pine bush/oak savanna habitat is to protect not only the beautiful wild lupine that cloaks it in blue each June but with it the complex ecosystem in which lupine plays a key role. The loss of one component in this natural environment could have a cascading effect on the associated species as well.

Water-lily Family

Fig. 682. The showy parts of the yellow pond-lily flower (*Nuphar variegata*) are the bright yellow sepals that form a cuplike structure surrounding the other floral parts. The round stigmatic disk in the center of the flower sits atop the pistil and has several stigmatic rays that are receptive to pollen on the first day the flower is open. The many stamens surrounding the ovary, just below the surface of the disk, reflex outward to expose their anthers over the next few days.

Yellow Pond-lily

Fig. 683. *Nuphar variegata* flowers are held above the surface of the water on stout peduncles. Although this species has floating leaves, most of the leaves seen surrounding this flower belong to the white water-lily, *Nymphaea odorata*.

Yellow Pond-lily

Nuphar variegata Engelm. ex Durand
Water-lily Family (Nymphaeaceae)

Nymphaeaceae is a family of six genera with only two of those occurring in North America, *Nuphar* and *Nymphaea*. The family formerly included *Nelumbo* (treated elsewhere in this book), which was segregated from the Nymphaeaceae in 1997, based initially on plant chemistry and the morphology of the waxy cuticle, which suggested that *Nelumbo* had no affinity at all with the water-lily order (Nymphaeales) but rather was more closely related to the buttercup order (Ranunculales). Subsequent DNA research showed *Nelumbo* to belong in an entirely different order, the Proteales, which includes such Southern Hemisphere genera as *Banksia* and *Protea* that look vastly different from *Nelumbo*. The Angiosperm Phylogeny Group, the current authority on botanical taxonomy, states that the inclusion of *Nuphar* within the Nymphaeaceae still needs to be confirmed, but it remains there pending further investigation.

Habitat: Freshwater ponds, lakes, sluggish streams, swamps, and ditches.

Range: Maine to Delaware, west to Idaho (but not in Missouri or Wyoming); in all provinces and territories of Canada, plus the French archipelago of St. Pierre and Miquelon (just off the southeast coast of Newfoundland). *Nuphar variegata* tends to be a more northern species and has the widest geographical distribution of any of the *Nuphar* species in North America, ranging from Newfoundland well up into the Yukon and as far south as New Jersey and Nebraska.

Nuphar, or yellow pond-lily, has been described, rather cruelly, as "the Nymph's ugly cousin" (by none other than Dr. Donald Padgett, author of the excellent monograph of *Nuphar*, when he compared *Nuphar* to the white water-lily, *Nymphaea*). I beg to differ with Dr. Padgett's tongue-in-cheek assessment of the attractiveness of *Nuphar*; while yellow pond-lilies might not have inspired the impressionist artist Claude Monet to dedicate the last decades of his life to completing nearly 250 paintings of the same subject (the water-lilies in his Giverny garden), yellow pond-lilies possess their own bolder, less ethereal, beauty.

Water-lily Family

The main subject of this account is *Nuphar variegata* (commonly known as yellow pond-lily, spatterdock, or cow-lily and at one time classified under the name *N. lutea*, a Eurasian species). The species has a checkered nomenclatural past, with five different synonyms listed on Missouri Botanical Garden's Tropicos.org website, two of which are in *Nymphaea*, the genus of the white water-lily. In 1753, Linnaeus originally included three entities in the genus *Nymphaea*: *Nymphaea*, *Nuphar*, and *Nelumbo*. As mentioned above, the taxonomy of the family and the species included in it is still confusing partly because of the intergrading of species (now numbering 11), the striking variation in morphology throughout a species' range, and natural hybridization between species. The genus *Nuphar*, though, is quite easy to identify. Members of *Nuphar* have a yellow petal-like calyx (with *N. variegata* usually having three outer, small green sepals and three inner yellow sepals that tend toward red or purple near the base of the interior surface); hypogynous flowers (all other floral parts attached below the ovary), several whorls of stamens; small, straplike petals having slightly raised nectaries on their lower surfaces; leaves that are primarily floating (but with some submerged leaves near the rhizome); and seeds without arils. Typical of the floating leaves of many aquatics, the stomata (openings used for gas exchange) of *N. variegata* are on the leaf's upper surface. The genus has been divided into two sections by Padgett, one comprising the Old World species; the other including species from the New World. Only one species, *Nuphar microphylla*, is found in both the Old and New Worlds (see discussion below).

Fig. 684. The flowers of our common water-lily (*Nymphaea odorata*) float on the surface of the water. They comprise four green sepals and many white petals.

Nuphar has an ancient lineage based on fossil seeds found in China that date from the early Eocene, and on a well-preserved North Dakota specimen from the Paleocene. The genus is evolutionarily at the base of the Nymphaeaceae family, itself an early, primitive family. The name *Nuphar* is of ancient Arabic or Persian origin.

The genus *Nuphar* differs from that of our white water-lilies, *Nymphaea*, in many ways—among them: (1) in place of water-lily's many delicate petals, the "showy" parts of the flower of *Nuphar* are the relatively few sepals that form a globular-shaped flower (fig. 682); (2) the flowers of yellow pond-lily are held above the water surface (fig. 683) rather than resting on the water as do the flowers of the white water-lily (*Nymphaea odorata*) (fig. 684); (3) while the leaves of both genera in our area float on the water (with the exception of those of *Nuphar advena* [fig. 685]), those of *Nuphar* are more elongated than the rounded leaves of the fragrant white water-lily (fig. 686); (4) although yellow pond-lily has petals, they are small and hidden by numerous series of stamens above them, with all these structures enclosed within the sepals (fig. 687); and (5) leaf venation in *Nuphar* is essentially pinnate with the veins forking near the leaf margins, while in *Nymphaea* venation is palmate. All members of this family have leaf blades attached to their petioles either at the center of the blade (peltate as in *Nymphaea*) or, in the case of *Nuphar*, at a point on the margin where the lobes are separated. In *Nuphar variegata* the petioles are flattened on the top with a raised medial ridge and are slightly winged where the petiole joins the blade. In contrast, the petioles of white water-lily are round in their entirety (as are those of *Nuphar advena*).

Generally, the plants of yellow pond-lily grow in water that is 0.5–2 meters deep; however, they can grow, but less successfully, in deeper water. The leaves have long petioles that arise from the tips of branching rhizomes (horizontal stems) on, or in, the substrate of the pond or other quiet water body. In like manner, the peduncles of yellow pond-lily flowers also grow from the rhizomes and extend even above the water's surface. Roots help to anchor the rhizomes in the substrate. Pond-lilies reproduce vegetatively by proliferation of the rhizomes, as well as by seed. Occasionally, a piece of rhizome will become broken off from the mass below and float to the top of the water to be carried elsewhere, perhaps lodging in shallow water where it may sprout

Yellow Pond-lily

Fig. 685. (Above) *Nuphar advena* (also called spatterdock, like *N. variegata*) has leaves that emerge from the water and usually have a deep V-shaped sinus, measuring up to 90° wide, between the basal lobes of the leaf blade.

Fig. 686. (Below) The mixture of floating leaves in this pond includes the oblong leaves of *Nuphar variegata*, which have rounded lobes that overlap each other and pinnate veins; and the round, "Pac-Man-shaped" leaves of white water-lily with palmate venation.

new leaves. The plant body has a system of lacunae (interconnected air spaces) that allow air to move throughout the plant, especially important for underwater portions to obtain the oxygen needed for metabolism; this is an adaptation found in most aquatic plants.

Nuphar is heterophyllus, meaning that it has two types of leaves, those that reach the water's surface and others, on shorter petioles, that remain submersed. The submersed leaves are similar in form to those that float on the surface of a pond, but they are usually wider and very membranous, rather than tough and leathery with a thick, waxy upper surface (cuticle). The membranous leaves never reach the surface, and they remain on the plants throughout the winter, while the surface leaves die back when cold temperatures arrive. The floating leaves of *N. variegata* have a sinus (gap) approximately one-third to one-half the length of the leaf's midrib; the lobes of the leaf are adjacent to each other or even

Fig. 687. One of the inner sepals of this flower of yellow pond-lily has been removed to show the many series of straplike yellow stamens, which have already arched outward from the pistil. The underside of the remaining upright stamens in this flower are reddish. The petals

are all but hidden beneath the stamens. Grains of small, light-colored pollen can be seen on the stigmatic disk, which is no longer receptive at this time. Note how the interior of the sepals is reddish-purple near the base.

Fig. 688. (Right) Leaf beetles in the genus *Donacia* are common visitors to the flowers and leaves of yellow pond-lily. They pollinate the flowers and make small feeding holes in the floating leaves through which they deposit their eggs, gluing them onto the underside of the leaves. The larvae feed on the submersed parts of the plants. Note that this beetle is covered with pollen. Two of these beetles may also be glimpsed lurking within the sepals of the flower in fig. 682. **Fig. 689.** (Far right) Small sweat (halictid) bees visit *Nuphar* flowers to consume pollen. They may also pollinate the flowers.

Water-lily Family

overlapping. Leaves of both types are eaten by herbivores. Among the consumers of *Nuphar* are members of two leaf beetle genera, *Donacia* (fig. 688), also an important pollinator of the flowers of some pond-lily species, and *Galerucella*, whose larvae and adults feed on only the upper surface of the floating (or emergent, in the case of *N. advena*) leaves, leaving shallow, irregular trenches. *Galerucella* spp. are also known to be effective pollinators (of *Nuphar advena*, in particular).

Donacia beetles are very fast-moving active fliers, moving from flower to leaf to flower, and disappearing quickly into the recesses of the flowers when alarmed. They are often covered with pollen (see fig. 688). Females lay their eggs by ovipositing through small (2–5 mm) holes they make when feeding on the surface leaves, gluing the eggs to the leaf's underside in concentric arcs. Their larvae spend two years as aquatic grubs, presumably feeding on the underwater, winter-hardy parts of the plant: the roots, rhizomes, and petioles. They attach themselves to the rhizomes or petioles with two respiratory hooks through which they are able to acquire oxygen from the internal air spaces (lacunae). The beetles also feed on leaves of other aquatic plants.

Other insects reported to visit *Nuphar* include small sweat (halictid) bees (fig. 689), and syrphid flies (fig. 690), both of which feed on pollen. Honeybees and even aphids have been observed on the flowers, perhaps contributing to their pollination. The plants have value to other wildlife as well: the larvae of certain moths feed on either rhizomes or leaves, various ducks and other waterfowl eat the seeds, and the entire plant, especially the rhizomes and lower parts of the petioles, provide food for mammals such as beaver and muskrat.

Yellow pond-lily flowers are open for four or five days. On the first day, the sepals open only enough to form a triangular window that exposes the flat stigmatic surface at the apex of the pistil. The stigmatic rays are moist with mucus and receptive at this time, and the nectaries on the petals are also secreting nectar; the flower emits a sweet aroma that emanates from the stigmatic rays, the nectaries, and the freshly opened anthers. If an insect, such as one of the above-mentioned beetles, crawls down into the sepals, it risks being trapped overnight when the flowers close in the evening. During the night the upper whorls of stamens begin to arch outward, exposing their paired, strap-shaped anthers, which are beginning to shed pollen. Thereby, the beetles may leave the flower already dusted with a load of pollen they carry to the next flower they visit on the following morning (they have also been found to have pollen in their gut, documenting that they consume it while trapped in the flowers). During this initial 24-hour period, the female and male phases of the flower overlap slightly, and it is possible for the flower to be self-pollinated by insect activity. The flower opens more widely on subsequent days, the mucus on the stigmatic disk dries up, and the aroma becomes somewhat unpleasant; the flower is now fully in its male phase. This daily sequence of additional stamens reflexing outward continues (although the flowers may no longer close completely at night) until all the stamens have released the pollen from their anthers, at which time the petals and stamens wither (fig. 691) and the ovary begins to ripen into a purplish-green, strongly ribbed fruit that, because it is filled with lacunae, floats on the top of the water (fig. 692); the sepals often remain attached to the fruit. After some days, the connection of the fruit with its peduncle weakens, and the fruit may break off and drift freely on the surface. Seed dispersal occurs in this manner, for once the fruit

Fig. 690. (Far left) Hover flies (syrphid flies) are frequent visitors to yellow pond-lily flowers as well. They also eat pollen and serve as pollinators. This one (the larger one on the left) is on a flower of *Nuphar ×rubrodisca*, a hybrid between *N. variegata* and *N. microphylla*. The distinctive bright red disk is scalloped along its margin. **Fig. 691.** (Left) An old flower of *Nuphar variegata* has been cut longitudinally to reveal the immature seeds inside its compound ovary. Most of the stamens have withered and fallen. Some petals are visible surrounding the bottom of the pistil.

Yellow Pond-lily

disintegrates, the seeds are released and soon sink to the bottom. By this time, the fruit may have traveled many meters from the mother plant.

Some botanists believe that *Nuphar*, having flowers that close overnight and sometimes trap beetles within, has coevolved with beetles as their main pollinators. This close partnership has occurred in other genera, including *Magnolia*, *Nelumbo*, and *Victoria* (the Amazon water-lily). Other researchers disagree with the beetle hypothesis, basing their objections on the observation that in Europe, bees and flies are the primary pollinators on indigenous species of *Nuphar*. Nevertheless, beetles seem to predominate in New World pond-lily interactions.

Hybridization between species of *Nuphar* is not uncommon and has been one cause of the taxonomic difficulty that led to the describing of many baseless "new species." One hybrid that has been clearly elucidated is that between *N. variegata* and *N. microphylla*, which, in this case, is a hybrid between species belonging to the two different sections of the genus. The result of this pairing, called *Nuphar* ×*rubrodisca*, is obviously intermediate between the two parents. The parent species are strikingly different in appearance: *N. variegata* is a robust plant with large leaves, while *N. microphylla*, as its name indicates, is a dwarf with small leaves. In the hybrid, leaves are, in all dimensions, intermediate in size between those of the parents. Intermediacy applies as well to the number of sepals; the shape and color of the stigmatic disk, and the number of its stigmatic rays; and the size of the fruit. Even the length of the anthers (about equal in length to the filaments in *N.* ×*rubrodisca*) (fig. 693) falls between the extremes of the two parents (*N. variegata* has anthers 1–2× *longer* than the filaments, and *N. microphylla* has anthers 5× *shorter* than the filaments). Further evidence is based on *N.* ×*rubrodisca* having genetic material of both purported parents, as shown by amplification of random segments of DNA from the two species and their hybrid.

The ranges of the two parent species overlap in North America, and all but three collections of the hybrid have been found within the bounds of the overlap. Populations of the hybrid are often sterile and reproduce only by the spread of their rhizomes, although in some areas, a low fruit set has been observed. The result of this cross, *Nuphar* ×*rubrodisca*, is a beautiful little plant distinguished by its bright red stigmatic disk with a scalloped margin (see fig. 690), features resembling

Fig. 692. A fruit of yellow pond-lily floats, still attached to its peduncle, on the surface of a lake. Leaves of *Nuphar* and *Nymphaea* surround it.

Fig. 693. The anthers of *Nuphar* ×*rubrodisca* are clearly visible as lighter yellow, parallel, sessile lines on the recurved strap-shaped stamens. The anthers are near equal in length to the filaments, intermediate between the anthers of *N. variegata*, which are longer than the filaments, and those of *N. microphylla*, which are shorter than the filaments.

those of its *N. microphylla* parent, and by its fruits, which differ in color but are similar in shape to those of *N. variegata*, lacking a pronounced constriction beneath the stigmatic disk as found in *N. microphylla*).

Interesting alkaloids isolated from *Nuphar* show potential for medical use. One *Nuphar* alkaloid, castoramine, is also found in the scent glands of beavers and is used by them to mark their territories. One hypothesis is that the beavers obtain this compound from feeding on *Nuphar*. Despite the presence of strong alkaloids, the rhizomes have been used as a starchy vegetable or ground into meal or flour by Native Americans, who also dried and roasted the seeds. Although yellow pond-lilies have been used ornamentally in lakes and ponds, they have a tendency to become weedy and are considered in most places an expensive nuisance that must be eliminated.

Glossary

Botany, as does any discipline, has its own terminology. In many cases, equivalent terms are lacking in everyday English, making explanation of some words necessary to enable the general reader to understand their meaning as used in this book. Many such terms are defined where they are used in the text. This glossary is meant to provide brief definitions of (mostly) botanical terms used in the text that might not be familiar to the general reader. The plural form is given in parentheses.

achene: a small, indehiscent, dry fruit with a single seed attached to the fruit wall at only one point and easily separated from it

adventitious roots: roots that arise from a part of the plant other than the primary root, e.g., from a stem or corm

aerenchyma: a soft plant tissue containing air spaces; common in aquatic plants

agamospermy: the formation of seeds without fertilization

alien: non-native; from a foreign land

alkaline: having a pH greater than 7; used in describing soil chemistry

alkaloid: an organic, nitrogen-based compound found in plants that is often toxic, but in some cases has important medicinal properties; probably serves as a botanical defensive compound against herbivores

allelopathic: producing and secreting compounds that interfere with the growth or survival of other plants growing in the same area

androecium: the collective male reproductive organs of a plant; the stamens

androgynophore: a stalk bearing the androecium and gynoecium above the perianth of a flower (e.g., *Passiflora*)

annual plant: a plant that completes its life cycle within one growing season; it does not persist into the following year

antenna (antennae): one of a pair of appendages on an insect's head; used as a sensory organ

anther: the pollen-bearing part of a stamen

anthocyanin: a water-soluble plant pigment that is red, blue, or purple in color

antihelminthic: a substance used to destroy parasitic worms (in medicine)

antioxidant: a substance that inhibits oxidation; often found in the fruits of plants (e.g., vitamin C or E)

APG: initials of the Angiosperm Phylogeny Group, an organization responsible for summarizing plant family classification based on current research

apical meristem: an area at the top of a plant that produces cells for plant growth in height, width, or length

appressed: close to another structure but not fused to it

apomixis: the ability to produce viable seeds without fertilization

aposematic: referring to coloration or markings that serve to warn predators of distastefulness or toxicity

aquatic: a plant growing in water

areole: in the cactus family, a small area on the stem of a plant that produces spines, glochids, leaves, or flowers

aril: a fleshy outgrowth of the funicle (the structure that attaches the seed to the fruit wall); often attractive to an animal that disperses the seed

arthropod: an invertebrate in the phylum Arthropoda, which includes insects, spiders, and crustaceans

asymmetrical flower: having two halves or sides that are unequal in size and/or shape

atropine: an alkaloid found in plants

autogamy: self-pollination

awn: a bristle, often at the apex of a leaf

backcross: the crossing of a hybrid with one of its parents in order to achieve offspring with a genetic makeup closer to that parent

basal rosette: the basal leaves of a plant growing from a central point and lying more or less flat on the ground in a circular pattern

BCE: abbreviation for Before the Common (Current) Era; equivalent to BC

biennial plant: having a two-year life cycle, with flowering taking place the second year; the plant then dies

bifid: divided in two, as in a stigma or leaf

biomimicry: the design of materials or structures modeled on biological entities or processes

bisexual flowers: having both male (stamens) and female (pistils) reproductive structures in the same flower

bog: an acidic freshwater wetland of spongy material comprising mainly sphagnum moss, which when decayed is referred to as peat

boreal: referring to regions of high northern latitude or to forests that grow in such regions

BP: a time scale indicating "before the present." The "present" has been set at 1950, the year when that time scale was adopted based on the new technology of radiocarbon dating

bulb: an underground storage organ comprising concentrically wrapped fleshy leaf bases

bulbil: a small, bulblike structure capable of producing a new plant asexually, and is usually formed in a leaf axil

bulblet: a small bulb arising as an offshoot from a larger bulb

buzz-pollination: a technique used by some bees, particularly bumblebees (*Bombus* spp.) to release pollen from poricidal anthers by vibrations; also called sonication

Glossary

bract: a modified leaf in an axil from which a flower or inflorescence stalk arises

bracteole: a small bract, especially one on a floral stem

bractlet: a bract on the stalk of an individual flower, which is itself on a main stalk of several flowers

calcareous: containing calcium carbonate (limestone); chalky

calciophile: a plant that prefers to grow in calcium-rich soils (having a higher pH)

calyptra: a hood- or caplike structure that covers a plant organ such as a flower bud and is pushed off by the growing of the organ

calyx (calices): a collective term for the sepals of a flower

capitulum (capitula): an inflorescence of sessile flowers born on a flattened and expanded portion of the inflorescence axis; used to refer to the inflorescence of a member of the aster family

capsule: a dry, dehiscent fruit that splits along one or many lines into many separate seed-bearing chambers

carbohydrate: a compound of carbon, hydrogen, and oxygen in the form of sugar or starch; the product of photosynthesis in a plant

cardenolide: a type of steroid found in plants; in the form of cardenolide glycosides it is often toxic and provides defense against herbivores

carminic acid: a red compound found in some scale insects, such as the cochineal, that feeds on *Opuntia* cactus; the insects are collected and crushed to produce a red dye

carnivorous: in botany, referring to plants that derive some or most of their nutrients from trapping and digesting insects and other animals

carpel: the primary unit of a pistil, consisting of an ovary, style, and stigma and containing ovules

caudicle: the slender, stalklike appendage of the pollen masses in orchids

CE: abbreviation for Common (Current) Era; equivalent to AD

center of diversity: a region where a particular group of organisms has its greatest genetic diversity; it is often the region where the group originated

chaff: dry, fine bracts or scales as found in flowers of the Asteraceae

chasmogamous: a flower open to pollinators

chlorophyll: the green pigment in plants that allows for photosynthesis to occur

chromosome: a threadlike structure found in the nucleus of most living cells that carries the organism's genetic information in the form of genes

chrysalis: a transitional state in which a caterpillar becomes a butterfly; also refers to the structure itself

circumarctic: found in the Arctic region surrounding the North Pole

circumboreal: found through the North American and Eurasian boreal region in northern part of the Northern Hemisphere, south of the Arctic

circumnutate: movement of the growing portions of a plant (e.g., a tendril or stem) to form spirals, or ellipses

circumpolar: surrounding or found in the vicinity of either terrestrial pole

cirque: a half-open, steep-sided hollow at the head of a valley, or on a mountainside, formed by glacial erosion

CITES: an international treaty, Convention on International Trade in Endangered Species of Wild Fauna and Flora, to help conserve and protect endangered plants and animals

cladode: a stem that looks and functions like a leaf

claw: the narrowed base of a petal, sepal, or bract (e.g., the tepals of *Lilium philadelphicum* [wood lily])

cleistogamous: flowers that never open and self-fertilize within the closed buds

clone: a plant that is genetically identical to one or more other plants, having originated vegetatively from the same parent plant; may cover a large area with apparently separate plants that are attached underground

cochineal: a scale insect containing red carminic acid used to make a dye; also refers to the dye itself

coevolution: the influence of species upon each other throughout time

column: a structure in orchid flowers that contains both male and female parts

coma: silky fibers attached to a milkweed seed that allow it to be carried by wind currents (= floss)

composite: a common name for a member of the aster family

compound leaf: a leaf consisting of more than one division; comprising several leaflets

coniferous: referring to trees or shrubs having cones as the reproductive structures, as in gymnosperms (e.g., pines and spruces)

conjoined: united

connivant: separated at the base, but close at the top; often refers to anthers

contiguous: touching and continuing, but not fused

contorted: twisted

corbiculae: pollen-carrying modifications of the hind legs of some bees (pollen baskets)

corm: a fleshy subterranean storage organ comprising the base of a stem

corolla: the floral whorl of petals within the calyx; a collective name for the petals

corona: a modified floral structure located between the stamens and the corolla as in *Asclepias* (milkweed) or *Passiflora* (passion-flower)

corpusculum: a small organ that connects the two translator arms of milkweed pollinia

cotyledon: a leaf of an embryonic plant within a seed

crassulacean acid metabolism (CAM): a type of photosynthesis that developed in some plants as an

Glossary

adaptation to arid conditions; the stomata remain closed during the day to reduce water loss, but open at night to collect and allow CO_2 to diffuse into the mesophyll cells; it is stored as malate (a four-carbon acid) during the night, and then in daytime, transported to the chloroplasts where it is converted back to CO_2 and used during photosynthesis

crenate: toothed along the margin, the teeth rounded
crepuscular: appearing or active in twilight
cross-pollination: pollination of a flower with pollen from another plant of the same species
cryptically colored: colored or patterned so as to blend in with the environment
cultivar: a plant that has originated in cultivation by manipulating a normal plant by some means, often by hybridization
cushion plant: a compact, low-growing plant found in alpine, subalpine, and Arctic environments
cuticle: a waxy layer covering a leaf or stem
cypsela: a specialized achene that develops from an inferior ovary

deciduous: shedding foliage (or other organs) at the end of a growth period or season
decurrent: having an adnate wing or margin extending down a stem below the point of insertion, as in a leaf
decussate: leaves arranged oppositely on a stem and at right angles to the pair above and the pair below
dehisce: split open at maturity, allowing discharge of the contents, as with the seeds of a fruit or the pollen of an anther
diapause: a temporary pause in the development of an organism, especially in insects
diaspore: a unit of dispersal such as a seed or fruit
dimorphic: of two forms
dioecious: producing male and female flowers on separate individual plants
dipteran: referring to a fly
disjunct: the range of a species with populations that are widely separated geographically from others of the same species
disk flower: a small, tubular flower in the Asteraceae lacking an expanded corolla and usually occurring in the central portion of the capitulum (e.g., the yellow center of a daisy)
dispersal: the spreading of seeds or other propagules
distal: toward or at the tip or far end
distylous: having styles of two different lengths in flowers of different plants (e.g., as in *Mitchella repens*)
diurnal: functioning or active during the day
DNA: a molecule composed of two chains of nucleotides that form a double helix and carry genetic instructions for an organism
drupe: a fleshy fruit with a firm endocarp that permanently encloses the usually solitary seed (e.g., a peach)

druse: a rounded cluster of calcium oxalate crystals found in some plant cells
dust seeds: very fine seeds such as those of orchids or Indian pipe (*Monotropa uniflora*) that are transported by wind currents

ectomycorrhizal relationship: a symbiotic relationship that occurs between a fungal symbiont and the roots of various plant species; the fungi do not penetrate the host's cell walls, but instead form an intercellular interface
elytra: the forewings of most beetles, which are hardened and serve to protect the body and membranous hind wings; they are not used for flying
embryo: a young plant in a seed
emergent: used to describe an aquatic plant with parts that extend above the water's surface
endemic: confined to a specific geographical area
endosperm: the food storage tissue of a seed
ephemeral: transient, short-lived
epicalyx: a set of bracts, located closely below the calyx, appearing like a second calyx (as found in *Hibiscus*)
epichile: the apical portion of an orchid's labellum
epiphyte: a plant without connection to the soil, growing upon another plant but not deriving its food or water from it
epithet, specific: the second part of a species name in binomial nomenclature
erose: slightly jagged along the margin
exoskeleton: the rigid external covering of the body of an arthropod that provides support and protection
extirpate: to cause to become extinct from an area
exudate: a substance secreted by a plant

fall: the sepal of an iris flower
family: a level of taxonomic classification made up of genera with similar characteristics; related families are classified into orders
fasciated: having abnormal fusion of parts or organs, resulting in a flattened ribbonlike structure
fen: a wetland with neutral or calcium-rich soils
fertilization: in botany, the union between a sperm cell released by a pollen grain deposited on a receptive stigma, and an egg cell in an ovule of the flower
filament: the anther-bearing stalk of a stamen
floret: a little flower; an individual flower of a cluster as in the head of a composite (Asteraceae)
floss: another term for the coma of milkweed seeds
follicle: a fruit that opens only along one line of dehiscence as in milkweed (*Asclepias*), which differentiates it from a pod that opens along two lines of dehiscence, such as a pea pod
form: anglicized version of the Latin *forma*; a level of classification within a species that recognizes minor differences from the species; may be abbreviated as f. (e.g., *Claytonia virginica* f. *lutea*) or fo.

Glossary

formic acid: an irritant volatile acid present in the fluid emitted by some ants

frass: the excrement of insects

fructose: a hexose sugar found especially in honey and fruit

fungus (fungi): a spore-producing organism, including mushrooms, molds, and yeast, that feeds on organic matter

frazil ice: soft, amorphous ice formed by the accumulation of ice crystals in water that is too turbulent to freeze solid

funicle: a cordlike structure that attaches an ovule or seed to the interior of a fruit

gall: a mass of plant tissue that proliferates at the site where an insect has inserted its egg; the larva will live in the gall and feed on its interior until ready to pupate

gas chromatography: a technique that dissolves and vaporizes a sample solution in order to separate and analyze individual organic compounds in the gas phase

genome: the hereditary information encoded in the DNA of an organism

genotype: the set of genes in the DNA of an organism that is responsible for a particular trait

genus (genera): a level of taxonomic classification that includes species with similar characteristics; related genera are classified into families

gestalt: a structure (such as a plant) that, when considered as a whole, has qualities that are more than the sum of parts

gilled mushroom: a fungus having thin, papery ribs under its cap, which open to allow the spores to disperse

germination: emergence from a dormant state, as when a seed begins to develop into a plant

glochid (glochids or **glochidia):** small barbed hair, as found at the nodes of cacti

glucose: a simple sugar made mainly by plants

glucoside: a molecule commonly found in plants that contains a sugar group bonded to a nonsugar group, which when hydrolyzed, breaks off the sugar group and activates the nonsugar group, often a poison

grub: the larva of an insect, especially a beetle

guild: in ecology, a group of species that exploit the same resources

guttation: expelling of droplets of liquid from hydathodes along the leaf margin and apices (evident in the leaves of strawberry)

gynophore: a stalked ovary

gynostegium: a column of fused male (stamens) and female (pistils) reproductive structures, as in milkweeds (*Asclepias*)

habitat: a particular type of environment regarded as the home of an organism

haustorium (haustoria): the part of a parasitic plant that invades a host species and joins with the host's vascular tissue in order to procure nutrients from it

head: equal to the capitulum in members of the Asteraceae

heath: a dwarf shrub with small leathery leaves in the Ericaceae

hemiparasitic: a plant that photosynthesizes but obtains some of its water and/or nutrients from a host plant (e.g., lousewort [*Pedicularis*]); compare with holoparasite

herbaceous: nonwoody, soft-stemmed

herbarium (herbaria): a collection of dried and mounted plant specimens

herbicide: a substance used to kill plants

herbivore: an animal that eats plants

herbivory: the act of feeding on plants

heterophyllous: having leaves of more than one type

holoparasite: an organism that is completely reliant on a host species for its nutrients (e.g., squawroot); compare with hemiparasite

homologous: in biology, similar in position, structure, and evolutionary origin

homostylous: having flowers with styles of uniform length

honeydew: a liquid waste product produced by aphids and relished by ants

host plant: a plant that is the target plant from which an insect or a parasitic plant derives its nutrition

hover fly: also called flower fly and syrphid fly; a fly belonging to the Syrphidae family of flies

hummingbird syndrome: flowers having certain characteristics that attract hummingbirds (e.g., often red tubular corolla, scentless flowers, lack of a landing platform, thickened basal corolla walls, generally anthers at the dorsal part of the corolla, and copious, thin nectar)

hummock: a small mound, as in a small area of land raised above the level of water in a swamp or bog

hybrid: the offspring of the sexual union of two different species

hybridize: crossbreed two different species or varieties

hydathode: a pore of a leaf that is modified to exude water and sometimes dissolved minerals

hydrophobic: tending to repel water, as when water beads up on a leaf

hyoscyamine: an alkaloid found in plants

hypanthium (hypanthia): a ring or cup around the ovary, formed either by expansion of the receptacle or by the union of the lower parts of the calyx, corolla, and androecium

hypha (hyphae): each of the branching filaments that make up the mycelium of a fungus

hypochile: the basal portion of the labellum of an orchid

hypogeal germination: a seed germination process whereby the cotyledons remain underground

hypogynous: referring to flowers in which all other floral parts are attached below the ovary

Glossary

ichneumon wasp: a slender parasitic wasp of the family Ichneumonidae that deposits eggs in, on, or near the larvae of other insects by means of a long ovipositor

in vitro: a process taking place in a test tube or culture dish, not in a living organism

indehiscent: not opening at maturity, as in a fruit

inferior ovary: an ovary located beneath the other floral parts

inflorescence: an arrangement or grouping of flowers on a plant

instar: in insects, a phase between each molt until sexual maturity is reached

intergeneric hybrid: an offspring of parents belonging to two different genera

introduced: a non-native plant intentionally or unintentionally brought into an area; it is not necessarily invasive

invasive: a non-native plant that reproduces prolifically to the detriment of the native species in its area

involucre: the collective bracts subtending a flower or a collection of flowers, as with the inflorescences of the aster family (Asteraceae)

krummholz: a vegetation type characterized by stunted trees, usually at higher elevations in mountains; from the German for twisted wood

labellum (labella): the median petal of an orchid; it serves as a landing platform for insects, attracting them by virtue of (usually) being larger and variously embellished in structure and color; also called the lip

lacuna (lacunae): an empty space within a tissue, as found in many aquatic plants

larva (larvae): the active, immature form of an insect (e.g., a caterpillar or grub)

latex: milky sap, as found in milkweeds and some members of the Asteraceae

layering: propagation that triggers root production on stems that are still attached to the parent plant; usually done by securing a portion of the stem to the ground and covering it with soil; the newly rooted plants may then be separated and planted

leaflet: a small leaflike segment of a compound leaf

legume: a general name for a member of the pea family; a type of fruit, usually dry, that opens along two sutures

liana: a climbing woody vine

lignin: a complex chemical component of wood and of all vascular plant cell walls providing mechanical strength

ligulate head: a head of a member of the Asteraceae bearing only ray flowers

lithophytic: a plant capable of growing on rock

limy: having a high pH due to calcium carbonate in the soil

Linnaeus, Carl: an eighteenth-century Swedish botanist, zoologist, and physician who is credited with the creation of the system of binomial nomenclature used for naming all living things; in 1753 Linnaeus published *Species Plantarum* in which names were given to all plants then known to him

lip: an alternate name for one petal of an orchid that is usually showier than the other two; it may be in the bottom position in resupinate orchids or at the uppermost position in nonresupinate orchids; also used to refer to a distinct portion of other asymmetrical flowers such as members of the mint family

lipid: a fatty or waxy organic compound that is not solvent in water, and whose main biological function involves energy storage, being a structural component of a cell membrane, and cell signaling

locule: a chamber within an ovary or fruit

loculicidal: opening along the midrib of each locule

lotus effect: self-cleaning properties that are the result of ultrahydrophobicity; dirt particles are picked up by water droplets due to the nanoscopic architecture on the surface (e.g., the leaves of *Nelumbo* spp.)

lump: to group a number of formerly separate taxa into one taxon

maggot: the larval form of a fly

mandibles: the mouthparts of an arthropod

mano: a stone used for grinding grains or other things, such as dyestuffs, by hand in a metate

marcescent: withering without falling off

mass spectrometry: a method for identifying the chemical constitution of a substance by means of the separation of gaseous ions according to their differing mass and charge

mericarp: an individual carpel of a schizocarp, as in the fruits of Apiaceae and Malvaceae

meristem: an area of a plant that produces cells for plant growth

mesic: descriptive of a habitat that has a moderate amount of moisture

metamorphosis: a marked developmental change in the form or structure of an animal (such as a butterfly transforming from caterpillar to chrysalis, to butterfly)

metate: a stone with a concave upper surface used as the lower millstone for grinding grains or other things, such as dyestuffs

midge: a small fly

midrib: the main, central vein of a leaf

mixed mating system: a system in which plants in a population use more than one method of reproducing (e.g., selfing and outcrossing)

mixotrophic: capable of deriving nourishment from autotrophic and heterotrophic means, as in hemiparasites that can produce some of their own carbohydrates through photosynthesis, but also derive some nutrients from a host plant

Glossary

monocarpic: having a lifestyle in which a plant flowers and fruits once in its lifetime and then dies

monocot: a flowering plant whose embryo generally has only one seed leaf (cotyledon), has leaves usually with parallel venation, and flower parts in threes or multiples of three

monotypic: referring to a family or a genus with only one species

mordant: a chemical that fixes a dye in a substance by combining with it; the type of mordant may affect the color of the dye

morph: a plant or animal with a visual difference from others in the same population of a species

morphology: the form and structure of an organism and the study of the same

mucilage: a thick, slimy, or gelatinous substance inside of some plants (e.g., some Malvaceae and aloe)

mucro: a short, stiff, sharp point, as on a leaf apex

mutualistic relationship: one in which two organisms of different species interact so that each organism benefits; such relationships generally coevolve over time

mycorrhizal: descriptive of the symbiotic union between a fungus and the roots of a plant

mycoheterotroph: a plant that has an obligate relationship with mycorrhizal fungi in the soil in order to obtain water and nutrients

mycologist: a scientist who studies fungi

mycorrhizal fungi: fungi having a symbiotic union with the roots of a plant

nanometer: one-billionth of a meter

nanotechnology: a branch of technology that manipulates and manufactures materials and devices on the scale of atoms or small groups of atoms between 1 and 100 nanometers

naturalized: introduced from another area but spread widely and naturally throughout a new landscape

nectar: (noun) a sugary liquid produced within flowers to attract pollinators or by other parts of the plant to attract insects that subsequently protect the plant from herbivores (see nectary, extrafloral)

nectar: (verb) the imbibing of nectar by a butterfly or moth

nectary: a tissue or structure that produces nectar

nectar, extrafloral: nectar that is produced in a part of the plant other than the flower

New World: includes all the Americas

nocturnal: blooming or active at night

node: an area on a stem where a leaf or branch is attached

nodule: in botany, referring to a rounded swelling on the roots of plants, primarily legumes, that forms a symbiosis with nitrogen-fixing bacteria; such plants are used as cover crops to enrich the soil of cultivated fields

nonresupinate: as in an orchid that does not rotate 180° from its original position in bud as it opens; also occurs in some other usually resupinate flowers in other families

Northeast: referring to the northeastern portion of the United States; in a broad sense (as used in this book), this term would include the northeastern quarter of the country; in a more narrow sense, it refers to New England, New York, Pennsylvania, and New Jersey

obligate: dependent on certain conditions, such as the presence of a host plant for a parasite

Old World: includes Europe, Africa, and Asia

opposite leaves: leaves that arise across from each other from the same node on a stem

orchidologist: one who studies orchids

ornamental: an attractive plant used in gardens

osmophore: a tissue or gland, usually associated with flowers, from which aromas emanate

outcross: reproduction by individuals of the same species that are not closely related genetically

ovary: the part of a flowering plant that contains the ovules; the fertilized ovary becomes a fruit and the ovules become the seeds

ovipositor: a tubular organ through which a female insect deposits eggs

ovule: a young or undeveloped seed containing an egg cell inside the ovary

oxycodone: an opioid capable of causing narcotic effects (found in some nectars)

pahoehoe: lava that cools to form smooth, undulating or ropy masses

palmate: with three or more lobes, veins, or leaflets arising from a common point

palynologist: one who studies pollen in both living and fossil form

papillae: short, rounded, blunt projections

pappus (pappi): the modified calyx crowning the ovary (and achene) in members of the Asteraceae; pappus may consist of hairs, scales, bristles, or a mixture of these

parasite: an organism that lives in or on an organism of another species (its host) and derives nutrients from it, often at the other's expense

parasitoid: an insect, especially a wasp or fly, that completes its larval development within the body of another insect, eventually killing it; it is free-living as an adult

pathogen: a causative agent of disease (e.g., a bacterium or virus)

pedicel: the stalk of an individual flower

peduncle: the stalk of a solitary flower or of an inflorescence of more than one flower (each of which will have its own pedicel)

peltate: descriptive of a leaf blade attached to its petiole at a point within its lower surface rather than at the leaf blade's margin

pepo: technically a berry with a hard, thick rind (e.g., a pumpkin)

Glossary

perennial: a plant that lives for longer than two growing seasons and continues to grow after it has reproduced

perianth: the collective name for the calyx and the corolla

petal: a unit of the corolla (the floral whorl between the calyx and the stamens); often serves to attract pollinators

petiole: a leaf stalk

pH: a unit of measurement of the acidity or alkalinity of a solution or of soil

pheromone: a chemical substance usually produced by an animal that serves especially as a stimulus to other individuals of the same species for one or more behavioral responses (e.g., mating)

phloem: the vascular tissue in plants that conducts sugars and other metabolic products downward from the leaves

photosynthesis: the process by which green plants produce carbohydrates from carbon dioxide and water using the sun's energy; oxygen is a byproduct

phytoestrogen: an estrogen occurring naturally in plants, especially legumes

phyllary: an involucral bract of an Asteraceae

phylloclade: a portion of the stem having the general form and function of a leaf; same as cladophyll

pin flower: a type of heterostylic flower that has a relatively long style and short stamens

pine barrens: also called pine bush or sand plains; a plant community on nutrient-poor, often sandy soil dominated by pines (especially Jack pine and pitch pine), oaks, shrubs (especially members of the heath family), and herbaceous plants that have adapted to this dry, acidic habitat

pinnate: with two rows of lateral branches or appendages, or parts along an axis, as in the venation of some leaves (e.g., birch)

pistil: the female part of a flower composed of the ovary, style, and stigma

pistillate: descriptive of a flower that has only functional female organs

placenta (plancentae): in botany, the tissue to which seeds are attached

placentation: the arrangement of the placenta or plancentae in the ovary of a flower

Pleistocene: the first epoch of the Quaternary period

plica (plicae): in botany, the folds of corolla tissue between lobes (as in some gentians)

pollen: the microspores (male reproductive structures) of seed plants contained in the anthers

pollinarium: in milkweeds (*Asclepias*), two pollinia, two translator arms, and the corpusculum that attaches them

pollinium (pollinia): a coherent cluster of many pollen grains, transported as a unit during pollination, as in many orchids and milkweeds (in milkweeds, part of a pollinarium)

pollination: in flowering plants, the deposition of pollen on the stigma of the same or a different flower

pollinator: an animal that serves as a vector to transport pollen from the anthers of one flower to the stigma of another (or within the same flower)

polyploidy: containing more than two homologous sets of chromosomes

poricidal anther: an anther that opens by at pore at its tip

pouch: a term for a specialized lip (labellum) in a lady-slipper orchid

proboscis (proboscises): a long tongue of an insect, such as a butterfly or moth, used for sucking nectar from flowers

protandrous: said of a flower in which the male reproductive organs function before the female organs are receptive; this helps prevent self-pollination

protogynous: a flower having its stigma receptive before the anthers release their pollen

pseudanthium (pseudanthia): a group of flowers that appear to be one flower, as in dogwood (*Cornus*) or sunflowers (*Helianthus*); a false flower

pubescent: bearing hairs (trichomes) of any sort

pulvinus (pulvini): the swollen base of a petiole or petiolule, which may govern the attitude of the leaf or leaflet

pupa (pupae): an insect in its inactive immature form between its larval stage and its adult form (e.g., a chrysalis)

pupate: become a pupa

raceme: an inflorescence with a single axis, the stalked flowers of which bloom from the bottom upward

radiocarbon dating: a technique developed in the 1940s for measuring the approximate age of organic materials by comparing the known rate of decay of carbon-14 (^{14}C) with the stable amount of carbon-12 (^{12}C) in the same material

ramet: an individual member of a clone

range: the area in which a plant grows

ray flower: a strap-shaped ligule flower of an Asteraceae, which is tubular at its base; one branch of an umbel

receptacle: the expanded apex of a pedicel that bears the floral parts

receptive: descriptive of a stigma when at a stage when pollination can occur

reflexed: arched sharply downward or backward

refugium (refugia): area of relatively unaltered climate inhabited by plants and animals during a period of continental climatic change (such as glaciation)

regular symmetry: having similar parts regularly arranged around a central axis, as in a daisy head

reproductive structures: the structures of (male) stamens and (female) pistils in a flower

resupinate: referring to the position of the lip of an orchid flower that has turned 180° while still in bud, resulting in a flower with the lip in the lowermost position in the flower

rhipidium: a fan-shaped inflorescence in which the branches lie in the same plane

Glossary

rhizobia: soil bacteria capable of forming symbiotic nodules on the roots of leguminous plants and fixing atmospheric nitrogen

rhizome: a prostrate stem on or below the ground that sends off roots, vertical stems, and/or leaves

saprophyte: a fungus or microorganism that lives on dead or decaying organic matter; formerly used in place of mycoheterotroph to describe plants such as Indian pipe, which actually get their nutrients via a mycorrhizal network between the plant and the trees in its vicinity

scape: the leafless stalk of a flower or inflorescence

schizocarp: a fruit derived from a two- to many-carpellate pistil that splits at maturity into individual segments (mericarps) as found in Apiaceae

scientific name: the two-word name of an organism—the binomial, based on Latin or the Latinized version of words in other languages, which includes first the genus (always capitalized) followed by the specific epithet (never capitalized), both always italicized; each plant has a scientific name that distinguishes it from all others

sclerenchyma tissue: a supportive tissue of vascular plants, consisting of thick-walled, usually lignified cells

scopolamine: an alkaloid found in plants

scopa (scopae): a small, brushlike tuft of hairs on some insects, especially that on which pollen collects on the leg of a bee

seed coat: the layer of tissue covering a seed, derived from the integuments surrounding the ovule; the testa

self-compatible: in a plant, when pollen from the same plant lands on the stigma of the same flower or another flower on the same plant and results in fertilization

self-pollination: also called selfing; pollination of a plant with pollen from the same plant or another plant within a clone; this does not necessarily result in fertilization

self-sterile: in a plant, incapable of self-fertilization if pollen from a flower lands on the stigma of the same flower or another flower on the same plant

sepal: one unit of the calyx, the outer whorl of the perianth of a flower

septum (septa): in an ovary, a partition formed by connate walls of adjacent carpels

serpentine-rich: pertaining to soil rich in a green mineral, the presence of which limits the ability of most species to grow because of relatively high concentrations of certain metals such as nickel, manganese, and chromium

serrate: in a leaf, toothed along the margin, the teeth sharp and pointed forward

sessile: without a stalk, as in leaves or flowers

social wasp: a wasp that lives as part of a colony in a nest of its own making

soil remediator: a plant, microorganism, or other biological entity capable of improving soil polluted by human-caused activities

sonication: see buzz-pollination

snowbank community: a group of alpine plants that assemble in sheltered sites above tree line where late-lying snow provides insulation from late-season frosts

species: (singular [sp.] and plural [spp.]) the basic rank of classification within a genus; may be further subdivided into subspecies, varieties, or forms; a plant's name consists of the genus name and the specific epithet

Species Plantarum: a 1753 publication by Carl Linnaeus that is the basis for all validly published plant names at that time; the foundation for the binomial system of nomenclature

sperm nuclei: the cells in the nucleus of a pollen grain that function in the fertilization of the ovules of a seed plant

sphinx moth: a member of the moth family Sphingidae; large-bodied, fast-flying moths that are sometimes diurnal; a hawk moth

spicule: in botany, a small spike; often tiny, as in those along the margins of the leaves of some lilies

spiracle: a respiratory opening, especially the pores on the body of an insect

stamen: the male part of a flower comprising the filament and pollen-bearing anther

staminal ring: a ringlike arrangement made up of stamens fused at least at their bases

staminal tube: a tube formed from connivant stamens, which often open to the inside of the tube to release their pollen

staminode (staminodia): a sterile stamen

standard: the uppermost petal in a papilionaceous flower such as a pea flower (the banner) or an upright petal of an iris

stigma: pollen-receptive tissue located at the apex of the style (or directly on the apex of the carpel if the style is lacking)

stipule: one of a pair of basal appendages found in association with the leaves of many species

stoma (stomata): a special opening in a plant part (especially the leaves), the opening of which is controlled by two surrounding guard cells; it functions for gas exchange; a stomate

stratification: the process of pretreating seeds to simulate natural conditions that a seed must endure before it can germinate, such as a cold period

stylar arms: the separate portions of a style that divides into branches

style: the part of the pistil between the ovary and the stigma

subfamily: a taxonomic category that ranks below family and above genus, usually ending in -oideae in botany

subshrub: a dwarf shrub, especially one that is woody only at its base

stellate: starlike; used to refer to radiating tufts of hairs on a plant

subspecies: a population within a species that differs somewhat from the recognized species, but not enough to be classified as a separate species; abbreviated subsp. or ssp.

subterranean: underground

Glossary

sucrose: a sweet crystalline sugar composed of fructose and glucose; it is produced naturally in plants and refined for table sugar, especially from sugarcane and sugar beets

suffrutescent: partially shrubby, or dying back to a perennial woody base

superhydrophobic: highly hydrophobic, such that water beads up and rolls off, as in the leaves of *Nelumbo*

superior ovary: an ovary located above the other floral parts

symbiotic: denoting a mutually beneficial relationship between different organisms

taproot: a root system in which the primary root is of much greater diameter than any lateral roots; it often serves as a storage organ

tarn: a small mountain lake

taxon (taxa): a named group at any level of classification, for example, a species, a genus, a family

taxonomist: one who studies taxonomy

taxonomy: the study and description of variation in the natural world and the subsequent classification and naming thereof

tendril: a twining stem, partial stem, leaf, or leaflet that helps a plant attach to neighboring plants or other supports as it climbs

tepal: the unit of the perianth of a flower in which the sepals and petals are similar in appearance (e.g., lilies)

tepui: a table-top mountain or mesa in the Guiana Highlands of South America

terrestrial: of plants, growing on land rather than in water or on trees or rocks

thigmotropism: growth or change in orientation in response to touch

thorax: in entomology, the midsection of an insect's body to which the head, legs, wings, and abdomen are attached

thrum flower: a type of heterostylic flower that has a relatively short style and longer stamens

tissue culture: a collection of techniques used to maintain or grow plant cells, tissues, or organs on a nutrient culture medium; widely used to produce clones of a plant

translator: a narrow connector between the two pollen sacs (pollinia) from different anthers in a milkweed flower (*Asclepias*)

trebuchet: (Fr.) a medieval machine used for hurling large stones or other missiles

tree line: the altitude on a mountain above which trees cannot grow

trichome: a hairlike outgrowth from the epidermis of a plant part

trinitrotouline (TNT): a chemical used as an explosive

tuber: an underground swollen stem that serves to store food for the plant

tubular flower: a flower shaped like a tube or narrow funnel

tundra: a level or rolling treeless plain characteristic of Arctic and subarctic regions dominated by mosses, lichens, herbs, and small shrubs

turgor pressure: the force within a plant cell that pushes the plasma membrane against the cell wall; if turgor pressure drops, the cell or plant structure will wilt

type specimen: a specimen upon which a species name is based and from which at least a portion of the species description was made (other specimens may also be used for providing descriptive information)

ultraviolet or UV pattern: markings as on a flower discernible only under ultraviolet light or to insects able to see colors in this wavelength

umbel: an inflorescence with flowers or branches borne from a common point, presenting flowers on a rounded or flat plane at the top; may be simple or compound

umbellet: one of the ultimate umbellate flower clusters in a compound umbel

USDA PLANTS Database: a website of the United States Department of Agriculture with information about all the plants growing in the United States, with maps of their ranges and their native or introduced status

variety: a population within a species that has minor differences that are more significant than those that denote a form, but less significant than those used to determine subspecies (the delineation of the various ranks of classification is often subjective); abbreviated var.

venation: the arrangement of the vascular system (veins) of a leaf

verrucose: warty, covered with wartlike projections

versatile: in stamens, having the anther attached to the filament at its midpoint, thus allowing it to move easily when touched

viscidium: a sticky structure attached to an orchid pollinium via stipes; the viscidium facilitates pollen transfer by sticking to pollinators

viviparous: in plants, sprouting or germinating on the parent plant, as the bulbils in the inflorescence of some plants (e.g., wild leek)

zygomorphic: in flowers, a flower that is capable of being bisected by only one plane into similar halves; bilaterally symmetric

References

The General References were consulted for background information and referred to for multiple essays. They are not included again in the references for the individual species.

General References

Clemants, Steven, and Carol Gracie. 2006. *Wildflowers in the Field and Forest: Field Guide to the Northeastern United States.* Oxford University Press, New York.

Fernald, Merritt Lyndon. 1950. *Gray's Manual of Botany*, 8th edition. American Book Company, New York.

Flora of North America Editorial Committee, eds. 1993+. *Flora of North America North of Mexico*, 16+ vols., http://www.fna.org/families (accessed online 2012–2019).

Foster, Steven, and James A. Duke. 1990. *Peterson Field Guides: Eastern/Central Medicinal Plants.* Houghton Mifflin, Boston.

Gleason, Henry A., and Arthur Cronquist. 2004. *Manual of the Vascular Plants of the Northeastern United States and Adjacent Canada*, 2nd edition. The New York Botanical Garden Press, Bronx.

Holmgren, Noel H. 1998. *Illustrated Companion to Gleason and Cronquist's Manual.* The New York Botanical Garden, Bronx.

Hyam, Roger, and Richard Pankhurst. 1995. *Plants and Their Names: A Concise Dictionary.* Oxford University Press, New York.

International Plant Names Index. 2015. "The International Plant Names Index," http://www.ipni.org (accessed 2012–2019).

Judd, Walter S., Christopher S. Campbell, E. A. Kellogg, P. F. Stevens, and M. J. Donoghue. 2008. *Plant Systematics: A Phylogenetic Approach*, 3rd edition. Sinauer Associates, Sunderland, MA.

Lamoureux, G. 2002. *Flore Printanière.* Fleurec Éditeur, Saint-Henri-de Lévis, Quebec.

Mabberley, David J. 2008. *Mabberley's Plant-Book*, 3rd edition. Cambridge University Press, Cambridge.

Mehrhoff, Leslie J., John A. Silander Jr., et al. n.d. "Invasive Plant Atlas of New England," https://www.invasive.org/weedcd/html/ipane.htm (accessed 2012–2019).

NatureServe. 2007. "NatureServe Explorer: An Online Encyclopedia of Life," Version 6.2, http://www.natureserve.org/explorer (accessed 2012–2019).

Pell, Susan K., and Bobbi Angell. 2016. *A Botanist's Vocabulary.* Timber Press, Portland, OR.

Radcliffe-Smith, A. 1998. *Three-language List of Botanical Name Components.* Whitstable Litho, Whitstable, Kent, UK.

Stevens, Peter F. (2001 onward). Angiosperm Phylogeny Website Version 14, July 2017 (and more or less continuously updated since), http://www.mobot.org/MOBOT/research/APweb/ (accessed 2012–2019).

The Plant List. 2013. Version 1.1. Published on the Internet; http://www.theplantlist.org/ (accessed 2013–2019).

Tropicos.org. Missouri Botanical Garden, http://www.tropicos.org (accessed 2012–2019).

USDA, NRCS. 2019. The PLANTS Database. National Plant Data Team, Greensboro, NC, http://plants.usda.gov/ (accessed 2012–2019).

Usher, George. 1996. *The Wordsworth Dictionary of Botany.* Wordsworth Editions, Ware, Hertfordshire, UK.

Alpine Wildflowers references

Aiken, S. G., M. J. Dallwitz, L. L. Consaul, et al. 1999 onward. Flora of the Canadian Arctic Archipelago: Descriptions, Illustrations, Identification, and Information Retrieval website, http://nature.ca/aaflora/data/index.htm (accessed 2013).

Appalachian Mountain Club. 2009. "The Climate Connection." The Mountain Watcher Newsletter, Spring Issue 2009, http://www.outdoors.org/conservation/mountainwatch/upload/MW_Newsletter_Spr09.pdf (accessed 2012).

Appalachian Mountain Club. 2012. "Mt. Washington Climate Trends," The Mountain Watcher Newsletter, Spring Issue 2012, http://www.outdoors.org/conservation/mountainwatch/upload/MW_2012Newsletter.pdf (accessed 2013).

"Back from the Brink: Ten Success Stories Celebrating the Endangered Species Act at 40: Robbins' Cinquefoil (*Potentilla robbinsiana*)," Endangered Species Coalition website, http://www.endangered.org/cms/assets/uploads/2013/06/2013-Back-from-the-Brink-Top Ten.pdf?utm_source=January+2014+Member+eNews&utm_campaign=January+2014+eNews&utm_medium=email (accessed 2014).

Bell, Katherine L., and L. C. Bliss. 1977. Overwinter Phenology of Plants in a Polar Semi-desert. *Arctic* 30(2): 118–121.

Bliss, Lawrence C., 1963. Alpine Plant Communities of the Presidential Range, New Hampshire. *Ecology* 44(4): 678–697.

Bliss, Lawrence C., 1963. *Alpine Zone of the Presidential Range.* Self-published, Edmonton, Canada.

Brown, Janice. 2006. "Found Only in New Hampshire: Robbins' Cinquefoil," New Hampshire's History Blog, http://cowhampshire.blogharbor.com/blog/_archives/2006/10/23/2292758.html (accessed 2014).

Brumback, William E., Doug M. Weihrauch, and Kenneth D. Kimball. 2003. Propagation and Transplanting of an Endangered Alpine Species: Robbin's Cinquefoil (*Potentilla robbinsiana*, Rosaceae). *Native Plants.* 5(1): 91–97. http://muse.jhu.edu/login?auth=0&type=summary&url=/journals/native_plants_journal/v005/5.1brumback.pdf (accessed online 2014).

Bucker, C. 2013. "Northeast Alpine Peaks," Summitpost.org website, http://www.summitpost.org/northeast-alpine-peaks/419290 (accessed 2013).

Butterfield, B. J., L. A. Cavieres, R. M. Callaway, et al. 2013. Alpine Cushion Plants Inhibit the Loss of Phylogenetic Diversity in Severe Environments. *Ecology Letters* 16(4): 478–486.

"*Rhododendron lapponicum*: Lapland Rosebay," Central Yukon Species Inventory Project website, http://www.flora.dempstercountry.org/0.Site.Folder/Species.Program/Species.php?species_id=Rhodo.lappo (accessed 2004).

Cogbill, Charles V. 1993. The Interplay of Botanists and *Potentilla robbinsiana*: Discovery, Systematics, Collection, and Stewardship of a Rare Species. *Rhodora* 95(881): 52–75.

Coker, P. D., and A. M. Coker. 1973. Biological Flora of the British Isles: *Phyllodoce caerulea* (L.) Bab. *Journal of Ecology* 61(3): 901–913.

Elberling, Heidi. 2001. Pollen Limitation of Reproduction

References

in a Subarctic-alpine Population of *Diapensia lapponica* (Diapensiaceae). *Nordic Journal of Botany* 21: 277–282.

Fitzgerald, B. T., K. D. Kimball, C. V. Cogbill, and T. D. Lee. 1990. "The Biology and Management of *Potentilla robbinsiana*, an Endemic from New Hampshire's White Mountains," *in* Mitchell, Richard S., Charles J. Sheviak, and Donald J. Leopold (eds.). Proceedings of the 15th Annual Natural Areas Conference. Ecosystem Management: Rare Species and Significant Habitats. *New York State Museum Bulletin* 471: 163–166.

Grimaldi, David. 1988. Bee Flies and Bluets: *Bombylius* (Diptera: Bombyliidae) flower-constant on the Distylous Species, *Hedyotis caerulea* (Rubiaceae), and the Manner of Foraging. *Journal of Natural History* 22(1): 1–10.

Hocking, Brian. 1968. Insect–Flower Associations in the High Arctic with Special Reference to Nectar. *Oikos* 19(2): 359–387.

Jansen, Steven, Martin R. Broadley, Elmar Robbrecht, and Erik Smets. 2002. Aluminum Hyperaccumulation in Angiosperms: A Review of Its Phylogenetic Significance. *Botanical Review* 68(2): 235–269.

Jonasson, Sven. 1995. Resource Allocation in Relation to Leaf Retention Time of the Wintergreen *Rhododendron lapponicum*. *Ecology* 76(2): 475–485.

Karlsson, P. Staffan. 1992. Leaf Longevity in Evergreen Shrubs: Variation Within and Among European Species. *Oecologia* 91(3): 346–349.

Karlsson, P. Staffan. 1994. The Significance of Internal Nutrient Cycling in Branches for Growth and Reproduction of *Rhododendron lapponicum*. *Oikos* 70(2): 191–200.

Kimball, Kenneth D., and Douglas M. Weihrauch. 2000. Alpine Vegetation Communities and the Alpine-Treeline Ecotone Boundary in New England as Biomonitors for Climate Change. *USDA Forest Service Proceedings* RMRS-P-15 (3): 93–101.

Kubitzki, Klaus. 2004. *The Families and Genera of Vascular Plants*, Vol. 6. Springer-Verlag, Berlin.

Kudo, Gaku. 1991. Effects of Snow-free Periods on the Phenology of Alpine Plants Inhabiting Snow Patches. *Arctic and Alpine Research* 23(4): 436–443.

Larson, B. M. H., P. H. Kevan, and D. W. Inouye. 2001. Flies and Flowers: Taxonomic Diversity of Anthophiles and Pollinators. *Canadian Entomologist* 133(4): 438–465.

LaRue, Diane, and Brad Toms. 2012. Establishing a Foundation for the Recovery of Eastern Mountain Avens (*Geum peckii*). Mersey Tobeatic Research Institute website, http://www.merseytobeatic.ca/userfiles/file/Science%20Conference%202012/Presentations/LaRue_D_Geum_MTRI_Scie_Conf_2012.pdf (accessed 2013).

Levesque, Christine M., and John F. Burger. 1982. Insects (Diptera, Hymenoptera) Associated with *Minuartia groenlandica* (Caryophyllaceae) on Mount Washington, New Hampshire, U.S.A., and Their Possible Role as Pollinators. *Arctic and Alpine Research* 14(2): 117–124.

Marren, Peter. 1999. *Britain's Rare Flowers*. Academic Press, London: 60.

Masters, Jeff. 2010. "A New World Record Wind Gust: 253 mph in Australia's Tropical Cyclone Olivia," Wunderground.com. Dr. Jeff Masters' WunderBlog, http://www.wunderground.com/blog/JeffMasters/a-new-world-record-wind-gust-253-mph-in-australias-tropical-cyclone- (accessed 2013).

McFarland, Kent P. 2003. "Conservation Assessment of Two Endemic Butterflies (White Mountain Arctic, *Oeneis melissa semidea*, and White Mountain Fritillary, *Boloria titania montinus*) in the Presidential Range Alpine Zone, White Mountains, New Hampshire," Vermont Institute of Natural Science website, http://www.vtecostudies.org/PDF/wmalpinebuttrep03.pdf (accessed 2014).

Molau, Ulf. 1997. Age-related Growth and Reproduction in *Diapensia lapponica*, an Arctic-alpine Cushion Plant. *Nordic Journal of Botany* 17(3): 225–234.

Mount Washington Observatory website, http://www.mountwashington.org/about/visitor/surviving.php (accessed 2013).

Murphy, T., and James W. Hardin. 1976. A New and Unique Venation Pattern in the Diapensiaceae. *Bulletin of the Torrey Botanical Club* 103(4): 177–179.

Nature Gate. 2013. "Creeping Azalea," Nature Gate website, http://www.luontoportti.com/suomi/en/kukkakasvit/creeping-azalea (accessed 2013).

Nature Gate. 2013. "Mountain Heath," Nature Gate website, http://www.luontoportti.com/suomi/en/kukkakasvit/mountain-heath (accessed 2014).

Nature Serve. 2013. "*Geum peckii*," Nature Serve Explorer website: An Encyclopedia of Life, http://www.natureserve.org/explorer/servlet/NatureServe?searchName=Geum+peckii (accessed 2014).

Oliver, Meghan. 2013. "Top Flowers: Adaptations for Living on the Alpine Edge," Northern Woods website, http://northernwoodlands.org/articles/artilce/flowers-adaptions-alpine (accessed 2013).

Pendergast, B. A., and D. A. Boag. 1970. Seasonal Changes in Diet of Spruce Grouse in Central Alberta. *Journal of Wildlife Management* 34(3): 605–611.

Pendergast, B. A., and D. A. Boag. 1971. Nutritional Aspects of the Diet of Spruce Grouse in Central Alberta. *The Condor* 73(4): 437–443.

Scott, Peter J., and Robin T. Day. 1983. Diapensiaceae: A Review of the Taxonomy. *Taxon* 32(3): 417–423.

Slack, Nancy G. and Allison W. Bell. 1995. *Field Guide to the New England Alpine Summits*. Appalachian Mountain Club, Boston.

Stevens, P. F. 1970. *Calluna*, *Cassiope* and *Harrimanella*: A Taxonomic and Evolutionary Problem. *New Phytologist* 69(4): 1131–1148.

Wallner, Jeff, and Mario J. DiGregorio. *New England's Mountain Flowers: A High Country Heritage*. 1997. Mountain Press Publishing, Missoula, MT.

Wikipedia contributors, "Double-flowered," *Wikipedia, The Free Encyclopedia*, https://en.wikipedia.org/w/index.php?title=Double-flowered&oldid=883998672 (accessed 2013).

Wikipedia contributors, "Grayanotoxin," *Wikipedia, The Free Encyclopedia*, https://en.wikipedia.org/w/index.php?title=Grayanotoxin&oldid=903385713 (accessed 2014).

Wikipedia contributors, "*Potentilla robbinsiana*," *Wikipedia, The Free Encyclopedia*, https://en.wikipedia.org/w/index.php?title=Potentilla_robbinsiana&oldid=865931086 (accessed 2014)

Wikipedia contributors, "Rhododendron," *Wikipedia, The Free Encyclopedia*, https://en.wikipedia.org/w/index.php?title=Rhododendron&oldid=903482425 (accessed 2013).

Zwinger, Ann H., and Beatrice E. Willard, 1972. *Land Above the Trees: A Guide to American Alpine Tundra*. Harper & Row, New York.

American Cranberry references

Bayot, Sarah Mags. "Heritage Cranberry: A Thanksgiving Tradition," Fashion & Philosophers blog, http://fashionandphilosophers.com/2013_11_29/cranberries/ (accessed 2013; no longer accessible).

Brooks, Rebecca Beatrice. "The History of the First Thanksgiving," History of Massachusetts Blog, https://historyofmassachusetts.org/the-first-thanksgiving/ (accessed 2019).

Conner, Brenda. 2010. "Cranberry Culture" (lecture, Pine Barrens Short Course, Burlington County College, Pemberton, New Jersey, March 20).

Costea, Mihai, and Francois J. Tardif. 2005. The Biology of Canadian Weeds. 133. *Cuscuta campestris* Yuncker, *C. gronovii* Willd. Ex Schult., *C. umbrosa* Beyr. Ex Hook., *C. epithymum* (L.) L. and *C. epilinum* Weihe. *Canadian Journal of Plant Science*. www.nrcresearchpress.com (accessed 2019).

Cumo, Christopher Martin, ed. 2013. *Encyclopedia of Cultivated Plants: From Acacia to Zinnia* (3 volumes), Vol. 1: A–F. ABC-CLIO, Santa Barbara, CA.

Fadul, Clair, and Daniel Mosquin, "*Vaccinium macrocarpon* cultivar," Botany Photo of the Day website, University of British Columbia Botanical Garden, https://botanyphoto.botanicalgarden.ubc.ca/2010/11/vaccinium-macrocarpon-cultivar/, posted November 25, 2010 (accessed 2010).

Gervais, Amélie, Madeleine Chagnon, and Valerie Fournier. 2018. Diversity and Pollen Loads of Flower Flies (Diptera: Syrpidae) in Cranberry Crops. *Annals of the Entomological Society of America* 111(6): 326–334.

Iowa State University, Agricultural Marketing Resource Center. 2019. "Cranberries," https://www.agmrc.org/commodities-products/fruits/cranberries (accessed 2019).

Kloet, Sam P. Vander. 2009. *Vaccinium macrocarpon*, in Flora of North America Editorial Committee, eds. 1993+. *Flora of North America North of Mexico*, vol. 8, pp. 515–520. New York.

MacKenzie, K. E. 1994. The Foraging Behaviour of Honey Bees (*Apis mellifera* L.) and Bumble Bees (*Bombus* spp.) on Cranberry (*Vaccinium macrocarpon* Ait.) *Apidologia* 25: 375–383.

Oudemans, P. V., J. J. Polashock, and B. T. Vinyard. 2008. Fairy Ring Disease of Cranberry: Assessment of Crop Losses and Impact on Cultivar Genotype. *Plant Disease* 92(4): 616–622.

Whitman-Salkin, Sarah. "Cranberries, A Thanksgiving Staple, Were a Native American Superfood," *National Geographic News*, November 28, 2012. https://news.nationalgeographic.com/news/2013/11/131127-cranberries-thanksgiving-native-americans-indians-food-history/ (accessed 2019).

Wikipedia contributors, "Cranberry," *Wikipedia, The Free Encyclopedia*, https://en.wikipedia.org/w/index.php?title=Cranberry&oldid=897682191 (accessed 2011).

Wisconsin Historical Society. 1996–2019. "Cranberry Harvesting Rake," Historical Essay. https://www.wisconsinhistory.org/Records/Article/CS2724 (accessed 2019).

American Ginseng references

Anderson, Roger C., M. Rebecca Anderson, and Gregory Houseman. 2002. Wild American Ginseng. *Native Plant Journal* 3(2): 93–105. http://npj.uwpress.org/content/3/2/93.full.pdf+html?sid=6753dc4d-1b62-4bd8-a863-21262da0cf19 (accessed online 2019).

Beattie-Moss, Melissa. "Roots and Regulations: The Unfolding Story of Pennsylvania Ginseng," *Penn State News*, June 19, 2006, https://news.psu.edu/story/141751/2006/06/19/research/roots-and-regulations (accessed online 2019).

Beyfuss, Robert. 2002. "Growing Ginseng in Your Woodlot," *The Natural Farmer*, Special Supplement on AgroForestry. Northeast Organic Farming Association: 1–5.

Bolgiano, Chris. 2000. "Gold in the Woods." *American Forests*, Winter 2000: 7–9.

Bonnabeaux, Maddy. 2016. "Ginseng: The Black Market Herb of the Appalachian." *Technician* (online version). North Carolina State University. March 15, 2016. http://www.technicianonline.com/arts_entertainment/article_771f9b02-eb2b-11e5-936c-132b98d780d8.html (accessed 2019).

Bourne, Joel. 2000. "On the Trail of the 'Sang Poachers." *Audubon* 102(March-April): 84–90.

Fenton, William N. 1974–2019. "Lafitau, Joseph-François," *Dictionary of Canadian Biography*. University of Toronto/Université Laval, http://www.biographi.ca/en/bio/lafitau_joseph_francois_3E.html (accessed 2019).

Furedi, Mary Ann, and James B. McGraw. 2004. White-tailed Deer: Dispersers or Predators of American Ginseng Seeds? *American Midland Naturalist* 152(2): 268–276.

Hruska, Amy M., Sara Souther, and James B. McGraw. 2014. Songbird Dispersal of American Ginseng (*Panax quinquefolius*). *Ecoscience* 21(1): 46–55.

Jiménez-Mejías, P., and R. F. C. Naczi. 2017. Araliaceae, the Ginseng Family. Online edition. R. F. C. Naczi, J. R. Abbott, and Collaborators, *New Manual of Vascular Plants of Northeastern United States and Adjacent Canada*, compiled in 2016, 2017. NYBG Press, New York.

Johannsen, Kristin. 2006. *Ginseng Dreams*. University of Kentucky Press, Lexington.

Juel, H. O., and John W. Harshberger. 1929. New Light on the Collection of North American Plants Made by Peter Kalm. *Proceedings of the Academy of Natural Sciences of Philadelphia* 81: 297–303.

Krochmal, Arnold, and Connie Krochmal. 1978. Ginseng, Panacea of Five Leaves. *Garden Magazine* (2): 25–28.

Macera, Lucile P. 1982. "The Wonder Root," *The Conservationist*. State of New York, Department of Environmental Conservation. May–June: 31–33.

Memorial Sloan Kettering Cancer Center. "Ginseng (Asian)," Integrative Medicine, About Herbs, Botanicals, & Other Products, For Healthcare Professionals: Clinical Summary, https://www.mskcc.org/cancer-care/integrative-medicine/herbs/ginseng-asian#references-25 (accessed 2019).

Mitchell, Paul D., and Lan Cheng. 2009. "Status of the Wisconsin Ginseng Industry." Department of Agricultural and Applied Economics, University of Wisconsin–Madison.

McMullen, Joseph M. 2010. American Ginseng (*Panax quinquefolius*)—Facts and Folklore. *New York Flora Association Quarterly Newsletter* 21(2): 1–5.

Schmid, Doug. 2014. American Ginseng. *New York State Conservationist*, June 2014: 28–29.

References

Taylor, David A. 2006. *Ginseng, the Divine Root.* Algonquin Books of Chapel Hill, Chapel Hill, NC.

Tennis, Joe. 2018. "Seeking Your Roots: Ginseng Hunting Season is Here, But There Are Rules You Must Follow." *Bristol Herald Courier*, Bristol, VA. September 10, 2018.

American Lotus references

Baier, Wilhelm Richard, and Georg Heinrich. 1993. Gas-Liquid Chromatographic Analyses of the Water-soluble Compounds of Plant Latices. *Phyton* 33(1): 77–85.

Bhushan, Bharat. 2012. *Biomimetics: Bioinspired Hierachial-structured Surfaces for Green Science and Technology.* Springer, Heidelberg. Available at Google Books. http://books.google.com/books (accessed 2013).

Bug Guide. 2003–2011. Iowa State University Entomology website. http://bugguide.net/node/view/15740 (accessed 2011).

Cox, Paul Alan, and Steven King. 2013. "Bioprospecting," *in Encyclopedia of Biodiversity*, 2nd ed., vol. 1, pp. 588–599. Academic Press, Waltham, MA.

Crow, Garrett E., and C. Barre Hellquist. 2000. *Aquatic and Wetland Plants of Northeastern North America*, vol. 1, pp. 44, 49. University of Wisconsin Press, Madison.

Estrada-Ruiz, Emilio, Gilbert R. Upchurch Jr., J. A. Wolfe, and Sergio R. S. Cevallos-Ferriz. 2011. Comparative Morphology of Fossil and Extant Leaves of Nelumbonaceae, Including a New Genus from the Late Cretaceous of Western North America. *Systematic Botany* 36(2): 337–351.

Farwell, Oliver A. 1936. The Color of the Flowers of *Nelumbo pentapetala*. *Rhodora* 38(445): 272.

Forbes, Peter. 2008. Self-Cleaning Materials. *Scientific American* 299(2): 88–95.

Gandolfo, Maria A., and Ruben N. Cuneo. 2005. Fossil Nelumbonaceae from the La Colonia Formation (Campanian–Maastrichtian, Upper Cretaceous), Chubut, Patagonia, Argentina. *Review of Palaeobotany & Palynology* 133(3–4): 169–178.

Hall, Thomas F., and William T. Penfound. 1944. The Biology of the American Lotus, *Nelumbo lutea* (Wild.) Pers. *American Midland Naturalist* 31(3): 744–758. https://www.jstor.org/stable/2421417?read-now=1&seq=1#metadata_info_tab_contents (accessed 2012).

Hayes, Virginia, Edward L. Schneider, and Sherwin Carlquist. 2000. Floral Development of *Nelumbo nucifera* (Nelumbonaceae). *International Journal of Plant Sciences* 161 (S6 Current Perspectives on Basal Angiosperms): S183–S191.

Heritage, Benjamin. 1895. Preliminary Notes on *Nelumbo lutea*. *Bulletin of the Torrey Botanical Club* 22(6): 266–271.

Kaufman, Rachel. 2012. "32,000-year-old Plant Brought Back to Life—Oldest Yet." *National Geographic News*, February 23, 2012, http://news.nationalgeographic.com/news/2012/02/120221-oldest-seeds-regenerated-plants-science/ (accessed 2013).

Kuo-Huang, Ling-Long, Chiou-Rong Sheue, Yuen-Po Yang, and Shu-Hwa Tsai Chiang. 1994. Calcium Oxalate Crystals in Some Aquatic Angiosperms of Taiwan. *Botanical Bulletin of the Academy Sinica* 33(2): 179–188.

Kupchan, S. M., B. Dasgupta, E. Fugita, M. L. King. 1963. The Alkaloids of American Lotus, *Nelumbo lutea*. *Tetrahedron* 19(1): 227–232.

Marshall, Steven A. 2006. *Insects: Their Natural History and Diversity: With a Photographic Guide to Insects of Eastern North America*. Firefly Books, Buffalo, p. 355.

Masuda, Jun-ichiro, Yukio Ozaki, and Hiroshi Okubo. 2007. Rhizome Transition to Storage Organ is Under Phytochrome Control in Lotus (*Nelumbo nucifera*). *Planta* 226(4): 909–915.

Minard, Anne. 2008. "'Methuselah' Tree Grew from 2,000-year-old Seed." *National Geographic News*, Oct. 28, 2008, http://news.nationalgeographic.com/news/2008/06/080612-oldest-tree.html (accessed 2013).

"The Lotus Symbol in Buddhism." Religion Facts website. http://www.religionfacts.com/buddhism/symbols/lotus.htm (accessed 2012).

Sassafras River Water Trail: Along the River. "American Lotus." Sultana Projects: Preservation through Education. http://srwt.org/along-the-sassafras (accessed 2013).

Schneider, Edward L., and John D. Buchanan. 1980. Morphological Studies of the Nymphaeaceae. XI. The Floral Biology of *Nelumbo pentapetala*. *American Journal of Botany* 67(2): 182–193.

Shen-Miller, J., et al. 1995. Exceptional Seed Longevity and Robust Growth: Ancient Sacred Lotus from China. *American Journal of Botany* 82(11): 1367–1380.

Sohmer, S. H., and D. F. Sefton. 1978. The Reproductive Biology of *Nelumbo pentapetala* (Nelumbonaceae) on the Upper Mississippi River. II. The Insects Associated with the Transfer of Pollen. *Brittonia* 30(3): 355–364.

Swan, Daniel C. 2012. The North American Lotus (*Nelumbo lutea* Willd.): Sacred Food of the Osage People. *Ethnobotany & Research & Applications* 8: 249–253. http://lib-ojs3.lib.sfu.ca:8114/index.php/era/article/viewFile/496/297 (accessed 2012).

Tenaglia, Dan. n.d. "*Nelumbo lutea* Willd." MissouriPlants.com. http://www.missouriplants.com/Yellowalt/Nelumbo_lutea_page.html (accessed 2011).

The Official Website of the Osage Nation Historic Preservation. The Osage Nation © 2006. http://www.osagetribe.com/historicpreservation/info_sub_page.aspx?subpage_id=14 (accessed 2013).

Thomas, Robert. 2009. "American Lotus, *Nelumbo lutea*: locally known as graine à voler." Loyola University, New Orleans; Center for Environmental Communication, http://www.loyno.edu/lucec/natural-history-writings/american-lotus-nelumbo-lutea-locally-known-graine-à-voler (accessed 2011).

Sto. StoLotusan Color. http://www.stoshop.co.uk/products/item/stolotusan-color (accessed 2012).

Vogel, Stefan. 2004. Contributions to the Functional Anatomy and Biology of *Nelumbo nucifera* (Nelumbonaceae) I. Pathways of air circulation. *Plant Systematics and Evolution* 249(1): 9–25.

Vogel, Stefan. 2004. Contributions to the Functional Anatomy and Biology of *Nelumbo nucifera* (Nelumbonaceae) III. An ecological reappraisal of floral organs. *Plant Systematics and Evolution* 249(3–4): 173–189.

Wikipedia contributors, "Mass Suicide," *Wikipedia, The Free Encyclopedia*, https://en.wikipedia.org/w/index.php?title=Mass_suicide&oldid=574069152 (accessed 2013).

Wikipedia contributors, "Lotus Effect," *Wikipedia, The Free Encyclopedia*, https://en.wikipedia.org/w/index.php?title=Lotus_effect&oldid=890565461 (accessed 2012).

Wikipedia contributors, "*Nelumbo lutea*," *Wikipedia, The

Free Encyclopedia, https://en.wikipedia.org/w/index.php?title=Nelumbo_lutea&oldid=863446433 (accessed 2011).

Wikipedia contributors, "Nelumbonaceae," *Wikipedia, The Free Encyclopedia*, https://en.wikipedia.org/w/index.php?title=Nelumbo&oldid=898387425 (accessed 2012).

Zorba, Vassilia, et al. 2008. Biomimetic Artificial Surfaces Quantitatively Reproduce the Water Repellency of a Lotus Leaf. *Advanced Materials* 20(21): 4049–4054.

Aster references

Beadle, David, and Seabrooke Leckie. 2012. *Peterson Field Guide to Moths of Northeastern North America*. Houghton Mifflin Harcourt, New York.

DeBarros, Nelson. 2010. "Star of September's Glory—New England Aster," Nature Northeast Blogspot, http://naturenortheast.blogspot.com/2012/10/symphyotrichum-novae-angliae.html (accessed 2012).

Gray, Jane Loring (ed.) 1883. "Letters of Asa Gray." Houghton, Mifflin, Boston. Electronic version on Internet Archive, http://archive.org/stream/lettersofasagray02gray#page/726/mode/2up (accessed 2012).

Haines, Arthur. 2001. "Clarifying the Generic Concepts of Aster Sensu Lato in New England," *Botanical Notes*, no. 7. Topsham, ME. http://www.arthurhaines.com/botanical_notes/BotNotes_N7.prd (accessed 2012).

Perry, Leonard. n.d. "Asters and Goldenrod: The Mythology," The Green Mountain Gardener, University of Vermont Extension, Department of Plant and Soil Science, http://www.uvm.edu/pss/ppp/articles/asters.html (accessed 2012).

Poole, Robert W. "Nocturidae—Cuculliinae: *Cucullia asteroides* Guenée," Nearctica: A Natural History of North America website, http://www.nearctica.com/moths/noctuid/cucullia/cucullia_asteroides.htm (accessed 2012).

Semple, John C., Stephen B. Heard, and ChunSheng Xiang. 1998. *The Asters of Ontario* (Compositae: Astereae): *Diplactis* Raf., *Oclemena* E. L. Greene, *Dollingeria* Nees, and *Aster* L. (including *Canadanthus* Nesom, *Symphyotrichum* Nees, and *Virgulus* Raf.). University of Waterloo Biology Series 38. Waterloo, Ontario.

Willmer, Pat. 2011. *Pollination and Floral Ecology*. Princeton University Press, Princeton, NJ.

Yatshievych, George, and Richard W. Spellenberg. 1993. Chapter 10. Plant Conservation in the Flora of North America Region, in *Flora of North America North of Mexico*, vol. 1: Introduction. Oxford University Press, New York.

Beechdrops references

Abbate, Anthony P., and Joshua W. Campbell. 2013. Parasitic Beechdrops (*Epifagus virginiana*): A Possible Ant-Pollinated Plant. *Southeastern Naturalist* 12(3): 661–665.

Brooks, A. E. 1960. A Preliminary Morphological Study of *Epifagus virginiana* (L.) Bart. *Proceedings of the Indiana Academy of Science* (70): 73–78.

Grafton, Emily. 2005. "Beech Drops," Notes of the Pennsylvania Native Plant Society Newsletter (8)4: 1,5.

Heide-Jørgensen, Henning S. 2008. *Parasitic Flowering Plants*. Koninklijke Brill, The Netherlands.

Kuijt, Job. 1969. *The Biology of Parasitic Flowering Plants*. University of California Press, Berkeley.

Musselman, Lytton J. 1982. The Orobanchaceae of Virginia. *Castanea* 47(3): 266–275.

Thieret, John W. 1969. Notes on *Epifagus. Castanea* 34(4): 397–402.

Tsai, Yi-Hsin Erica, and Paul S. Manos. 2010. Host Density Drives the Postglacial Migration of the Tree Parasite, *Epifagus virginiana. Proceedings of the National Academy of Sciences* 107(39): 17035–17040.

Wesche, Spencer L., Elizabeth L. Barker, and Alice L. Heikens. 2016. Population Ecology Study of *Epifagus virginiana* (L.) W. P. C. Barton (Beechdrops) in Central Indiana. *Proceedings of the Indiana Academy of Science* 125(1): 69–74.

Wikipedia contributors, "*Epifagus*," *Wikipedia, The Free Encyclopedia*, https://en.wikipedia.org/w/index.php?title=Epifagus&oldid=884662303 (accessed 2019).

Willmer, Pat. 2011. *Pollination and Floral Ecology*. Princeton University Press, Princeton, NJ.

Blackberry-lily references

Brown, William H. 1938. The Bearing of Nectaries on the Phylogeny of Flowering Plants. *Proceedings of the American Philosophical Society* 79(4): 549–595.

Chimphamba, Brown B. 1973. Intergeneric Hybridization between *Iris dichotoma*, Pall. and *Belamcanda chinensis*, Leman. *Cytologia* 38(3): 539–547.

Garofalo, Joseph F. 2002. "Blackberry-lily, a Flowering Perennial for South Florida." Fact Sheet No. 25, Miami-Dade County/University of Florida Cooperative Extension Service.

Goldblatt, Peter. 2000. Phylogeny and Classification of the Iridaceae and the Relationships of *Iris. Annali di Botanica* 58(0): 13–28.

Goldblatt, Peter. 2002. *Belamcanda* Adason, *in* Flora of North America Editorial Committee, eds. 1993+. *Flora of North America North of Mexico*. vol. 26, pp. 395–396. Oxford University Press, New York.

Goldblatt, Peter, and David John Mabberly. 2005. *Belamcanda* Included in *Iris*, and the New Combination *I. domestica* (Iridaceae: Iridaea). *Novon* 15(1): 129–132.

Harborne, J. B., and C. A. Williams. 2000. The Phytochemical Richness of the Iridaceae and its Systematic Significance. *Annali di Botanica* 58(0): 43–50.

Hyam, Roger, and Richard Pankhurst. 1995. *Plants and Their Names*. Oxford University Press, Oxford.

Plants for a Future. 2012. "*Belamcanda chinensis* (L.) DC." http://pfaf.org/user/Plant.aspx?LatinName=Belamcanda+chinensis (accessed 2017).

Mavrodiev, Evgeny V., Mario Martínez-Azorin, Peter Dranishnikov, and Manuel B. Crespo. 2014. At least 23 Genera Instead of One: The Case of *Iris* L. s.l. (Iridaceae). *PLOS ONE* 9(8): e106459. http://journals.plos.org/plosone/article?id=10.1371/journal.pone.0106459 (accessed 2017).

Morrissey, Colm, Jasmin Bektic, Barbara Spengler, et al. 2004. Phytoestrogens Derived from *Belamcanda chinensis* Have an Antiproliferative Effect on Prostate Cancer Cells in Vitro. *Journal of Urology* 172(6): 2426–2433.

Redouté, Pierre Joseph. 1807. *Belamcanda. Les Liliacees* 3(21), pl. 121. Didot Jeune, Paris.

Thelen, Paul, Thomas Peter, Anika Hünermund, et al. 2007. Phytoestrogens from *Belamcanda chinensis* Regulate the Expression of Steroid Receptors and Related Cofactors in

References

LNCaP Prostate Cancer Cells. *BJU International* 100(1): 199–203. http://onlinelibrary.wiley.com/doi/10.1111/j.1464-410X.2007.06924.x/full (accessed 2017).

Thomas Jefferson Monticello website: "Blackberry-lily." http://www.monticello.org/site/house-and-gardens/blackberry-lily (accessed 2012).

Tillie, Nico, Mark W. Chase, and Tony Hall. 2000. Molecular Studies in the Genus *Iris* L.: A Preliminary Study, *in* M. A. Colasante and Paula J. Rudall (eds.), *Irises and Iridaceae: Biodiversity and Systematics*. Annali di Botanica (Rome), nuova serie 1(58): 105–122.

Vitali, Maria de Jesus, João Clovis Stanzani Dutra, and Vera Lígia Letízio Macado. 1995. Entomofauna Visitante de *Belamcanda chinensis* (L.) DC (Iridaceae) Durante o Período de Floração. *Revista Brasiliera Zoologia* 12(2): 239–250.

Wikipedia contributors, "*Iris domestica*," *Wikipedia, The Free Encyclopedia*, https://en.wikipedia.org/w/index.php?title=Iris_domestica&oldid=877634893 (accessed 2017).

Wikipedia contributors, "Bernard McMahon," *Wikipedia, The Free Encyclopedia*, https://en.wikipedia.org/w/index.php?title=Bernard_McMahon&oldid=897166538 (accessed 2017).

Willmer, Pat. 2011. *Pollination and Floral Ecology*. Princeton University Press, Princeton, NJ.

Wilson, Carol A. 2011. Subgeneric Classification in *Iris* Re-examined Using Chloroplast Sequence Data. *Taxon* 60(1): 27–35.

Wulf, Andrea. 2011. *Founding Gardeners: The Revolutionary Generation, Nature, and the Shaping of the American Nation*. Random House, New York.

Bog Orchids references

Baily, W. W. 1880. *Botanical Gazette* 5(7): 79. University of Chicago Press, Chicago.

House, Homer D. 1906. *Wildflowers of New York*, Part I, Memoir 15. University of the State of New York, Albany.

Johnson, Charles W. 1985. *Bogs of the Northeast*. University Press of New England, Hanover, NH.

NatureServe Explorer: An Online Encyclopedia of Life. "*Arethusa bulbosa* L." http://explorer.natureserve.org/servlet/NatureServe?searchName=Arethusa+bulbosa (accessed 2015).

Nelson, Tom, and Eric Lamont. 2012. *Orchids of New England & New York*. Kollath and Stenaas, Duluth, MN.

North American Pollinator Protection Campaign. 2015. Pollinator Partnership. http://www.pollinator.org/orchidposter_orc.htm (accessed 2015).

Pennsylvania Natural Heritage Fact Sheet. Swamp Pink (*Arethusa bulbosa*) adapted from Felbaum, Mitchell, et al. Endangered and Threatened Species of Pennsylvania. http://www.naturalheritage.state.pa.us/factsheets/15418.pdf (accessed 2015).

Percival, Mary S. 1965. *Floral Biology*. Pergamon Press, Elmsford, NY.

Poem Hunter. "Arethusa—Poem by Percy Bysshe Shelley". http://www.poemhunter.com/percy-bysshe-shelley/ (accessed 2015).

Poem Hunter. "Rose Pogonias—Poem by Robert Frost." http://www.poemhunter.com/poem/rose-pogonias/ (accessed 2015).

Robert W. Freckmann Herbarium. "*Arethusa bulbosa* L." http://www.botany.wisc.edu/orchids/Arethusa.html (accessed 2015).

Rorres, Chris. n.d. Coins of Syracuse: "Coins of Arethusa." http://www.math.nyu.edu/~crorres/Archimedes/Coins/Arethusa.html (accessed 2015).

Sheviak, Chuck, and Steve Young. 2012. Orchids of New York. *New York State Conservationist*: 2–7.

Smith, W. B., 1993. *Orchids of Minnesota*. University of Minnesota Press, Minneapolis.

Thien, Leonard B., and Brian G. Marcks. 1972. The Floral Biology of *Arethusa bulbosa, Calopogon tuberosus*, and *Pogonia ophioglossoides* (Orchidaceae). *Canadian Journal of Botany* 50(11): 2319–2325.

Tucson Botanical Garden. 2015. "Orchids: Mad About Orchids." http://www.tucsonbotanical.org/gardening/more-orchids/ (accessed 2015).

Weston, Peter H., Andrew J. Perkins, and Timothy J. Entwisle. 2005. More than Symbioses: Orchid Ecology, with Examples from the Sydney Region. *Cunninghamia* 9(1): 1–15.

Wikipedia contributors, "*Rhizanthella gardneri*," *Wikipedia, The Free Encyclopedia*, https://en.wikipedia.org/w/index.php?title=Rhizanthella_gardneri&oldid=898950311 (accessed 2015).

Willmer, Pat. 2011. *Pollination and Floral Biology*. Princeton University Press, Princeton, NJ.

Helleborine references

Amonin, Gérard G., Marcel Bournérias, Michel Démares, et al. 1998. *Les Orchideés de France, Belgique et Luxembourg*. Collection Parthénope, Paris.

Baskin, Carol C., and Jerry M. Baskin. 2000. *Seeds: Ecology, Biogeography, and Evolution of Dormancy and Germination*. Academic Press, Cambridge, MA.

Brodmann, Jennifer. 2010. Pollinator Attraction in Wasp-flowers. Summary of Dissertation zur Erlängung des Doktorgrades Dr. rer. Nat. der Fakultät für Naturwissenschaften der Universität Ulm. Germany.

Cameron, Duncan D., Irene Johnson, David J. Read, and Jonathan R. Leake. 2008. Giving and Receiving: Measuring the Carbon Cost of Mycorrhizas in the Green Orchid, *Goodyera repens*. *New Phytologist* 180(1): 176–184.

Dearnaley, John. 2007. Further Advances in Orchid Mycorrhizal Research. *Mycorrhiza* 17(6): 475–486.

Delforge, Pierre. 1995. *Collins Photo Guide: Orchids of Britain & Europe*. HarperCollins, London.

Darwin, Charles. 1869. XVI—Notes on the Fertilization of Orchids. *Annals and Magazine of Natural History* 4(21): 141–159. Biodiversity Heritage Library https://www.biodiversitylibrary.org/page/27734968#page/157/mode/1up (accessed 2017).

Dressler, Robert L., 1981. *The Orchids: Natural History and Classification*. Harvard University Press, Cambridge, MA.

Ehlers, B. K., J. M. Olesen, and J. Ågren. 2002. Floral Morphology and Reproductive Success in the Orchid *Epipactis helleborine*: Regional and Local Across-habitat Variation. *Plant Systematics and Evolution* 236(1–2): 19–32.

Flora of Wisconsin. 2015 on. Online Virtual Flora of Wisconsin. *Epipactis helleborine* (L.) Cranz. Wisconsin State Herbarium, UW–Madison. http://wisflora.herbarium.wisc.edu/taxa/index.php?taxon=3529 (accessed 2018).

Jacquemyn, Hans, Marijke Lenaerts, Daniel Tyteca, and Bart

Lievens. 2013. Microbial Diversity in the Floral Nectar of Seven *Epipactis* (Orchidaceae) Species. *Microbiology Open* 2(4): 244–258.

Jakubska, Anna, Daniel Przado, M. Steiniger, Jadwiga Aniol-Kwiatkowska, and Marcin Kadej. 2005. Why Do Pollinators Become "Sluggish"? Nectar Chemical Constituents from *Epipactis helleborine* (L.) Crantz (Orchidaceae). *Applied Ecology and Environmental Research* 3(2): 29–38.

Keenan, Philip E., 1998. *Wild Orchids Across North America: A Botanical Travelogue*. Timber Press, Portland, OR.

Mayr, Hubert. 1998. *Orchid Names and Their Meanings*. A. R. G. Gantner Verlag K.-G., Vaduz, Germany.

McCormick, Melissa. 2017. Picky Partners: Why Native Orchids Grow Where They Grow. Smithsonian Environmental Research Center, Edgewater, MD. https://serc.si.edu/sites/default/files/EveningLectures/mccormick_2017_orchid_optimism_web.pdf (accessed 2018).

Nelson, Tom, and Eric Lamont. 2012. *Orchids of New England & New York*. Kollath & Stensaas, Duluth, MN.

Paul Smith's College, Visitor Interpretive Center (VIC). Adirondack Wildflowers: Helleborine Orchid (*Epipactis helleborine*). http://www.adirondackvic.org/Adirondack-Wildflowers-Helleborine-Orchid-Epipactis-helleborine.html (accessed 2017).

Pennisi, Elizabeth. 2004. The Secret Life of Fungi. *Science* 304(5677): 11620–11622.

Selosse, Marc-André, and Duncan D. Cameron. 2010. Editorial. Introduction to a "Virtual Special Issue" on Mycoheterotrophy: *New Phytologist* Sheds Light on Non-green Plants. *New Phytologist* 185(3): 591–593.

Smith, Welby R., *Native Orchids of Minnesota*. University of Minnesota Press, Minneapolis.

Swearingen, Jil M., and Charles T. Bargeron. 2016. Invasive Plant Atlas of the United States. University of Georgia Center for Invasive Species and Ecosystem Health. http://www.invasiveplantatlas.org/ (accessed 2018).

Talalaj, Izabela, and Emilia Brzosko. 2008. Selfing Potential in *Epipactis palustris*, *E. helleborine*, and *E. atrorubens* (Orchidaceae). *Plant Systematics and Evolution* 276(1): 21–29.

Těšitelová, Tamara, Jakub Těšitel, Jana Jersáková, Gabriela Říhova, and Marc-André Selosse. 2012. Symbiotic Germination Capability of Four *Epipactis* Species (Orchidaceae) is Broader than Expected from Adult Ecology. *American Journal of Botany* 99(6): 1020–1032.

Wikipedia contributors, "Epipactis helleborine," *Wikipedia, The Free Encyclopedia*, https://en.wikipedia.org/w/index.php?titel=Epipactis_helleborine&oldid=880719229 (accessed 2017).

Williams, John G., and Andrew E. Williams. 1983. *Field Guide to Orchids of North America from Alaska, Greenland, and the Arctic, South to the Mexican Border*. Universe Books, New York.

Buckbean references

GRIN Taxonomy for Plants. "Taxon: *Menyanthes trifoliata* L.)." USDA Germplasm Resources Information Network (GRIN) website. http://www.ars-grin.gov/cgi-bin/npgs/html/taxon.pl?400088 (accessed 2014).

Hewett, D. G. 1964. *Menyanthes trifoliata* L. *Journal of Ecology* 52(3): 723–735.

Huang, C., H. Tunon, and L. Bohlin. 1995. Anti-inflammatory Compounds Isolated from *Menyanthes trifoliata* L. *Acta Pharmaceutica Sinica* 30(8): 621–626.

Landscape Architect's Pages. "Plant of the Week: *Menyanthes trifoliata*." Davis Landscape Architecture on Planting & Landscape Architect's Projects. http://davisla.wordpress.com/2012/05/18/plant-of-the-week-menyanthes-trifoliata/ (accessed 2014).

Nic Lughadha, Eimear M., and John A. N. Parnell. 1989. Heterostyly and Gene-flow in *Menyanthes trifoliata* L. (Menyanthaceae). *Botanical Journal of the Linnaean Society* 100(4): 337–354.

Olesen, Jens Mogens. 1986. Heterostyly, Homostyly, and Long-distance Dispersal of *Menyanthes trifoliata* to Greenland. *Canadian Journal of Botany* 65(7): 1509–1513.

Plants for a Future. 1996–2012. "*Menyanthes trifoliata* L." Plants for a Future. http://www.pfaf.org/user/Plant.aspx?LatinName=Menyanthes+trifoliata (accessed 2014).

Tippery, Nicholas P., Donald H. Les, Donald J. Padgett, and Surrey W. L. Jacobs. 2008. Generic Circumscription in Menyanthaceae: A Phylogenetic Evaluation. *Systematic Botany* 33(3): 598–612.

Thompson, Faye L., Luise A. Hermanutz, and David J. Innes. 1998. The Reproductive Ecology of Island Populations of Distylous *Menyanthes trifoliata* (Menyanthaceae). *Canadian Journal of Botany* 76(5): 818–828.

Wildscreen. 2003–2013. "Bogbean (*Menyanthes trifoliata*)." Arkive webpage. http://www.arkive.org/bogbean/menyanthes-trifoliata/ (accessed 2014).

Bunchberry references

Barrett, Spencer C., and Kaius Helenurm. 1987. The Reproductive Biology of Boreal Forest Herbs. I. Breeding Systems and Pollination. *Canadian Journal of Botany* 65(10): 2036–2046.

Eyde, Richard H., 1988. Comprehending *Cornus*: Puzzles and Progress in the Systematics of the Dogwoods. *Botanical Review* 54(3): 233–351.

Edwards, Joan, Dwight Whitaker, Sarah Kilonsky, and Marta J. Laskowski. 2005. A Record-breaking Pollen Catapult. *Nature* 435: 164.

Gucker, Corey L. 2012. "*Cornus canadensis*," in Fire Effects Information System, U.S. Department of Agriculture, Forest Service, Rocky Mountain Research Station, Fire Sciences Laboratory website. http://www.fs.fed.us/database/feis/ (accessed 2014).

Murrell, Zack E., 1994. Dwarf Dogwoods: Intermediacy and the Morphological Landscape. *Systematic Botany* 19(4): 539–556.

"Tale of the Dogwood." 2005. Williams College Biology Department website. http://web.williams.edu/Biology/explodingflower/ (accessed 2014).

Wikipedia contributors, "Cornus canadensis," *Wikipedia, The Free Encyclopedia*, https://en.wikipedia.org/w/index.php?title=Cornus_canadensis&oldid=877779848 (accessed 2014).

Cardinal Flower references

Brown, James H., and Astrid Kodric-Brown. 1979. Convergence, Competition, and Mimicry in a Temperate Community of Hummingbird-pollinated Flowers. *Ecology* 60(5): 1022–1035.

Burroughs, John. 1904. *The Writings of John Burroughs Vol. IX: Riverby*. Houghton Mifflin, Boston: 13.

References

Burroughs, John. 2009. "The Cardinal Flower." *Bird and Bough*. Reprint of 1906 original. Houghton Mifflin, Boston: 50.

Chittka, Lars, and Nickolas M. Waser. 1996. Why Red Flowers Are Not Visible to Bees. *Israel Journal of Plant Sciences* 45(2–3): 169–283.

Devlin, B., and A. G. Stephenson. 1985. Sex Differential, Floral Longevity, Nectar Secretion, and Pollinator Foraging in a Protandrous Species. *American Journal of Botany* 72(2): 303–310.

Dickinson, Emily. 1873. "Indian Pipe—The Most Amazing Flower," in Poems, http://emilydickinsonsgarden.wordpress.com/2009/04/18/indian-pipe-the-most-amazing-flower/ (accessed 2012).

Dussourd, David E. 2005. In the Trenches: Bioprospecting with a Caterpillar Probe. *Wings* 28(2): 20–24.

Felpin, François-Xavier, and Jacques Lebreton. 2001. "History, Chemistry, and Biology of Alkaloids from *Lobelia inflata*," Tetrahedron Report Number 694, Available online September 21, 2004, Elsevier, smartshop.nazwa.pl/coffeshop/lobelia_inflata.pdf (accessed 2013).

Hawthorne, Nathaniel. 1842. *Passages from the American Notebooks, Vol. 2*. Released as Project Gutenberg Ebook #8089 in 2005, http://www.gutenberg.org/files/8089/8089-h/8089-h.htm (accessed 2013).

Hilty, John. 2002–2012. "Great Blue Lobelia," Illinois Wildflowers website, http://www.illinoiswildflowers.info/wetland/plants/gb_lobeliax.htm (accessed 2013).

Johnston, Mark O. 1991. Limitation of Female Reproduction in *Lobelia cardinalis* and *L. siphilitica*. *Ecology* 72(4): 1500–1503.

Johnston, Mark O. 1991b. Natural Selection of Floral Traits in Two Species of *Lobelia* with Different Pollinators. *Evolution* 45(6): 1468–1479.

Lady Bird Johnson Wildflower Center. 2013. "Native Plant Database: *Lobelia cardinalis*." University of Texas website, http://www.wildflower.org/plants/result.php?id_plant=LOCA2 (accessed 2013).

Noonan, James-Charles, Jr. 1996. *The Church Visible: The Ceremonial Life and Protocol of the Roman Catholic Church*. Viking, New York: 191.

Peterson, Roger Tory, and Margaret McKenny. 1968. *A Field Guide to Wildflowers of Northeastern and North-central North America*. Houghton Mifflin, Boston: 216.

Stine, Anne. 2013. "Lobelias and Pollinators." Hubbard Fellowship Blog, The Prairie Ecologist website, http://prairieecologist.com/2013/09/24/hubbard-fellowship-blog-lobelias-and-pollinators/ (accessed 2013).

Willmer, Pat. 2011. *Pollination and Floral Ecology*. Princeton University Press, Princeton, NJ.

Chicory references

Barth, Friedrich G. 1991. *Insects and Flowers: The Biology of a Partnership*. Princeton University Press, Princeton, NJ.

Bently, Barbara, and Thomas Elias, eds. 1983. *The Biology of Nectaries*. Columbia University Press, New York.

Cichan, Michael, A. 1983. Self Fertility in Wild Populations of *Cichorium intybus* L. *Bulletin of the Torrey Botanical Club* 110(3): 316–323.

Del Tredici, Peter. "Forests of the Future." Lecture, Cary Institute for Ecosystem Studies, Millbrook, New York. January 25, 2013.

Duke, James. "*Cichorium intybus* L. in Handbook of Energy Crops" (unpublished manuscript, 1983). https://hort.purdue.edu/newcrop/duke_energy/Cichorium_intybus.html (accessed 2019).

Eastman, John. 2003. *The Book of Field and Roadside: Open-country Weeds, Trees, and Wildflowers of Eastern North America*. Stackpole Books, Mechanicsburg, PA.

Faegri, K., and L. van der Pijl. 1979. *The Principles of Pollination Ecology: Third Revised Edition*. Pergamon Press, Elmsford, NY.

Foote, Knowlton. "Chicory (*Cichorium intybus* L.)." *New York Flora Association Newsletter*, Spring 2005: 4–7.

Gardiner, Brian G. 2007. "The Linnean Tercentenary: Some Aspects of Linnaeus' Life, 4. Linnaeus' Floral Clock," http://www.linnean.org/fileadmin/images/The_Linnean_-_Tercentenary/4-Floral_Clock.pdf (accessed 2011).

Koch, Minna Frotscher. 1930. Studies in the Anatomy and Morphology of the Composite Flower I. The Corolla. *American Journal of Botany* 17(9): 938–952.

Mahr, Susan. 2012. "Chicory, *Cichorium intybus*," Master Gardener Program, Division of Extension. University of Wisconsin–Madison, https://wimastergardener.org/article/chicory-cichorium-intybus/ (accessed 2019).

McCadden, Helen Matzke. "Chicory—Sponsored by Presidents." *Garden Journal*, March/April 1969: 35–36.

Smith, K. Annabelle. 2014. "The History of the Chicory Coffee Mix that New Orleans Made Its Own," Smithsonian.com, https://www.smithsonianmag.com/arts-culture/chicory-coffee-mix-new-orleans-made-own-comes-180949950/ (accessed 2019).

Stokes, Donald, and Lillian Stokes. 1985. *A Guide to Enjoying Wildflowers*. Little, Brown and Company, Boston.

Torres, C., and Leonardo Galetto. 2002. Are Nectar Sugar Composition and Corolla Tube Length Related to the Diversity of Insects that Visit Asteraceae Flowers? *Plant Biology* 4(3): 360–366.

Wikipedia contributors, "Chicory," *Wikipedia, The Free Encyclopedia*, https://en.wikipedia.org/w/index.php?title=Chicory&oldid=886636847 (accessed 2019).

Wikipedia contributors, "Linnaeus's flower clock," *Wikipedia, The Free Encyclopedia*, https://en.wikipedia.org/w/index.php?title=Linnaeus%27s_flower_clock&oldid=881792305 (accessed 2019).

Common Milkweed references

Agrawal, Anurag. 2017. *Monarchs and Milkweed: A Migrating Butterfly, A Poisonous Plant, and Their Remarkable Story of Coevolution*. Princeton University Press, Princeton, NJ.

Borders, Brianna, and Matthew Shepherd. 2011. Milkweeds: Not Just for Monarchs. *Wings* 34(1): 14–18.

Camazine, Scott. 1986. Milkweed Pollination: An Insect Pony Express. *Cornell Plantations* 42(2): 19–23.

Fordyce, James A., and Stephen B. Malcolm. 2000. Specialist Weevil, *Rhyssomatus lineaticollis*, Does Not Spatially Avoid Cardenolide Defenses of Common Milkweed by Ovipositing into Pith Tissue. *Journal of Chemical Ecology* 26(12): 2857–2874.

Gomez, Tony. "Is Tropical Milkweed Killing Monarch Butterflies?," Monarch Butterfly Garden, https://monarchbutterflygarden.net/is-tropical-milkweed-killing-monarch-butterflies/ (accessed 2018).

Graham, Ada, and Frank Graham. 1976. *The Milkweed and Its World of Animals*. Doubleday, Garden City, NY.

Hristov, Nickolay L., and William E. Conner. 2005. Effectiveness of Tiger Moth (Lepidoptera, Arctiidae) Chemical Defenses Against an Insectivorous Bat (*Eptesicus fuscus*). *Chemoecology* 15(2): 105–113.

Hilty, John. 2002–2017. "Butterfly Milkweed," Illinois Wildflowers website, http://www.illinoiswildflowers.info/prairie/plantx/btf_milkweedx.htm (accessed 2018).

Hilty, John. 2002–2017. "Whorled Milkweed," Illinois Wildflowers website, http://www.illinoiswildflowers.info/prairie/plantx/wh_milkweedx.htm (accessed 2018).

Ordish, George. 1975. *The Year of the Butterfly*. Charles Scribner's Sons, New York.

Morse, Douglass H. 1981. Prey Capture by the Crab Spider *Misumena vatia* (Clerck) (Thomisidae) on Three Common Native Flowers. *American Midland Naturalist* 105(2): 358–367.

Morse, Douglass H. 1985. Milkweeds and Their Visitors. *Scientific American* 253(1): 112–119.

Morse, Douglass H. 1993. The Twinning of Follicles by Common Milkweed (*Asclepias syriaca*). *American Midland Naturalist* 130(1): 56–61.

North American Monarch Conservation Plan. 2008. Communications Department, Commission for Environmental Cooperation, Montreal.

Schwartz, David M. 1987. Underachiever of the Plant World. *Audubon* 89(5): 47–51.

USDA, NRCS. 2019. The PLANTS Database, "Common Milkweed. *Asclepias syriaca* L.," https://plants.usda.gov/core/profile?symbol=ASSY (accessed 2018).

Wagner, David L. 2005. *Caterpillars of Eastern North America*. Princeton University Press, Princeton, NJ.

Wikipedia contributors, "*Asclepias*," *Wikipedia, The Free Encyclopedia*, https://en.wikipedia.org/wiki/Asclepias (accessed 2018).

Wikipedia contributors, "*Euchaetes egle*," *Wikipedia, The Free Encyclopedia*, https://en.wikipedia.org/wiki/Euchaetes_engle (accessed 2018).

Common Mullein references

Alba, Christina, and Ruth Hufbauer. 2012. Exploring the Potential for Climatic Factors, Herbivory, and Co-occurring Vegetation to Shape Performance in Native and Introduced Populations of *Verbascum thapsus*. *Biological Invasions* 14: 2505–2518.

Darlington, H. T. 1931. The 50-Year Period for Dr. Beal's Seed Viability Experiment. *American Journal of Botany* 18(4): 262–265.

Donnelly, Sarah E., Christopher J. Lortie, and Lonnie W. Aarssen. 1998. Pollination in *Verbascum thapsus* (Scrophulariaceae): The Advantage of Being Tall. *American Journal of Botany* 85(11): 1618–1625.

Eisenmann, Charley. "Mullein Weevils," Bug Tracks blog, 06-09-2014. https://bugtracks.wordpress.com/2014/06/09/mullein-weevils/ (accessed 2014).

Gross, Katherine L. 1980. Colonization by *Verbascum thapsus* (Mullein) of an Old Field in Michigan: Experiments on the Effects of Vegetation. *Journal of Ecology* 68(3): 919–927.

Gross, Katherine L.. 1981. Predictions of Fate from Rosette Size in Four "Biennial" Plant Species: *Verbascum thapsus*, *Oenothera biennis*, *Daucus carota*, and *Tragopogon dubius*. *Oecologia* 48(2): 209–213.

Gross, Katherine L., and Patricia A. Werner. 1978. The Biology of Canadian Weeds, 28. *Verbascum thapsus* L. and *V. blattaria* L. *Canadian Journal of Plant Science* 58: 401–413.

Gross, Katherine L., and Patricia A. Werner. 1982. Colonizing Abilities of 'Biennial' Plant Species in Relation to Ground Cover: Implications for Their Distributions in a Successional Sere. *Ecology* 63(4): 921–931.

Gucker, Corey L. 2008. *Verbascum thapsus*, in Fire Effects Information System, U.S. Department of Agriculture, Forest Service, Rocky Mountain Research Station, Fire Sciences Laboratory. https://www.fs.fed.us/database/feis/plants/forb/vertha/all.html (accessed 2019).

Jiménez-Mejías, P. 2018. Scrophulariaceae, the Figwort Family, *in* R. F. C. Naczi and Collaborators, *New Manual of Vascular Plants of Northeastern United States and Adjacent Canada*, online edition 2016 onward. NYBG Press, New York.

Parker, Ingrid M., Joseph Rodriguez, and Michael E. Loik. 2003. An Evolutionary Approach to Understanding the Biology of Invasions: Local Adaptation and General-Purpose Genotypes in the Weed *Verbascum thapsus*. *Conservation Biology* 17(1): 59–72.

Telewski, Frank W., and Jan A. D. Zeevaart. 2002. The 120-Yr Period for Dr. Beal's Seed Viability Experiment. *American Journal of Botany* 89(8): 1285–1288.

Wikipedia contributors, "*Rhinusa tetra*," *Wikipedia, The Free Encyclopedia*, https://en.wikipedia.org/w/index.php?title=Rhinusa_tetra&oldid=835385716 (accessed 2019).

Wikipedia contributors, "Thapsus," *Wikipedia, The Free Encyclopedia*, https://en.wikipedia.org/w/index.php?title=Thapsus&oldid=880367759 (accessed 2019).

Evening-primrose references

Clare, John. 1935. "The Evening Primrose," in *The Rural Muse*. Accessed on Google Books. http://books.google.com/books/about/The_rural_muse_poems.html?id=G-x9vNL0EkIC (accessed 2014).

Cleland, R. E., 1964. The Evolutionary History of the North American Evening Primroses of the "*biennis*" Group. *Proceedings of the American Philosophical Society* 108(2): 88–98.

Cox, R. M. 1984. Sensitivity of Forest Plant Reproduction to Long Range Transported Air Pollutants: In Vitro and In Vivo sensitivity of *Oenothera parviflora* L. Pollen to Simulated Acid Rain. *New Phytologist* 97(1): 63–70.

Cruden, Robert William, and Kenneth G. Jensen. 1979. Viscin Threads, Pollination Efficiency and Low Pollen-Ovule Ratios. *American Journal of Botany* 66(8): 875–879.

Deng, Yu-Chen, Hui-Ming Hua, Li Jun, and Peter Lapinskas. 1991. Studies on the Cultivation and Uses of Evening Primrose (*Oenothera* spp.) in China. *Economic Botany* 55(1): 83–92.

Doorn, Wouter G. van, and Uulke van Meeteren. 2003. Flower Opening and Closure: A Review. *Journal of Experimental Botany* 54(389): 1801–1812.

Drugs and Supplements: Evening Primrose (*Oenothera* spp.). Mayo Clinic. http://www.mayoclinic.org/drugs-supplements/evening-primrose/evidence/hrb-20059889 (accessed 2014).

References

Grieve, M. 1995–2014. "Primrose, Evening." Botanical.com: A Modern Herbal. http://botanical.com/botanical/mgmh/p/primro70.html (accessed 2014).

Gross, Katherine L. 1981. Predictions of Fate from Rosette Size in Four "Biennial" Plant Species: *Verbascum thapsus, Oenothera biennis, Daucus carota,* and *Tragopogon dubius*. Oecologia 48 (2): 209–213.

Johnson, Marc T. J., and Anurag A. Agrawal. 2005. Plant Genotype and Environment Interact to Shape a Diverse Arthropod Community on Evening Primrose (*Oenothera biennis*). Ecology 86(4): 874–885.

Linalool Guide. 2014. GoodGuide.com. https://www.goodguide.com/ingredients/116460-linalool (accessed 2014).

"Oenothera." 2013. Wikimedia Commons. http://en.wikipedia.org/wiki/Evening_primrose (accessed 2014).

"Oenothera biennis L." 1996–2012. Plants For a Future. http://www.pfaf.org/user/Plant.aspx?LatinName=Oenothera+biennis (accessed 2014).

Raguso, Robert A., and Eran Pichersky. 1995. Floral volatiles from *Clarkia breweri* and *C. concinna* (Onagraceae): Recent Evolution of Floral Scent and Moth Pollination. *Plant Systematics and Evolution* 194(1–2): 55–67.

Raguso, Robert A., and Eran Pichersky. 1999. A Day in the Life of a Linalool Molecule: Chemical Communication in a Plant-Pollinator System. Part 1: Linalool Biosynthesis in Flowering Plants. *Plant Species Biology* 14(2): 95–120.

Wagner, David L. 2005. *Caterpillars of Eastern North America*. Princeton University Press, Princeton, NJ.

Wagner, W. L., and P. C. Hoch. 2005–2019. Onagraceae—The Evening Primrose Family website. http://botany.si.edu/onagraceae/index.cfm (accessed 2014).

Fringed Gentian references

A Celebration of Women Writers. "L. M. Montgomery. *The Alpine Path: The Story of My Career*." Originally published in installments in *Everywoman's World*, 1917. http://digital.library.upenn.edu/women/montgomery/alpine/alpine.html (accessed 2015).

Adkins, Leonard, and Joe Cook. 2005. *Wildflowers of the Blue Ridge and Great Smoky Mountains*. Menasha Ridge Press, Birmingham, AL.

Bernardo, Ana, Joana Gregório, and Laura Tammiste. *Gentiana clusii* and *Gentiana acaulis*: Relationship between Geology, Plants, and Insects. http://www.sjf.ch/wp-content/uploads/2012/10/Gentiana-clusii-and-Gentiana-acaulis.pdf (accessed 2016).

Cullina, William. 2000. *The New England Wild Flower Society Guide to Growing and Propagating Wildflowers of the United States and Canada*. Houghton Mifflin, Boston.

Frost, Robert. 1942. "The Quest of the Purple Fringed" in *A Witness Tree*. Henry Holt, New York. Poetry Nook website. http://www.poetrynook.com/poem/quest-purple-fringed (accessed 2015).

Georgia Department of Natural Resources: Wildlife Resources Division. http://georgiawildlife.com/sites/default/files/uploads/wildlife/nongame/pdf/accounts/plants/gentianopsis_crinita.pdf (accessed 2015).

Hobbs, Christopher. Gentian: "A Bitter Pill to Swallow." Christopher Hobbs webpage. http://www.christopherhobbs.com/library/articles-on-herbs-and-health/gentian-a-bitter-pill-to-swallow/ (accessed 2015).

Hilty, John. 2002–2016. "Stiff Gentian (*Gentianella quinquefolia*) Gentian family (Gentianaceae)." Illinois Wildflowers website. http://www.illinoiswildflowers.info/savanna/plants/stiff_gentian.htm (accessed 2016).

Iltis, Hugh H. 1965. The Genus *Gentianopsis* (Gentianaceae): Transfers and Phytogeographic Comments. Sida, Contributions to Botany 2(2): 129–153.

Luteyn, J. L. 1999. *Páramos: A Checklist of Plant Diversity, Geographic Distribution and Botanical Literature*. New York Botanical Garden Press, New York.

Prairie Moon Nursery: *Gentianella quinquefolia* (Stiff Gentian), https://www.prairiemoon.com/seeds/wildflowers-forbs/gentianella-quinquefolia-stiff-gentian.html (accessed 2016).

Struwe, Lena. 2002. Gentian Research Network website. Rutgers University. http://gentian.rutgers.edu/overview.htm (accessed 2015).

Struwe, Lena. 2002. Gentianales (Coffees, Dogbanes, Gentians and Milkweeds). *Encyclopedia of Life Sciences*. Wiley Online Library. http://onlinelibrary.wiley.com/doi/10.1038/npg.els.0003732/abstract (accessed 2015).

Red List. The IUCN Red List of Threatened Species. 2016–1. *Gentiana punctata*. http://www.iucnredlist.org/details/203221/0 (accessed 2016).

The L. M. Montgomery Literary Society. http://lmmontgomeryliterarysociety.weebly.com/about-the-lm-montgomery-literary-society.html (accessed 2015).

Thiers, Barbara. (continuously updated). *Index Herbariorum: A Global Directory of Public Herbaria and Associated Staff*. New York Botanical Garden's Virtual Herbarium. http://sweetgum.nybg.org/science/ih/ (accessed 2015).

USDA, NRCS. 2019. The PLANTS Database, "*Gentianella quinquefolia* (L.) Small subsp. *quinquefolia* (agueweed)," http://plants.usda.gov/core/profile?symbol=gequq (accessed 2016).

Weaver, Richard E. 1973. In Search of Tropical Gentians. *Arnoldia* 33(3): 189–198.

Wiener, Ayana, Marina Shudler, Anat Levit, and Masha Y. Niv. 2012. BitterDB: A Database of Bitter Compounds. *Nucleic Acids Research* 40(D1): D413–419.

Wikipedia contributors, "*Gentiana lutea*," *Wikipedia, The Free Encyclopedia*, https://en.wikipedia.org/w/index.php?title=Gentiana_lutea&oldid=878357667 (accessed 2016).

Wikipedia contributors, "*Gentiana verna*," *Wikipedia, The Free Encyclopedia*, https://en.wikipedia.org/w/index.php?title=Gentiana_verna&oldid=881392380 (accessed 2016).

Wikipedia contributors, "Illyria," *Wikipedia, The Free Encyclopedia*, https://en.wikipedia.org/w/index.php?title=Illyria&oldid=897598975 (accessed 2015).

Wikipedia contributors, "Moxie," *Wikipedia, The Free Encyclopedia*, https://en.wikipedia.org/w/index.php?title+Moxie&oldid=898784645 (accessed 2016).

Wikipedia contributors, "Tyrian Purple," *Wikipedia, The Free Encyclopeida*, https://en.wikipedia.org/w/index.php?title=Tyrian_purple&oldid=898452714 (accessed 2015).

Fringed Orchids references

Argue, Charles L. 2012. Chapter 11: Lacera Group, in *The Pollination Biology of North American Orchids: Volume 1*, pp. 143–148. Springer, New York.

Brown, Paul Martin. 1992. *Platanthera pallida* (Orchidaceae), a New Species of Fringed Orchis from Long Island, New York, U.S.A. *Novon* 2(4): 308–311.

Brown, Paul Martin. 2008. *Platanthera pallida*—Fifteen Years of Comparisons. *North American Native Orchid Journal* 14(2): 158–167.

Cole, F. Russell, and David H. Firmage. 1984. The Floral Ecology of *Platanthera blephariglottis*. *American Journal of Botany* 71(5): 700–710.

Coleman, Ronald A. 2002. *The Wild Orchids of California*. Cornell University Press, Ithaca, NY.

Evans, John Richard. 2006. Identification and Comparison of the Pollinators for the Purple-fringed Orchids *Platanthera psycodes* and *P. grandiflora*. Honors Thesis Project, University of Tennessee, Knoxville. http://trace.tennessee.edu/utk_chanhonoproj/953

Laroche, Vincent, Stéphanie Pellerin, and Luc Brouillet. 2012. White Fringed Orchid as Indicator of Sphagnum Bog Integrity. *Ecological Indicators* 14(1): 50–55.

Little, Karen J., Gregg Dieringer, and Michael Romano. 2005. Pollination Ecology, Genetic Diversity, and Selection on Nectar Spur Length in *Platanthera lacera* (Orchidaceae). *Plant Species Biology* 20(3): 183–190.

McGrath, Robert T. 2008. Contributions to the Status and Morphology of *Platanthera pallida*, Pale Fringed Orchis. *North American Native Orchid Journal* 14(2): 150–156.

Nelson, Tom, and Eric Lamont. 2012. *Orchids of New England & New York*. Kollath and Stenaas, Duluth, MN.

Sharp, Penelope C. 2004. *Platanthera ciliaris* (L.) Lindl. (Yellow-fringed Orchis) Conservation and Research Plan for New England. New England Wild Flower Society, Framingham, MA.

Smith, Gordon R., and Gerald E. Snow. 1976. Pollination Ecology of *Platanthera* (*Habenaria*) *ciliaris* and *P. blephariglottis* (Orchidaceae). *Botanical Gazette* 137(2): 133–140.

Stoutamire, Warren P. 1974. Relationships of the Purple-fringed Orchids *Platanthera psycodes* and *P. grandiflora*. *Brittonia* 26(1): 42–58.

Goldenrod references

Abhilasha, Dipti, Naira Quintana, Jorge Vivanco, and Jasmin Joshi. 2008. Do Allelopathic Compounds in Invasive *Solidago canadensis* s.l. Restrain the Native European Flora? *Journal of Ecology* 96(5): 993–1001.

Abrahamson, Warren G. n.d. "The *Solidago/Erosta* Gall Homepage." http://www.facstaff.bucknell.edu/abrahmsn/solidago/eurosta.html (accessed 2012).

Abrahamson, Warren G., Kenneth D. McCrea, and Stephen S. Anderson. 1989. Host Preference and Recognition by the Goldenrod Ball Gallmaker *Eurosta solidaginis* (Diptera: Tephritidae). *American Midland Naturalist* 121(2): 322–330.

Apáti, Pál György. 2003. Antioxidant Constituents in *Solidago canadensis* L. and its Traditional Phytopharmaceuticals. PhD thesis. Semmelweis University Department of Pharmacognosy. Budapest, Hungary.

Beadle, David, and Seabrooke Leckie. 2012. *Peterson Field Guide to Moths of Northeastern North America*. Houghton Mifflin Harcourt, New York.

Biography of Thomas Alva Edison. Thomas A. Edison Papers (PP). NYBG Mertz Library: Archives & Manuscripts. http://www.nybg.org/library/finding_guide/archv/edison_ppb.html (accessed 2013).

Buchanan, R. A., I. M. Cull, G. H. Otey, and C. R. Russell. 1978. Hydorcarbon- and Rubber-Producing Crops: Evaluation of U.S. Plant Species. *Economic Botany* 32(2): 131–145.

Collicutt, Doug. "Bug-sicles in the Class Room: Demonstrating Freeze Tolerance with Larvae of the Goldenrod Gall Fly." Manitoba Online Nature Magazine. *Nature North*. http://www.naturenorth.com/winter/gallfly/gflycvr.html (accessed 2012).

Gibbs, Jessica. 2008. "The Investigation of Competition Between *Eurosta solidaginsis* (Fitch) and *Rhopalomyia solidaginsis* (Loew), Two Gall-makers of *Solidago altissima* (Asteraceae)." *Essai* 5, Article 5: 62–65. The Berkley Electronic Press. http://dc.cod.edu/essai (accessed 2012).

Grimstad, P. R., and G. R. DeFoliart. 1974. Nectar Sources of Wisconsin Mosquitoes. *Journal of Medical Entomology* 11(3): 331–341.

Gugler, D. 2002. "More Information on Goldenrod Galls." http://www.deanswildflowers.com/DG/goldenrod.html (accessed 2012).

Hunt, Willis R. 1927. Miscellaneous Collections of North American Rusts. *Mycologia* 19(5): 286–288.

Kabuce, Nora. 2007–2010. "*Solidago canadensis*." NOBANIS–Invasive Alien Species Fact Sheet. http://www.nobanis.org/files/factsheets/Solidago%20canadensis.pdf (accessed 2012).

Langenhelm, Jean H., and Kenneth V. Thimann. 1982. *Plant Biology and Its Relation to Human Affairs*. John Wiley & Sons, New York.

Mooibrock, H., and K. Cornish. 2000. Alternative Sources of Natural Rubber. *Applied Microbiology and Biotechnology* 53(4): 355–365.

Penn State Extension website. "*Solidago* Diseases," http://extension.psu.edu/plant-disease-factsheets/all-fact-sheets/goldenrod-solidago-diseases (accessed 2012).

Perry, Leonard. n.d. "Asters and Goldenrod: The Mythology." The Green Mountain Gardener. University of Vermont Extension, Department of Plant and Soil Science. http://www.uvm.edu/pss/ppp/articles/asters.html (accessed 2012).

Plants for a Future. 1996–2012. "*Solidago virgaurea* L." https://pfaf.org/user/Plant.aspx?LatinName=Solidago+canadensis (accessed 2012).

"Science: Goldenrod Rubber." Time Magazine. Dec. 16, 1929. http://content.time.com/time/subscriber/article/0,33009,881890,00.html (accessed 2013).

Semple, John C., Gordon S. Ringius, and Jie Jay Zhang. 1999. *The Goldenrods of Ontario*: *Solidago* L. and *Euthamia* Nutt. 3rd ed. University of Waterloo Biology Series, Number 39. Waterloo, Ontario.

Storey, Janet M. n.d. "Frozen Alive!" Nature North. http://www.naturenorth.com/winter/frozen/Ffrozen.html (accessed 2012)

"The Goldenrod and Aster." n.d. (adapted from Cook's "Nature Myths"). The Baldwin Project, Bringing Yesterday's Classics to Today's Children. http://www.mainlesson.com/display.php?author=bailey&book=hour&story=goldenrod (accessed 2012).

Thomas A. Edison Papers. New York Botanical Garden Mertz Library Archives. http://www.nybg.org/library/finding_guide/archv/edison_ppb.html (accessed 2013).

References

Weber, Ewald. 2001. Current and Potential Ranges of Three Exotic Goldenrods (*Solidago*) in Europe. *Conservation Biology* 15(1): 122–128.

Wikipedia contributors, 2012. "*Solidago virauera*." *Wikipedia, The Free Encyclopedia,* http://en.wikipedia.org/wiki/Solidago_virgaurea (accessed 2012).

Wu Wei. "Obligatory Greek Food Blog." http://kosmyryk.typepad.com/wu_wei/2005/05/obligatory_food.html (accessed 2012).

Yatshievych, George, and Richard W. Spellenberg. 1993. Chapter 10. Plant Conservation in the Flora of North America Region, in *Flora of North America North of Mexico*, vol. 1: Introduction. Oxford University Press, New York.

Grass-of-Parnassus references

Anderson, William A. 1943. A Fen in Northwestern Lowa [sic]. *American Midland Naturalist* 29(3): 787–791.

Chambers, Henrietta L. 2003. "A New Plant Family for Oregon." Oregon Flora Newsletter 9(1): 1–2. http://www.oregonflora.org/ofn/OFNv9n1.pdf (accessed 2013).

Cronk, Julie K., and M. Siobhan Fennessy. 2001. *Wetlands Plants: Biology and Ecology*. CRC Press, Boca Raton, FL.

Crow, Garrett E., and C. Barre Hellquist. 2000. *Aquatic and Wetland Plants of Northeastern North America*. University of Wisconsin Press, Madison.

Eastman, John. 1995. *The Book of Swamp and Bog: Trees, Shrubs, and Wildflowers of Eastern Freshwater Wetlands*. Stakepole Books, Mechanicsburg, PA.

Kamstra, Klaas, and José van Son. 2006. Greek Mountain Flora. http://www.greekmountainflora.info/index.htm (accessed 2013; no longer accessible).

Ladd, Doug. 2001. *North Woods Wildflowers: A Field Guide to Wildflowers of the Northeastern United States and Southeastern Canada*. Falcon Publishing, Helena, MT.

Pace, Lula. 1912. *Parnassia* and Some Allied Genera. *Botanical Gazette* 54(4): 306329.

Sandvik, Sylvi M., and Ørjan Totland. 2003. Quantitative Importance of Staminodes for Female Reproductive Success in *Parnassia palustris* under Contrasting Environmental Conditions. *Canadian Journal of Botany* 81(1): 49–56.

Sharp, A. J., and Ailsie Baker. 1964. First and Interesting Reports of Flowering Plants in Tennessee. *Castanea* 29(4): 178–185.

Simmons, M. P. 2004. "Parnassiaceae," pp. 291–292 in *The Families and Genera of Vascular Plants. Volume VI. Flowering Plants: Dicotyledons: Celastrales, Oxalidales, Rosales, Cornales, Ericales*. Springer-Verlag, Berlin.

Swales, Dorothy E. 1979. Nectaries of Certain Arctic and Sub-arctic Plants with Notes on Pollination. *Rhodora* 81(823): 363–407.

The Home Bug Gardener. 2010. "Wildflower Wednesday: Broken Rocks, Mountain Grass, and Truthful Deceit." The Home Bug Garden website. http://homebuggarden.blogspot.com/2010/08/wildflower-wednesday-broken-rocks.html (accessed 2011).

USDA, NRCS. 2019. The PLANTS Database. National Plant Data Team, Greensboro, NC, http://plants.usda.gov/ (accessed 2019).

Indian Pipe references

Dickinson, Emily. 1873. "Indian Pipe—The Most Amazing Flower," in *Poems.* http://emilydickinsonsgarden.wordpress.com/2009/04/18/indian-pipe-the-most-amazing-flower/ (accessed 2012).

Harper, J. L., P. H. Lovell, and K. G. Moore. 1970. The Shapes and Sizes of Seeds. *Annual Review of Ecology and Systematics* 1(1): 327–356.

Ianson, David, and Jeff Smeenk, 2012. Mycorrhizae in the Alaska Landscape. Cooperative Extension Service. University of Alaska Fairbanks. www.uaf.edu/files.ces/publications-db/catalog/anr/HGA-00026pdf (accessed 2012).

Klooster, Matthew R., and Theresa M. Culley. 2009. Comparative Analysis of the Reproductive Ecology of *Monotropa* and *Monotropsis*: Two Mycoheterotrophic Genera in the Monotropoidae (Ericaceae). *American Journal of Botany* 96(7): 1337–1347.

Leake, Jonathan R. The Biology of Mico-heterotrophic ('saprophytic') Plants. 1994. *New Phytologist* 127(2): 171–216.

Neyland, Ray, and Melissa K. Hennigan. 2004. A Cladistic Analysis of *Monotropa uniflora* (Ericaceae) Inferred from Large Ribosomal Subunit (26S) rRNA Gene Sequences. *Castanea* 69(4): 265–271.

Olson, A. Randall. 1980. Seed Morphology of *Monotropa uniflora* L. (Ericaceae). *American Journal of Botany* 67(6): 968–974.

Olson, A. Randall. 1991. Gynoecial Pathway for Pollen Tube Growth in the Genus *Monotropa*. *Botanical Gazette* 152(2): 154–163.

Olson, A. Randall. 1991. Postfertilization Changes in Ovules of *Monotropa uniflora* L. (Monotropaceae). *American Journal of Botany* 78(1): 99–107.

Primack, Richard B. 1985. Longevity of Individual Flowers. *Annual Review of Ecology and Systematics* 16: 15–37.

Tedersoo, Leho, Prune Pellet, Urmas Kõljalg, and Marc-André Selosse. 2007. Parallel Evolutionary Paths to Mycoheterotrophy in Understory Ericaceae and Orchidaceae: Ecological Evidence for Mixotrophy in Pyroleae. *Oecologia* 151(2): 206–217.

Wallace, Gary D. 1977. Studies of the Monotropoideae (Ericaceae). Floral Nectaries: Anatomy and Function in Pollination Ecology. *American Journal of Botany* 64(2): 199–206.

Jewelweed references

Darwin, Erasmus. 1991. "The Loves of Plants" (originally published anonymously in 1769 as "Botanic Garden, part 2"). Woodstock Books, New York.

Day, Peter D., Jaume Pellicer, and Ralf G. Kynast. 2012. Orange Balsam (*Impatiens capensis* Meerb., Balsaminaceae): A Re-evaluation by Chromosome Number and Genome Size. *Journal of the Torrey Botanical Society* 139(1): 26–33.

Foote, Knowlton. 2007. Orange Jewelweed (*Impatiens capensis* Meerburgh). *New York Flora Association Quarterly Newsletter* 18(2): 10–17.

Groves, Katie. 2019. "Plant Dye Recipe and Tutorial: How to Make a Jewelweed Dye." Katie Grove Studios. https://katiegrovestudios.com/2015/10/12/plant-dye-recipes-tutorials-how-to-make-a-jewelweed-dye/ (accessed 2019).

Hayashi, Marika, Kara L. Felich, David J. Ellerby. 2009. The Mechanics of Explosive Seed Dispersal in Orange Jewelweed (*Impatiens capensis*). *Journal of Experimental Botany* 60(7): 2045–2053.

Hilty, John. 2002–2017. *Illinois Wildflowers: Wetlands Plants*.

"Orange Jewelweed *Impatiens capensis*." http://www.illinoiswildflowers.info/wetland/plants/or_jewelweed.htm (accessed 2019).

Leck, Mary Allessio. 1979. Germination Behavior of *Impatiens capensis* Meerb. (Balsaminaceae). *Bartonia* 46: 1–14. http://www.jstor.org/stable/41610369.

McMullen, Joseph M. 2012. Wetland Indicator Status Rankings: What Do They Mean and Why Do We Care? *New York Flora Association Quarterly Newsletter* 23(1): 1–5.

Rattink, Bruce. 2015. "Thirsty Plants!" Tryon Naturalist Notes: Blog Archives. Tryon Creek State Natural Area, Oregon State Parks, posted June 29, 2015. https://tryoncreek.wordpress.com/2015/06/29/thirsty-plants/#comments (accessed 2019).

Rust, Richard W. 1977. Pollination in *Impatiens capensis* and *Impatiens pallida* (Balsaminaceae). *Bulletin of the Torrey Botanical Club* 104(4): 361–367.

Rust, Richard W. 1979. Pollination of *Impatiens capensis*: Pollinators and Nectar Robbers. *Journal of the Kansas Entomological Society* 52(2): 297–308.

Schemske, Douglas W. 1978. Evolution of Reproductive Characteristics in *Impatiens* (Balsaminaceae): The Significance of Cleistogamy and Chasmogamy. *Ecology* 59(3): 596–613.

Schmidt, Richard J. 1994–2019. "Balsaminaceae (Balsam Family)." Botanical Dermatology Database. http://www.botanical-dermatology-database.info/BotDermFolder/BALS.html (accessed 2019).

Singh, Har Bhajan, and Kumar Avinash Bharati, eds. 2014. *Handbook of Natural Dyes and Pigments*, Chapter 6, Enumeration of Dyes: 295–299. Woodhead Publishing, India.

Steets, Janette A., Tiffany M. Knight, and Tia-Lynn Ashman. 2007. The Interactive Effects of Herbivory and Mixed Mating for the Population Dynamics of *Impatiens capensis*. *American Naturalist* 170(1): 113–127.

Waller, Donald M. 1982. Jewelweed's Sexual Skills. *Natural History Magazine* 91(5): 32–38.

Wikipedia contributors, "Henna," *Wikipedia, The Free Encyclopedia,* https://en.wikipedia.org/w/index.php?title=Henna&oldid=889880002 (accessed 2019).

Wikipedia contributors, "*Impatiens niamniamensis*," *Wikipedia, The Free Encyclopedia,* https://en.wikipedia.org/w/index.php?title=Impatiens_niamniamensis&oldid=831634297 (accessed 2019).

Zika, Peter F. 2006. Lectotypification of the Names of Two Color Forms of *Impatiens capensis* (Balsaminaceae). *Rhodora* 108(933): 62–64.

Jimsonweed references

Adler, Lynn S., and Judith L. Bronstein. 2004. Attracting Antagonists: Does Floral Nectar Increase Leaf Herbivory? *Ecology* 85(6): 1519–1526.

Beadle, David, and Seabrooke Leckie. 2012. *Peterson Field Guide to Moths of Northeastern North America*. Houghton Mifflin Harcourt, New York.

Burnside, Orvin C., Robert G. Wilson, Sanford Weisberg, and Kenneth G. Hubbard. 1996. Seed Longevity of 41 Weed Species Buried 17 Years in Eastern and Western Nebraska; Reviewed Works. *Weed Science* 44(1): 74–86.

Bolt, H. J., Letter to the editor. *New York Times*. February 7, 1915.

Editorial. 1915. "Twilight Sleep": the *Dämmerschlaf* of the Germans. *Canadian Medical Association Journal* 5(9): 805–808.

Erowid. 2012. "*Datura*–Legal Status," The Vault of Erowid, http://www.erowid.org/plants/datura/datura_law.shtml (accessed 2012).

Faegri, K., and L. van der Pijl. 1979. *The Principles of Pollination Ecology: Third Revised Edition*. Pergamon Press, Oxford.

Friedman, Mendel. 2004. Analysis of Biologically Active Compounds in Potatoes (*Solanum tuberosum*), Tomatoes (*Lycopersicon esculentum*), and Jimson Weed (*Datura stramonium*) Seeds. *Journal of Chromatography A*, 1054(1–2): 143–155.

Friedman, Mendel, and Carol E. Levin. 1989. Composition of Jimson Weed (*Datura stramonium*) Seeds. *Journal of Agricultural and Food Chemistry* 37(4): 998–1005.

Hilty, John. 2003–2012. "Jimsonweed," Illinois Wildflowers website, http://www.illinoiswildflowers.info/weeds/plants/jimsonweed.htm (accessed 2012).

Kogan, Marcos, and Richard D. Goeden. 1970. The Biology of *Lema trilineata daturaphila*, (Coleoptera: Chrysomelidae) with Notes on Efficiency of Food Utilization by Larvae. *Annals of the Entomological Society of America* 63(2): 537–546.

Lucero, M. E., W. Mueller, J. Hubstenberger, G. C. Phillips, and M. A. O'Connell. Tolerance to Nitrogenous Explosives and Metabolism of TNT by Cell Suspensions of *Datura innoxia*. *In Vitro Cellular & Developmental Biology. Plant* 35(6): 480–486.

Motten, A. F., and J. L. Stone. 2000. Heritability of Stigma Position and the Effect of Stigma-Anther Separation on Outcrossing in a Predominantly Self-fertilizing Weed, *Datura stramonium* (Solanaceae). *American Journal of Botany* 87(3): 339–347.

Weaver, Susan E., and Suzanne I. Warwick. 1984. The Biology of Canadian Weeds. 64. *Datura stramonium* L. *Canadian Journal of Plant Science* 64(4): 979–991.

Wikipedia contributors, "*Atropa belladonna*," *Wikipedia, The Free Encyclopedia,* https://en.wikipedia.org/w/index.php?title=Atropa_belladonna&oldid=90124424 (accessed 2012).

Wikipedia contributors, "*Datura stramonium*," *Wikipedia, The Free Encyclopedia,* https://en.wikipedia.org/w/index.php?title=Datura_stramonium&oldid=902120130 (accessed 2011).

Wikipedia contributors, "Hyoscine," *Wikipedia, The Free Encyclopedia,* https://en.wikipedia.org/w/index.php?title=Hyoscine&oldid=902829619 (accessed 2019).

Wikipedia contributors, "Twilight sleep," *Wikipedia, The Free Encyclopedia,* https://en.wikipedia.org/w/index.php?title=Twilight_sleep&oldid=884430260 (accessed 2012).

Willmer, Pat. 2011. *Pollination and Floral Design*. Princeton: Princeton University Press.

Lily references

Bernhardt, Peter. 1989. *Wily Violets and Underground Orchids*. Random House, New York.

Bernhardt, Peter. 1999. *The Rose's Kiss*. University of Chicago Press, Chicago.

Blanchan, Neltje. 1900. *Nature's Garden*. Garden City Publishing, Garden City, NY.

Boufford, D. E., and S. A. Songberg. 1983. Eastern Asian—North American Phytogeographical Relationships—A History from

References

the Time of Linnaeus to the Twentieth Century. *Annals of the Missouri Botanical Garden* 70: 423–439.

Burroughs, John. 1904. *The Writings of John Burroughs, Volume IX: Riverby*. Houghton Mifflin, New York.

Casagrande, Richard A., and Marc Kenis. 2004. "Evaluation of Lily Leaf Beetle Parasitoids for North American Introduction," in Driesche, R. G. Van, T. Murray, and R. Reardon (Eds.). *Assessing Host Ranges for Parasitoids and Predators*, Chapter 10, USDA Forest Health Technology Enterprise Team, Morgantown, WV, http://forestpestbiocontrol.info/fact_sheets/documents/wholehostrangebookpdf.pdf#page=127 (accessed 2019).

Chambers, Lannie. 2019. Hummingbirds.net website, http://www.hummingbirds.net/species.html (accessed 2019).

Doorn, Wouter G. van, and Uulke van Meeteren. 2003. Flower Opening and Closure: A Review. *Journal of Experimental Botany* 54(389): 1801–1812.

Editors of *Encyclopaedia Britannica*. "Orange Order: Irish Political Society." *Encyclopaedia Britannica* website, https://www.britannica.com/topic/Orange-Order (accessed 2019).

Edwards, Joan, and James R. Jordan. 1992. Reversible Anther Opening in *Lilium philadelphicum* (Liliaceae): A Possible Means of Enhancing Male Fitness. *American Journal of Botany* 79(2): 144–148.

Fletcher, J. Darl, Lisa A. Shipley, William J. McShea, and Durland L. Shumway. 2001. Wildlife Herbivory and Rare Plants: The Effects of White-tailed Deer, Rodents, and Insects on Growth and Survival of Turk's-cap Lily. *Biological Conservation* 101(2): 229–238.

Heus, Peter. 2003. Propagation Protocol for Canada Lily (*Lilium canadense*). *Native Plants Journal* 4(2): 107–109.

Hilty, John. 2014–2018. "Canada Lily," Illinois Wildflowers website, http://www.illinoiswildflowers.info/savanna/plants/canada_lily.htm (accessed 2019).

Hilty, John. 2014–2018. "Michigan Lily," Illinois Wildflowers website, http://www.illinoiswildflowers.info/prairie/plantx/mich_lilyx.htm (accessed 2019).

Hilty, John. 2014–2018. "Prairie Lily," Illinois Wildflowers website, http://www.illinoiswildflowers.info/prairie/plantx/pr_lily.html (accessed 2019).

Hilty, John. 2014–2018. "Turk's Cap Lily," Illinois Wildflowers website, http://www.illinoiswildflowers.info/savanna/plants/turkcap_lily.htm (accessed 2019).

Howe, Thomas D., and Susan Power Bratton. 1976. Winter Rooting Activity of the European Wild Boar in the Great Smoky Mountains National Park. *Castanea* 41(3): 256–264.

Majka, Christopher G., and Laurent LeSage. 2008. Introduced Leaf Beetles of the Maritime Provinces, 5: The Lily Leaf Beetle, *Lilioceris lilii* (Scopoli) (Coleoptera: Chrysomelidae). *Proceedings of the Entomological Society of Washington* 110(1): 186–195.

Meredith, Leda. 2013. "How to Sustainably Harvest and Eat Delicious Daylilies," Mother Earth News: The Original Guide to Living Wisely website https://www.motherearthnews.com/organic-gardening/how-to-sustainably-harvest-daylilies-zbcz1307 (accessed 2019).

New England Wildflower Society. 2011–2019. "*Lilium*," Go Botany website, https://gobotany.newenglandwild.org/search/?q=lilium (accessed 2019).

PetMD LLC. 1999–2019. "Lily Poisoning in Cats," PetMD website, https://www.petmd.com/cat/emergency/poisoning-toxicity/e_ct_lily_poisoning (accessed 2019).

Primack, Richard B. 2014. *Walden Warming: Climate Change Comes to Thoreau's Woods*. University of Chicago Press, Chicago.

Showalter, A. M. 1964. Allelic Segregation in Interspecific Hybrids. *Journal of Heredity* 55(4): 183.

Stack, Philip A., Eleanor Groden, and Lois Berg Stack. 2009. "Lily Leaf Beetle," Cooperative Extension Publications, Bulletin #2450. University of Maine, https://extension.umaine.edu/publications/2450e/ (accessed 2016).

Staff reporter. 2018. "Man Who Drove over Orange Lilies Convicted of Hate Crime." *Belfast Telegraph* digital, https://www.belfasttelegraph.co.uk/news/northern-ireland/man-who-drove-over-orange-lilies-convicted-of-hate-crime-36988030.html.

Wherry, Edgar T. 1942. The Relationship of *Lilium michiganense*. *Rhodora* 44(528): 453–456.

Wikipedia contributors, "*Lilium*," *Wikipedia, The Free Encyclopedia*, https://en.wikipedia.org/w/index.php?title=Lilium&oldid=885217987 (accessed 2019).

Wikipedia contributors. 2018. "*Lilium lancifolium*," *Wikipedia, The Free Encyclopedia*, https://en.wikipedia.org/wiki/Lilium_lancifolium (accessed 2019).

Wikipedia contributors. 2019. "*Lilium candidum*," *Wikipedia, The Free Encyclopedia*, https://en.wikipedia.org/wiki/Lilium_candidum (accessed 2019).

Wikipedia contributors. 2019. "Orange Order," *Wikipedia, The Free Encyclopedia*, https://en.wikipedia.org/wiki/Orange_Order (accessed 2019).

Partridge-berry references

Bierzychudek, Paulette. 1982. Life Histories and Demography of Shade-Tolerant Temperate Forest Herbs: A Review. *New Phytologist* 90(4): 757–776.

Burkle, Laura A., and Barry A. Logan. 2003. Seasonal Acclimation of Photosynthesis in Eastern Hemlock and Partridgeberry in Different Light Environments. *Northeastern Naturalist* 10(1): 1–16.

Darwin, Charles. 1877. *The Different Forms of Flowers on Plants of the Same Species, Part 5*. John Murray, London: 125–127.

Dorman, John Frederick, and James F. Lewis. Doctor John Mitchell, F.R.S.: Native Virginian. *The Virginia Magazine of History and Biography* 76(4): 437–440.

Ganders, Fred R. 1975. Fecundity in Distylous and Self-incompatible Homostylous Plants of *Mitchella repens*, Rubiaceae. *Evolution* 29(1): 186–188.

Hayden, W. John. 2012. "Partridge Berry (*Mitchella repens*)," Virginia Native Plant Society website, http://vnps.org/wildflowers-of-the-year/2012-partridge-berry-mitchella-repens/ (accessed 2015).

Hicks, David J., Robert Wyatt, and Thomas R. Meagher. 1985. Reproductive Biology of Distylous Partridgeberry, *Mitchella repens*. *American Journal of Botany* 72(10): 1503–1514.

Huang, Wei-Ping, et al. 2013. Molecular Phylogenetics and Biogeography of the Eastern Asian–Eastern North American Disjunct *Mitchella* and its Close Relative *Damnacanthus* (Rubiaceae, Mitchelleae). *Botanical Journal of the Linnean Society* 171(2): 395–412.

Keegan, Christine R., Robert H. Voss, and Kamalijit S. Bawa. 1979. Heterostyly in *Mitchella repens* (Rubiaceae). *Rhodora* 81(828): 567–573.

North, Robert L. 2000. Benjamin Rush, MD: Assassin or Beloved Healer? *Baylor University Medical Center Proceedings* 13(1): 45–49.

Stritch, Larry. "Partridge Berry (*Mitchella repens* L.)," Plant of the Week," U.S. Forest Service: Celebrating Wildflowers website, http://www.fs.fed.us/wildflowers/plant-of-the-week/mitchella_repens.shtml (accessed 2015).

Wikipedia contributors, "*Rubia tinctorum*," *Wikipedia, The Free Encyclopedia,* https://en.wikipedia.org/w/index.php?title=Rubia_tinctorum&oldid=891780610 (accessed 2015).

Wu, Cheng-yi, Peter H. Raven, and De-yuang Hong (eds.). Flora of China Editorial Committee. 2011. *Flora of China* 19: Rubiaceae. Chen, Tao, and Charlotte M. Taylor, "*Mitchella*," http://flora.huh.harvard.edu/china/PDF/PDF19/Mitchella.pdf (accessed 2014).

Passion-flower references

Benson, Woodruff W., Keith S. Brown Jr., and Lawrence E. Gilbert. 1975. Coevolution of Plants and Herbivores: Passion Flower Butterflies. *Evolution* 29(4): 659–680.

Bowling, Andrew J., and Kevin C. Vaughn. 2009. Gelatinous Fibers are Widespread in Coiling Tendrils and Twining Vines. *American Journal of Botany* 96(4): 719–727.

CABI. 2019. *Passiflora tripartita* var. *mollissima*: Banana Passionfruit (original text by Ian Popay), *in* Invasive Species Compendium, CAB International, Wallingford, UK. https://www.cabi.org/isc/datasheet/38802 (accessed 2019).

Cech, Rick, and Guy Tudor. 2005. *Butterflies of the East Coast: An Observer's Guide.* Princeton University Press, Princeton, NJ.

Coppens d'Eeckenbrugge, Geo, Victoria E. Barney, Peter Møller Jørgensen, and John M. MacDougal. 2001. *Passiflora tarminiana*, a New Cultivated Species of *Passiflora* subgenus Tacsonia (Passifloraceae). *Novon* 11(1): 8–15.

Glassberg, Jeffery. 1999. *Butterflies Through Binoculars: The East.* Oxford University Press, New York.

Gerbode, Sharon J., Joshua R. Puzey, Andrew G. McCormick, and L. Mahadevan. 2012. How the Cucumber Tendril Coils and Overwinds. *Science*, New Series, 327(6098): 1087–1091.

Heenan, P. B., and W. R. Sykes. 2003. "*Passiflora* (Passifloraceae) in New Zealand: A Revised Key with Notes on Distribution." *New Zealand Journal of Botany* 41(2): 217–221.

Holland, J. Burks, and J. Lanza. 2008. Geographic Variation in the Pollination Biology of *Passiflora lutea* (Passifloraceae), *Journal of the Arkansas Academy of Science* 62: 32–36.

Trevisan Scorza, Livia Camilla, and Marcelo Carnier Domelas. 2011. Plants on the Move: Toward Common Mechanisms Governing Mechanically-induced Plant Movements. *Plant Signaling and Behavior* 6(12): 1979–1986.

Wagner, David L. 2005. *Caterpillars of Eastern North America.* Princeton University Press, Princeton, NJ.

Williams, Kathy S., and Lawrence E. Gilbert. 1981. Insects as Selective Agents on Plant Vegetative Morphology: Egg Mimicry Reduces Egg Laying by Butterflies. *Science*, New Series 212(4493): 467–469.

Willmer, Pat. 2011. *Pollination and Floral Ecology.* Princeton University Press, Princeton, NJ.

Pipsissewa and Related Species references

Blanchan, Nellie. 1900. *Nature's Garden.* Doubleday, New York.

De Luca, Paul A., and Mario Vallejo-Marín. 2013. What's the 'Buzz' About? The Ecology and Evolutionary Significance of Buzz-pollination. *Current Opinion in Plant Biology* 16(4): 629–435.

Eriksson, Ove, and Kent Kainulainen. 2011. The Evolutionary Ecology of Dust Seeds. *Perspectives in Plant Ecology, Evolution, and Systematics* 13(2): 73–87.

Haber, Erich, and James E. Cruise. 1974. Generic Limits in the Pyroloideae (Ericaceae). *Canadian Journal of Botany* 52(4): 877–883.

Hausen, B. M., and I. Schiedermair. 1988. The Sensitizing Capacity of Chimaphilin, a Naturally-occurring Quinone. *Contact Dermatitis* 19(3): 180–183.

Johansson, Veronika A., Gregor Muller, and Ove Eriksson. 2014. Dust Seed Production and Dispersal in Swedish Pyroleae Species. *Nordic Journal of Botany* 32(2): 209–214.

Moneses uniflora: One-flowered Wintergreen. 2011. Central Yukon Species Inventory Project website, http://www.flora.dempstercountry.org/0.Site.Folder/Species.Program/Species2.php?species_id=Mone.uni (accessed 2014).

Moneses uniflora: One-flowered Wintergreen. 2013. Encyclopedia of Life (EOL) website, http://eol.org/pages/486756/names/synonyms (accessed 2014).

Knudsen, Jette T., and Lars Tollsten. 1991. Floral Scent and Intrafloral Scent Differentiation in *Moneses* and *Pyrola* (Pyrolaceae). *Plant Systematics and Evolution* 177(1–2): 81–91.

Knudsen, Jette T., and Jens Mogens Olesen. 1993. Buzz-Pollination and Patterns in Sexual Traits in North European Pyrolaceae. *American Journal of Botany* 80(8): 900–913.

Saxena, Geeta, S. W. Farmer, R. E. W. Hancock, and G. H. N. Towers. 1996. Chlorochimaphilin: A New Antibiotic from *Moneses uniflora*. *Journal of Natural Products* 59(1): 62–65.

Tedersoo, Leho, Prune Pellet, Urmas Kõljalg, and Marc-André Selosse. 2007. Parallel Evolutionary Paths to Mycoheterotrophy in Understorey Ericaceae and Orchidaceae: Ecological Evidence for Mixotrophy in Pyroleae. *Oecologia* 151(2): 206–217.

Ursic, Ken, Todd Farrell, Margot Ursic, and Margaret Stalker. 2010. Recovery Strategy for the Spotted Wintergreen (*Chimaphila maculata*) in Ontario. Ontario Recovery Strategy Series. Prepared for the Ontario Ministry of Natural Resources, Peterborough. vi + 28 pp. https://www.ontario.ca/page/spotted-wintergreen-recovery-strategy (accessed online 2014).

Prickly-pear references

Braam, Janet. 2005. In Touch: Plant Responses to Mechanical Stimuli. *New Phytologist* 165(2): 373–389.

Chachere, Vickie. 2012. "Cactus a Natural Oil Dispersant," University of South Florida News website, http://news.usf.edu/article/templates/?a=4387&z=123 (accessed 2013).

Donck, Adriaen van der, with Charles T. Gehring, William A. Starna (eds.) and Diederik Willem Goedhuys (translator). 2008. *A Description of New Netherland (The Iroquoians and Their World).* University of Nebraska, Lincoln. 24.

Fact Sheet: Invasive Plants and Animals. 2013. "The Prickly Pear Story," Queensland Government, Biosecurity Queensland website, http://cactiguide.com/pdf_docs/IPA-Prickly-Pear-Story-PP62.pdf (accessed 2013).

References

Fitter, Julian, Daniel Fitter, and David Hosking. 2000. *Wildlife of the Galápagos*. Princeton University Press, Princeton, NJ.

Galápagos Conservancy. 2012. "The Galápagos Islands: Santa Fé," http://www.galapagos.org/about_galapagos/santa-fe/ (accessed 2013).

Garrett, Howard. 2013. "Cochineal," The Dirt Doctor website, http://www.dirtdoctor.com/Cochineal_vq1321.htm (accessed 2013).

"*Opuntia humifusa* (Raf.) Raf. (eastern prickly pear)," New England Wild Flower Society, Go Botany website, https://gobotany.newenglandwild.org/species/opuntia/humifusa/ (accessed 2013).

Hilty, John. 2002–2012. "Eastern Prickly Pear," Illinois Wildflowers website, http://www.illinoiswildflowers.info/prairie/plantx/prickly_pearx.htm (accessed 2012).

Jackson, Michael H. 1985. *Galápagos: A Natural History Guide*. University of Calgary Press, Calgary.

Maxwell, Phil. 1998. "The *Rhipsalis* Riddle—or the Day the Cacti Came Down from the Trees," *Rhipsalis, Lepismium, Hatiora, Schlumbergera* website, http://www.rhipsalis.com/maxwell.htm (accessed 2013).

McMullen, Conley K. 1999. *Flowering Plants of the Galápagos*. Cornell University Press, Ithaca, NY.

Nichols, William F. 2012. Three Non-native Plant Species Newly Documented for New Hampshire. *Rhodora* 114(958): 209–213.

Orozco-Segovia, A., J. Márquez-Guzmán, M. E. Sánchez-Coronado, et al. 2007. Seed Anatomy and Water Uptake in Relation to Seed Dormancy in *Opuntia tomentosa* (Cactaceae, Opuntioideae). *Annals of Botany* 99(4): 581–592.

Palumbo, Barbara, Yannis Efthimiou, Jorgos Stamatopoulos, et al. 2003. Prickly Pear Induces Upregulation of Liver LDL Binding in Familial Heterozygous Hypercholesterolemia. *Nuclear Medicine Review* (synopsis online), http://science.naturalnews.com/2003/5912984_Prickly_pear_induces_upregulation_of_liver_LDL_binding_in_familial.html (accessed 2013).

Pinkava, Donald J. 2003. *Opuntia*, in Flora of North America Editorial Committee, eds. 1993+. *Flora of North America North of Mexico*, vol. 4, p. 123. New York and Oxford.

Ríha, J., and R. Subík. 1981. *The Illustrated Encyclopedia of Cacti and Other Succulents*. Octopus Books, London.

Simons, Paul. 1992. *The Action Plant*. Blackwell, Oxford.

USDA, NRCS. 2019. The PLANTS Database, http://plants.usda.gov/ (accessed 2019).

Vigueras G., A. L., and L. Portillo. 2001. Uses of *Opuntia* species and the Potential Impact to *Cactoblastis cactorum* (Lepidoptera: Pyralidae) in Mexico. *Florida Entomologist* 84(4): 493–498.

Weiner, Jonathan. 1994. *The Beak of the Finch: A Story of Evolution in Our Time*. Alfred A. Knopf, New York.

Wiggins, Ira L., and Duncan M. Porter. 1971. *Flora of the Galápagos Islands*. Stanford University Press, Stanford, CA.

Wikipedia contributors, "Opuntia," *Wikipedia, The Free Encyclopedia*, https://en.wikipedia.org/w/index.php?title=Opuntia&oldid=901054157 (accessed 2013).

Purple Pitcher Plant—references

Atwater, Daniel Z., Jessica L. Butler, Aaron M. Ellison. 2006. Spatial Distribution and Impacts of Moth Herbivory on Northern Pitcher Plants. *Northeastern Naturalist* 13(1): 43–56.

Bradshaw, William E., and Robert A. Creelman. 1984. Mutualism between the Carnivorous Purple Pitcher Plant and its Inhabitants. *American Midland Naturalist* 112(2): 294–304.

Brown, Kate, Emily Burkett, Genevieve Johnson, and Michelle Plaxton. 1995. "The Relationship Between *Wyeomyia smithii* and *Metriocnemus knabi* Larvae and the Insectivorous Plant, *Sarracenia purpurea*," Deep Blue at the University of Michigan, http://hdl.handle.net/2027.42/54580 (accessed online 2011).

Catesby, Mark. 1731-1743. *The Natural History of Carolina, Florida, and the Bahama Islands; Containing the Figures of Birds, Beasts, Fishes, Serpents, Insects, and Plants; Particularly the Forest-trees, Shrubs, and Other Plants, not Hitherto Described, or Very Incorrectly Figured by Authors. Together with their Descriptions in English and French. To Which Are Added Observations on the Air, Soil, and Waters; with Remarks Upon Agriculture, Grain, Pulse, Roots, &c. To the Whole, is Prefixed a New and Correct Map of the Countries Treated Of/by the Late Mark Catesby; revised by Mr. Edwards*. Volume 2: 69, and Plate 70. Printed at the expense of the author, London.

Cheek, Martin. 1994. The Correct Names of the Subspecies of *Sarracenia purpurea* L. *Carnivorus Plant Newsletter* 23(3): 69–73.

Cheek, Martin, Don Schnell, James L. Reveal, and Jan Schlauer. 1997. Proposal to Conserve the Name *Sarracenia purpurea* (Sarraceniaceae) with a New Type. *Taxon* 46(4): 781–783.

Cipollini, Donald F. Jr., Sandra J. Newell, and Anthony J. Nastase. 1994. Total Carbohydrates in Nectar of *Sarracenia purpurea* L. (Northern Pitcher Plant). *American Midland Naturalist* 131(2): 374–377.

Charles Ho Consultants. "Help Manage Your Pain Through Knowledge," Beverly Pain Management website, http://www.pain-clinic.org/piriformissyndrome (accessed 2012).

Cresswell, James E. 1991. Capture Rates and Composition of Insect Prey of the Pitcher Plant *Sarracenia purpurea*. *American Midland Naturalist* 125(1): 1–9.

Dahlem, G. A., and Robert F. C. Naczi. 2006. Flesh Flies (Diptera: Sarcophagidae) Associated with North American Pitcher Plants (Sarraceniaceae), with Descriptions of Three New Species. *Annals of the Entomological Society of America* 99(2): 218–240.

Deppe, Jill L., et al. 2000. Diel Variation of Sugar Amount in Nectar from Pitchers of *Sarracenia purpurea* L. with and without Insect Visitors. *American Midland Naturalist* 144(1): 123–132.

Dussourd, David E. 2005. In the Trenches: Bioprospecting with a Caterpillar Probe. *Wings* 28(2): 20–24.

Ellison, Aaron M. 2005. Turning the Tables: Plants Bite Back. *Wings* 28(2): 25–29.

Ellison, Aaron M., Elena D. Butler, Emily Jean Hicks, Robert F. C. Naczi, Patrick J. Calie, et al. 2012. Phylogeny and Biogeography of the Carnivorous Plant Family Sarraceniaceae. *PLOS ONE* 7(6): e39291.

Fish, Durland, and Donald W. Hall. 1978. Succession and Stratification of Aquatic Insects Inhabiting the Leaves of the Insectivorous Pitcher Plant, *Sarracenia purpurea*. *American Midland Naturalist* 99: 172–183.

Folkerts, Debbie. 1999. Pitcher Plant Wetlands of the Southeastern United States: Arthropod Associates. Chapter 11, pp. 247–275, in Batzer, Darold P., Russell B. Rader, and Scott A. Wissinger (eds.). *Invertebrates in Freshwater Wetlands of North America: Ecology and Management*. Wiley, New York.

Forsyth, Adrian B., and Raleigh J. Robertson. 1975. K Reproductive Strategy and Larval Behavior of the Pitcher Plant Sarcophagid Fly *Blaesoxipha fletcheri*. *Canadian Journal of Zoology* 53(2): 174–177.

Gallie, Daniel R., and Su-Chih Chang. 1997. Signal Transduction in the Carnivorous Plant *Sarracenia purpurea*: Regulation of Secretory Hydrolase Expression During Development and in Response to Resources. *Plant Physiology* 115(4): 1461–1471.

Giberson, Donna, and M. Loretta Hardwick. 1999. Pitcher Plants (*Sarracenia purpurea*) Eastern Canadian Peatlands: Ecology and Conservation of the Invertebrate Inquilines. Chapter 18, pp. 401–422, *in* Batzer, Darold P., Russell B. Rader, and Scott A. Wissinger (eds.). *Invertebrates in Freshwater Wetlands of North America: Ecology and Management*. Wiley, New York.

Harris, Marjorie. 2003. *Botanica North America*. HarperCollins, New York.

Heard, Stephen B. 1998. Capture Rates of Invertebrate Prey by the Pitcher Plant, *Sarracenia purpurea* L. *American Midland Naturalist* 139(1): 79–89.

Manchikanti, Kavita N., Vidyasager Pampati, Kim S. Damon, and Carla D. McMannus. 2004. A Double-blind, Controlled Evaluation of the Value of Sarapin in Neural Blockade. *Pain Physician* 7(1): 59–62.

Manchikanti, Laxmiah, Vidyasager Pampati, Jose J. Rivera, Carla Beyer, et al. 2001. Caudal Epidural Injections with Sarapin or Steroids in Chronic Low Back Pain. *Pain Physician* 4(4): 322–335.

Miller, Thomas E., and Jamie M. Neitel. 2005. Inquiline Communities in Pitcher Plants as a Prototypical Metacommunity. Chapter 5, pp. 122–145, *in* Holyoak, Marcel, Mathew A. Leibold, and Robert D. Holt (eds.). *Metacommunities: Spatial Dynamics and Ecological Communities*, University of Chicago Press, Chicago.

Naczi, Robert F. C., Eric M. Soper, Frederick W. Case Jr., and Roberta B. Case. 1999. *Sarracenia rosea* (Sarraceniaceae), a New Species of Pitcher Plant from the Southeastern United States. *Sida* 18(4): 1183–1206.

Neeman, Gidi, Rina Neeman, and Aaron M. Ellison. 2006. Limits to Reproductive Success of *Sarracenia purpurea* (Sarraceniaceae). *American Journal of Botany* 93(11): 1660–1666.

Peterson, Raymond L., et al. 2000. Foreign Mosquito Survivorship in the Pitcher Plant *Sarracenia purpurea*: The Role of the Pitcher-plant Midge *Metriocnemus knabi*. *Hydrobiologia* 439(1–3): 13–19.

Rask, Michael R. 1984. The Omohyoideus Myofascial Pain Syndrome: Report of Four Patients. *Journal of Craniomandibular Practice* 2(3): 256.

Rice, Barry. "The Carnivorous Plant FAQ v. 11.5." Barry's website, http://www.sarracenia.com/faq/faq1250.html (accessed 2011).

Smith, Huron H. 1933. Ethnobotany of the Forest Potawatomi Indians. *Bulletin of the Public Museum of the City of Milwaukee* 7(1): 123.

Queen Anne's Lace references

Barlow, Jim. 2001. *Pollen from One Bt Corn Variety Reduced Growth Rates among Black Swallowtail*. University of Illinois. Illinois News Bureau. https://news.illinois.edu/view/6367/208201 (accessed 2018).

Borthwick, H. A., Mable Phillips, and W. W. Robbins. 1931. Floral Development in *Daucus carota*. *American Journal of Botany* 18(9): 784–796.

Bowling Green State University (BGSU) Libraries. n.d. Student Digital Gallery. "Writing Miss Newton: Mary Leslie Newton, 1874–1944." https://digitalgallery.bgsu.edu/student/exhibits/show/writing-miss-newton (accessed 2018).

Bhatnagar, Upendra. 1995. Postcoital Contraceptive Effects of an Alcoholic Extract of the *Daucus carota* Linn Seed in Rats. *Clinical Drug Investigation* 9(1): 33–36.

Clark, Larry, and J. Russell Mason. 1985. Use of Nest Material as Insecticidal and Anti-Pathogenic Agents by the European Starling. *Oecologia* 67(2): 169–176.

Clark, Larry, and J. Russell Mason. 1988. Effect of Biologically Active Plants Used as Nest Material and the Derived Benefit to Starling Nestlings. *Oecologia* 77(2): 174–180.

Dale, Hugh M. 1974. The Biology of Canadian Weeds. 5. *Daucus carota*. *Canadian Journal of Plant Science* 54: 673–685.

Eames, Edwin H. 1919. Another Exceptional Specimen of *Daucus carota*. *Rhodora* 21(248): 147–148.

Early Detection & Distribution Mapping System. "Queen Anne's lace, wild carrot (*Daucus carota* L.)," https://www.invasiveplantatlas.org/subject.html?sub=5514 (accessed 2019).

Elliott, Nancy B., and William M. Elliott. 1991. Effect of an Ambush Predator, *Phymata americana* Merlin, on Behavior of Insects Visiting *Daucus carota*. *American Midland Naturalist* 126(1): 198–202.

Feeny, Paul, William S. Blau, and Peter M. Kareiva. 1985. Larval Growth and Survivorship of the Black Swallowtail Butterfly in Central New York. *Ecological Monographs* 55(2): 167–187.

Goldman, Edward H. 1933. Comparisons of the Mouthparts of Adult Longhorn Beetles with Reference to Their Food (Coleoptera: Cerambycidae). *Transactions of the American Entomological Society* 59(2): 85–102.

Goulson, David, Kate McGuire, Emma E. Munro et al. 2009. Functional Significance of the Dark Central Floret of *Daucus carota* (Apiaceae) L.: Is It an Insect Mimic? *Plant Species Biology* 24(2): 77–82.

Hartigh, Cora den. 2014. "*Daucus carota*," Botany Photo of the Day website. Daniel Mosquin. http://botanyphoto.botanicalgarden.ubc.ca/2014/12/daucus-carota/ (accessed 2014).

Hilty, John. 2002–2017. "Weedy Wildflowers: Wild Carrot, *Daucus carota*," Illinois Wildflowers website, http://www.illinoiswildflowers.info/weeds/plants/wild_carrot.htm (accessed 2018).

Invasive Plant Atlas of the United States. 2018. "Queen Anne's lace, wild carrot," https://www.invasiveplantatlas.org/subject.html?sub=5514 (accessed 2019).

Jansen, Gabrielle Claire, and Hans Wohlmuth. 2014. Carrot Seed for Contraception: A Review. *Australian Journal of Herbal Medicine* 26(1): 10–17.

Koul, Pushpa, A.K. Koul, and I. A. Hamal. 1989. Reproductive Biology of Wild and Cultivated Carrot (*Daucus carota* L.). *New Phytologist* 112: 437–443.

Lacey, Elizabeth. 1980. The Influence of Hygroscopic Movement on Seed Dispersal in *Daucus carota* (Apiaceae). *Oecologia* 47(1): 110–114.

Lacey, Elizabeth. 1982. Timing of Seed Dispersal in *Daucus carota*. *Oikos* 39(1): 83–91.

References

Lacey, Elizabeth. 1986. The Genetic and Environmental Control of Reproductive Timing in a Short-lived Monocarpic Species, *Daucus carota* (Umbelliferae). *Journal of Ecology* 74(1): 73–86.

Lago, Paul K., and Michael O. Mann. 1987. Survey of Coleoptera Associated with Flowers of Wild Carrot (*Daucus carota* L.) (Apiaceae) in Northern Mississippi. *Coleopterists Bulletin* 41(1): 1–8.

Lamborn, E., and J. Ollerton. 2000. Experimental Assessment of the Functional Morphology of Inflorescences of *Daucus carota* (Apiaceae): Testing the "Fly catcher Effect." *Functional Ecology* 14(4): 445–454.

Lukes, Roy. 2000. "Queen Anne's Lace: Lovely Wildflower or Alien Invader?," Nature-Wise website, http://doorbell.net/lukes/a082804.htm (accessed 2004).

Mahr, Susan. 2004. "*Compsilura concinnata*, Parasitoid of Gypsy Moth," Biological Control News website, Department of Entomology, University of Wisconsin, http://www.entomology.wisc.edu/mbcn/kyf609.html (accessed 2018).

McDowell, William T. 2011. Diversity and Notes on the Reproductive Biology of Cerambycidae (Coleoptera) on *Hydrangea arborescens* L. and *Daucus carota* L. at LaRue-Pine Hills Research Natural Area in Southern Illinois, U.S.A. *Coleopterists Bulletin* 65(4): 422–416.

Meiners, Scott J., 2014. Functional Correlates of Allelopathic Potential in a Successional Plant Community. *Plant Ecology* 215(6): 661–672.

Newton, Mary Leslie. n.d. "Queen Anne's Lace," Poetry Nook: Poetry for Every Occasion website, https://www.poetrynook.com/poem/queen-annes-lace-0 (accessed 2018).

NJTODAY.net. 2016. "Black Swallowtail: New Jersey's Official Butterfly," CMD Media website, http://njtoday.net/2016/03/18/black-swallowtail-new-jerseys-official-butterfly/ (accessed 2018).

"List of U.S. State Butterflies," Obsession with Butterflies website, http://www.obsessionwithbutterflies.com/state-butterfly.html (accessed 2018).

Olsen, Lars, Chris Oostenbrink, Flemming Steen Jørgensen. 2015. Prediction of Cytochrome P450 Mediated Metabolism. *Advanced Drug Delivery Reviews* 86(5): 61–71.

Robinson, B. L. 1919. An Unusual *Daucus carota*. *Rhodora* 21(243): 70–71.

Ramen, Ryan. 2018. "Cilantro vs Coriander: What's the Difference?" Healthline Newsletter website. 2018. https://www.healthline.com/nutrition/cilantro-vs-coriander (accessed 2019).

Sifferlin, Alexandra. 2013. "Eat This Now: Rainbow Carrots," Time Healthland website, http://healthland.time.com/2013/08/20/eat-this-now-rainbow-carrots/ (accessed 2018).

Stachler, Jeff M., James J. Kells, and Donald Penner. 2000. Resistance of Wild Carrot (*Daucus carota*) to 2.4-D in Michigan. *Weed Technology* 14(4): 734–739. https://www.jstor.org/stable/3988662 (accessed online 2018).

Stolarczyk, John, and Jules Janick. 2011. Carrot: History and Iconography. *Chronica Horticulturae* 51(2): 13–18.

"The Wild Carrot," World Carrot Museum website, http://www.carrotmuseum.co.uk/wild.html (accessed 2018).

Wagner, David L. 2005. *Caterpillars of Eastern North America*. Princeton University Press. Princeton, NJ.

WeedUS—Database of Plants Invading Natural Areas in the United States, http://www.invasive.org/weedus/ (accessed 2019).

Williams, William Carlos. 1938. "Queen Anne's Lace." *The Collected Poems of William Carlos Williams, Volume 1, 1909–1939*," The Poetry Foundation website, https://www.dec.ny.gov/animals/113303.html (accessed 2018).

Wikipedia contributors. 2018. "*Daucus carota*," *Wikipedia, The Free Encyclopedia*, https://en.wikipedia.org/w/index.php?title=Daucus_carota&oldid=846499381 (accessed 2018).

Wikipedia contributors. 2018. "Anne, Queen of Great Britain," *Wikipedia, The Free Encyclopedia*, https://en.wikipedia.org/wiki/Anne,_Queen_of_Great Britain (accessed 2018).

Wikipedia contributors. 2018. "Anne of Denmark," *Wikipedia, The Free Encyclopedia*, https://en.wikipedia.org/wiki/Anne_of_Denmark (accessed 2018).

Wraight, C. Lydia, Arthur R. Zangerl, Mark J. Carroll, and May R. Berenbaum. 2000. Absence of Toxicity of *Baccillus thuringiensis* Pollen to Black Swallowtails under Field Conditions. *Proceedings of the National Academy of Sciences* 97(14): 7700–7703. http://www.pnas.org/content/97/14/7700 (accessed online 2018).

Showy Lady-slipper references

Argue, Charles L. 2011. *The Pollination Biology of North American Orchids: Volume I*, pp. 37–46. Springer, New York.

Bernhardt, Peter, and Retha Edens-Meier. 2010. What We Think We Know vs. What We Need to Know about Orchid Pollination and Conservation: *Cypripedium* L. as a Model Lineage. *Botanical Review* 76(2): 204–219.

Blanchan, Neltje. 1900. *Nature's Garden*. Garden City Publishing, Garden City, NY.

Botany Boy Plant Encyclopedia. "*Cypridium reginae*, the showy lady's slipper," http://botanyboy.org/cypripedium-reginae-the-showy-ladys-slipper/ (accessed 2015).

Burroughs, John. 1904. "Among the Wildflowers." *Riverby. The Writings of John Burroughs IX*, Riverby Edition, p. 9. Riverside Press, Cambridge, MA

"*Cypripedium reginae*," White Flower Farm website, http://www.whiteflowerfarm.com/27280-product.html (accessed 2015).

"*Cypripedium reginae*; Showy Hardy Ladyslipper Orchid," Plant Delights, Inc. website, http://www.plantdelights.com/Cypripedium-reginae-for-sale/Buy-Showy-Lady-Slipper-Orchid/ (accessed 2015).

"Cypripediums for Sale," Vermont Ladyslipper Company, Ltd. website, http://www.vtladyslipper.com/vtlscwebpg3REG.html (accessed 2015).

Edens-Meier, Retha, Mike Arduser, Eric Westhus, and Peter Bernhardt. 2011. Pollination Ecology of *Cypripedium reginae* Walter (Orchidaceae): Size Matters. *Telopea* 13(1–2): 327–340.

Gray, Asa. 1890. *The Manual of The Botany of the Northern United States Including the District East of the Mississippi and North of North Carolina and Tennessee*. Iveson, Blakeman, and Company, New York.

Keene, Philip. 1998. *Wild Orchids across North America*. Timber Press, Portland, OR.

Newton, Gilbert D., and Norris H. Williams. 1978. Pollen Morphology of the Cypripedioideae and the Apostasioideae (Orchidaceae). *Selbyana* 2(2/3): 169–182.

Owen, Wayne. "Showy Lady's Slipper (*Cypripedium reginae*)," USDA Forest Service. Plant of the Week, http://www.fs.fed.us/

wildflowers/plant-of-the-week/cypripedium_reginae.shtml (accessed 2015).

Swamp Rose-mallow references

Bernhardt, Peter. 1999. *The Rose's Kiss: A Natural History of Flowers.* University of Chicago Press, Chicago.

Borge, Mary Anne. 2016. "Exploring Nature's Connections: Swamp Rose Mallow—for Bees, Butterflies, Beetles, Birds and Beauty," The Natural Web website, https://the-natural-web.org/category/native-plants/ (accessed 2016).

Heslop-Harrison, J. 1968. Pollen Wall Development. *Science* New Series 161(3838): 230–237.

Anonymous. 2019. "Marshmallow," How Products Are Made, Volume 3 website, http://www.madehow.com/Volume-3/Marshmallow.html (accessed 2019).

Kingsolver, John M., Timothy J. Gibb, and Gary S. Pfaffenberger. 1989. Synopsis of the Bruchid Genus *Althaeus* Bridwell (Coleoptera) with Descriptions of Two New Species. *Transactions of the American Entomological Society* 115(1): 57–82.

Klips, Robert A., Patricia M. Sweeney, Elisabeth K. F. Bauman, and Allison A. Snow. 2005. Temporal and Geographic Variation in Predispersal Seed Predation on *Hibiscus moscheutos* L. (Malvaceae) in Ohio and Maryland, USA. *American Midland Naturalist* 154: 286–295.

Kudoh, Hiroshi, and Dennis F. Whigham. 1997. Microgeographic Genetic Structure and Gene Flow in *Hibiscus moscheutos* (Malvaceae) Populations. *American Journal of Botany* 84(9): 1285–1293.

Rust, Richard W. 1980. The Biology of *Ptilothrix bombiformis* (Hymenoptera: Anthophoridae). *Journal of the Kansas Entomological Society* 53(2): 427–436.

Sampson, B. J., C. T. Pounders, C. T. Werle, et al. 2016. Aggression Between Floral Specialist Bees Enhances Pollination of *Hibiscus* (Section Trionum: Malvaceae). *Journal of Pollination Ecology* 18(2): 7–12.

Anonymous. 2019. "*Hibiscus sabdariffa*," Science.gov: Your Gateway to U.S. Federal Science website, https://www.science.gov/topicpages/h/hibiscus+sabdariffa+linnaeus (accessed 2019).

Simpson, Melissa. "Pollinators: *Ptilothrix bombiformis*, the Rose-mallow Bee," USDA Forest Service Celebration Wildflowers website, https://www.fs.fed.us/wildflowers/pollinators/pollinator-of-the month/rosemallowbee.shtml (accessed 2019).

Spira, Timothy P., Allison A. Snow, Dennis F. Whigham, and Jen Leak. 1992. Flower Visitation, Pollen Deposition, and Pollen-tube Competition in *Hibiscus moscheutos* (Malvaceae). *American Journal of Botany* 79(4): 428–433.

Weiss, Harry B., and Edgar L. Dickerson. 1919. Insects of the Swamp Rose-mallow *Hibiscus moscheutos* L., in New Jersey. *Journal of the New York Entomological Society* 27(1): 39–68.

Wild Leek references

Core, Earl L. 1945. Ramps *Castanea* 10(4): 110–112.

Nault, Andre, and Daniel Gagnon. 1993. Ramet Demography of *Allium tricoccum*, a Spring Ephemeral, Perennial Forest Herb. *Journal of Ecology* 81(1): 101–119.

Herman-Antosiewicz, Anna, Anna A. Powolny, and Shivendra V. Singh. 2007. Molecular Targets of Cancer Chemoprevention by Garlic-derived Organosulfides. *Acta Pharmocologica Sinica* 28(9): 1355–1364.

Lukes, Roy. 2003. "Spring Has Sprung a Forest of Leeks." Nature-Wise website, originally published in *Door County Advocate*, May 9, 2003. http://www.doorbell.net/lukes/a050903.htm (accessed 2014).

Schriebstein, Jess. 2013. "The Salt: What's on Your Plate." National Public Radio blog. May 10, 2013. http://www.npr.org/blogs/thesalt/2013/05/08/182354602/in-the-land-of-wild-ramps-its-festival-time (accessed 2014).

Sen, Indrani. 2011. When Digging for Ramps Goes Too Deep. *New York Times*, April 19, 2011. http://www.nytimes.com/2011/04/20/dining/20forage.html?pagewanted=all&_r=0 (accessed 2014).

Štajner, D., N. Milić, J. Čanadanović-Brunet, A. Kapor, M. Štajner, and B. M. Popović. 2006. Exploring *Allium* Species as a Source of Potential Medicinal Agents. *Phytotherapy Research* 20(7): 581–584.

Wikipedia contributors. 2014. "*Allium tricoccum*." Wikipedia, The Free Encyclopedia, https://en.wikipedia.org/w/index.php?title=Allium_tricoccum&oldid=915649997 (accessed 2014).

Wild Lupine references

Albany Pine Bush Preserve Commission. 2012. "Discovery Center Now Open Daily Due to Popular Demand," Albany Pine Bush News. http://www.albanypinebush.org/discovery-center/discovery-center-now-open-daily-due-to-popular-demand (accessed 2016).

Albany Pine Bush Preserve Commission. n.d. Brochure: "Karner Blue Butterflies at the Albany Pine Bush: An Endangered Species in a Globally Rare, Nationally Significant, and Locally Distinct Ecosystem."

Anonymous. 2016. "Attributes of *Plebejus melissa*." Butterflies and Moths of North America website, http://www.butterfliesandmoths.org/species/plebejus-melissa (accessed 2016)

Bernhardt, Christopher E., Randall J. Mitchell, and Helen J. Michaels. 2008. Effects of Population Size and Density on Pollinator Visitation, Pollinator Behavior, and Pollen Tube Abundance in *Lupinus perennis*. *International Journal of Plant Sciences* 169(7): 944–953.

Black, Scott Hoffman, and D. Mace Vaughn. 2005. "Species Profile: Blues: Karner blue (*Lycaeides melissa samueli*)," Xerces Society for Invertebrate Conservation. http://www.xerces.org/karner-blue/ (accessed 2016).

Cech, Rick, and Guy Tudor. 2005. *Butterflies of the East Coast: An Observer's Guide*. Princeton University Press, Princeton, NJ.

Dirig, Robert. 1988. "Nabokov's Blue Snowflakes." *Natural History* 97(5): 68–69.

Forister, Matthew L., Zachariah Gompert, James A. Fordyce, and Chris C. Nice. 2010. "After 60 Years, an Answer to the Question: What Is the Karner Blue Butterfly?" *Biology Letters* (7): 399–402. http://rsbl.royalsocietypublishing.org (accessed online 2017).

Grundel, Ralph, Noel B. Pavlovic, and Christina L. Sulzman. 1998. The Effect of Canopy Cover and Seasonal Change on Host Plant Quality for the Endangered Karner Blue Butterfly (*Lycaeides melissa samuelis*). *Oecologia* 114(2): 243–250.

References

Grundel, Ralph, Noel B. Pavlovic, and Christina L. Sulzman.. 2000. Nectar Plant Selection by the Karner Blue Butterfly (*Lycaeides melissa samuelis*) at the Indiana Dunes National Lakeshore. *The American Midland Naturalist* 144(1): 1–10.

Lambers, Hans, Jon C. Clements, and Mattew N. Nelson. 2013. How a Phosphorus-acquisition Strategy Based on Carboxylate Exudation Powers the Success and Agronomic Potential of Lupines (*Lupinus*, Fabaceae). *American Journal of Botany* 100(2): 263–288.

Michigan Department of Natural Resources. 2016. "Karner Blue Butterfly (*Lycaeides melissa samuelis*)," http://www.michigan.gov/dnr/0,4570,7-153-10370_12145_12204-33007--,00.html (accessed 2016).

NatureServe Explorer: An Online Encyclopedia of Life. 2017. http://explorer.natureserve.org/servlet/NatureServe?searchSciOrCommonName=Lupinus%20perennis (accessed 2017).

New York State Department of Environmental Conservation. 2016. "Karner Blue Butterfly Fact Sheet," http://www.dec.ny.gov/animals/7118.html (accessed 2016).

New Zealand Plant Conservation Network. "*Lupinus polyphyllus*," http://nzpcn.org.nz/flora_details.aspx?ID=3144 (accessed 2016).

Plants & Soil Sciences eLibrary. Forms of Nitrogen in the Soil. http://passel.unl.edu/pages/informationmodule.php?idinformationmodule=1130447042&topicorder=2&maxto=8 (accessed 2016).

Rittner, Don. 1976. *Pine Bush: Albany's Last Frontier*. Pine Bush Historic Preservation Project, Albany, NY.

Royal Botanic Gardens Kew: Kew Science. "*Lupinus polyphyllus* (large-leaved lupin)." http://www.kew.org/science-conservation/plants-fungi/lupinus-polyphyllus-large-leaved-lupin (accessed 2017).

Seedaholic: "*Lupinus* x *russellii*," http://www.seedaholic.com/lupinus-x-russellii-the-pages.html (accessed 2017).

Shapiro, Arthur M. 1974. Partitioning of Resources among Lupine-feeding Lepidoptera. *American Midland Naturalist* 91(1): 243–248.

Shepherd, Matthew D. 2005. "Species Profile: Elfins: Frosted Elfin (*Callophrys irus*)," Xerces Society for Invertebrate Conservation website, http://www.xerces.org/frosted-elfin/ (accessed online 2016).

Stewart, Margaret M., and Claudia Ricci. 1988. "Death of the Blues." *Natural History* 97(5): 64–71.

Valtonen, Anu, Juha Jantunen, and Kimmo Saarinen. 2006. Flora and Lepidoptera Fauna Adversely Affected by Invasive *Lupinus polyphyllus* along Road Verges. *Biological Conservation* 133(3): 389–396.

Wikipedia contributors, "Karner blue," *Wikipedia, The Free Encyclopedia*, https://en.wikipedia.org/w/index.php?title=Karner_blue&oldid=899686032 (accessed 2016).

Wikipedia contributors, "*Lupinus polyphyllus*," *Wikipedia, The Free Encyclopedia*, https://en.wikipedia.org/w/index.php?title=Lupinus_polyphyllus&oldid=890710119 (accessed 2016).

Wikipedia contributors, "Pnin," *Wikipedia, The Free Encyclopedia*, https://en.wikipedia.org/w/index.php?title=Pnin&oldid=900164834 (accessed 2016).

Wikipedia contributors, "David Douglas (botanist)," *Wikipedia, The Free Encyclopedia*, https://en.wikipedia.org/w/index.php?title=David_Douglas_(botanist)&oldid=883209315 (accessed 2017).

Zimmer, Dieter E. 1996. "Excerpts from *A Guide to Nabokov's Butterflies and Moths*. Hamburg, Germany," https://www.libraries.psu.edu/nabokov/dzbutt4.htm (accessed 2016).

Yellow Pond-lily references

Borman, Susan, Robert Korth, and Jo Temte. 1997. *Through the Looking Glass: A Field Guide to Aquatic Plants*. Wisconsin Lakes Partnership. Stevens Point, WI.

Chen, Liu, Steven R. Manchester, and Zhiduan Chen. 2004. Anatomically Preserved Seeds of *Nuphar* (Nymphaeacea) from the Early Eocene of Wutu, Chandong Province, China. *American Journal of Botany* 91(8): 1265–1272.

Cronin, Greg, Katherine D. Wissing, and David M. Lodge. 1998. Comparative Feeding Selectivity of Herbivorous Insects on Water Lilies: Aquatic vs. Semi-terrestrial Insects and Submersed vs. Floating Leaves. *Freshwater Biology* 39: 243–257.

Crow, Garrett E., and C. Barre Hellquist. 2000. *Aquatic and Wetland Plants of Northeastern North America*, Vol. 1. University of Wisconsin Press, Madison.

Hilty, John. 2002–2014. "Wetland Wildflowers of Illinois," Illinois Wildflowers website, https://www.illinoiswildflowers.info/wetland/wetland_index.htm (accessed 2019).

Lippok, Barbara, Angela A. Gardine, Paula S. Williamson, and Suzanne S. Renner. 2000. Pollination by Flies, Bees, and Beetles of *Nuphar ozarkana* and *N. advena* (Nymphaeaceae). *American Journal of Botany* 87(6): 898–902.

Padgett, Donald J. 2007. A Monograph of *Nuphar* (Nymphaeaceae) 1. *Rhodora* 109(937): 1–95.

Padgett, Donald J., Donald H. Les, and Garrett E. Crow. 1998. Evidence for the Hybrid Origin of *Nuphar* ×*rubrodisca* (Nymphaceae). *American Journal of Botany* 85(10): 1468–1476.

Sohmer, S. H., and D. F. Sefton. 1978. The Reproductive Biology of *Nelumbo pentapetala* (Nelumbonaceae) on the Upper Mississippi River. II. The Insects Associated with the Transfer of Pollen. *Brittonia* 30(3): 355–364.

Weismera, John H., and C. Barre Hellquist. 1997. *Nuphar*, in Flora of North America Editorial Committee, eds. 1993+, vol. 3, pp. 67-71. *Flora of North America North of Mexico*. New York.

Index

This index includes scientific names of plants (with authors' names appended), common names of plants (both English and foreign), and names of animals (including people), fungi (with authors' names), and other organisms mentioned in the text.

Abelmoschus esculentus (L.) Moench, 305
Abies
    cephalonica Loudon, 193–194
    religiosa (Kunth) Schltdl. & Cham., 129
Abrahamson, Warren, 189
Abutilon theophrasti Medik., 312
Adiantum pedatum L., 26
Aegeus, King, 45
Agrawal, Anurag, 128, 135
Agrius cingulata, 224
Ailanthus altissima (Mill.) Swingle, 185
Alcea rosea L., 306–307
Alchemilla L., 37
alfalfa, 110, 325
Alliaceae Borkh., 317
Alliaria petiolata (M. Bieb.) Cavara & Grande, 320
alligator buttons, 41
Allium L., 318, 320
    tricoccum Aiton, 317
    ursinum L., 320
alpine heath/diapensia community, 5
alpine wildflowers, 3, 93
Althaea officinalis L., 307
Althaeus hibisci, 313
Ambrosia L., 184
    artemisiifolia L., 184
Anagrapha, 177
Angelica atropurpurea L., 295
angel's trumpet, 227
anise, 290
Anne, Queen of Denmark, 294
Anne, Queen of England, 294
ant(s), 46, 53–54, 66, 91, 112, 127, 165, 203, 213–214, 243, 253–254, 280, 294
Anthemurgus passiflorae, 249
Anthophora, 301
aphid(s), 113, 123, 126–127, 233 273, 314, 334
    oleander, 127
Aphis
    asclepiadis, 127
    nerii, 126–127
Apiaceae Lindl., 287, 290, 295–296
Apis, 301
    mellifica, 300
Apocynaceae Juss., 117, 126
Apocynum L., 119, 123, 134, 138, 184
    syriacum S.G. Gmel., 119
apple(s), 20, 24
apricot-vine, 247, 253
Aquilegia canadensis L., 100
Aralia nudicaulis L., 28–29
Araliaceae Juss., 27–29
Arctium L., 37
Arctostaphylos alpina (L.) Spreng., 14
Arenaria L., 11
Arethusa, 68
Arethusa L., 65–68
    bulbosa, 64–66
Asclepiadaceae Borkh., 117
Asclepiadoideae Burnett, 117, 131

Asclepias L., 118–120, 131, 170
    amplexicaulis Sm., 138
    curassavica L., 139
    exaltata L., 138
    incarnata L., 132, 134, 136
    lanceolata Walter, 136
    purpurescens L., 132, 137
    quadrifolia Jacq., 139
    rubra L., 139
    syriaca L., 117–118, 126, 134
    tuberosa L., 121, 128, 134, 136
    verticillata L., 138
    viridiflora Raf., 137
    viridis Walter, 128, 137
ash, 27
asphodel, bog, 67–68
aster(s), 42–49, 58, 111, 128, 163, 179–181, 186
    alpine, 45
    alpine leafybract, 45
    aromatic, 46
    lance-leaved, 46
    low showy, 43, 46
    New England, 42, 46–49, 128, 178
    New York, 43
    panicled, 46, 49
    purplestem, 45
    small-headed, 46
    stiff, 46
    white wood, 318
aster family, 43, 45–47, 66, 107, 112–113, 179, 183, 315
Aster L., 43–45, 107, 181
    alpinus L., 44
    ptarmicoides Torr. & A. Gray, 181
    tataricus L. f., 44
    tataricus 'Jindai', 44
Asteraceae Bercht. & J. Presl, 66, 107, 110, 112–113, 179, 186
Asterea, 45
Astor, John Jacob, 30
Atteva aurea, 185
Augochlorella aurata, 281
avens, mountain, 3, 13–14
    water, 301
azalea, 9, 17
    alpine, 2–3, 6–9

Bacillus thuringiensis, 296
Bacon, Nathaniel, 222
bacteria, 19, 160, 263, 280, 282–283, 324–325
Balsaminaceae A. Rich., 209
banana poka, 256–257
Banks, Sir Joseph, 20
Banksia L. f., 331
Baptisia tinctoria (L.) R. Br., 329
Barbour, Spider, 328
Barthlott, Wilhelm, 37
bat(s), 123, 226, 256
Bates, Henry Walter, 121
Beal, Dr. William, 147
bean family, 323–324
bear(s), 17, 94
bearberry, 14, 17
    alpine, 14
beard-tongue, 141
beaver(s), 30, 68, 334–335
bedstraw family, 241

bee(s), 10, 12, 18, 40, 53, 59–60, 62, 66–67, 69–71, 73, 78–79, 85, 91, 99, 102, 112–114, 118, 122, 132, 145, 159, 163–164, 166, 172–173, 176–177, 185, 197–198, 202–204, 213–214, 223, 243, 249, 252–253, 261–262, 267, 272, 283, 294, 300–301, 310, 312–313, 320, 328, 335
    andrenid, 249
    carpenter, 67, 164, 185, 249, 252–253, 267, 272
    chimney, 313
    Eastern carpenter, 184
    halictid, 28, 145, 267, 281, 333–334
    leaf-cutter, 73, 249, 300
    long-tongued, 167, 212, 185
    metallic green, 113
    rose-mallow, 310–312
    sweat, 333–334
beebalm, 100
beech, 27, 51, 54, 56, 201
    American, 51, 54, 56
beechdrops, 50–55, 142, 205
bee-fly(ies), 14
beetle(s), 12, 40, 46, 91, 123–125, 185, 189, 197, 233, 236–238, 267, 294–295, 301, 312–314, 333–334
    bruchid, 312–313
    four-eyed milkweed, 123–124, 126
    Japanese, 314
    ladybird, 126–127
    locust-borer, 185
    long-horned, 124, 185, 294–295 (*See also* cerambycids)
    milkweed, 124
    (red) lily leaf, 236–238
    rose-mallow, 309, 312–313
    spotted cucumber, 186
Belamcanda Adans., 60, 63
    chinensis (L.) Redouté, 57, 60–63, 320
belladonna, 227
bellflower family, 97
bellwort, 320
Benthamidia Spach, 95
Beverley, Robert, 222
bilberry(-ies), 25
    bog, 8, 10
    dwarf, 10
birch, white, 5
bird(s), 10, 16, 18–19, 28, 36, 62, 66, 86, 89, 94, 97, 99–102, 104, 108, 113, 121, 123, 125, 128–129, 159, 161, 171–174, 189, 219, 227, 238, 242, 250, 253–254, 256–257, 270, 272, 295, 297, 308, 327
bird's nest, 293, 296–297
bishop's lace, 293
bittersweet family, 193, 195
black-eyed Susan, 101
blackberry(-ies), 57, 61–62, 324
blackberry-lily, 56–63, 230
Blanchan, Neltje, 302
blazing star, 112–113
blue cohosh, 27
blue sailors, 108
blueberry(-ies), 5, 8, 10, 17, 19–20, 24–25, 163, 317
blue-eyed grass, 58
blue-flags, 58
bluestem, little, 191
bluets, alpine, 3, 13–15
boars, wild, 233
bogbean, 83

365

Index

bog-star, 199
Boloria titania montinus, 11
Boltonia asteroides (L.) L'Hér., 101
Bombacaceae, 305
Bombus, 40, 103, 163, 213, 267, 300, 310–311
    affinis, 281
    bimaculatus, 53
    impatiens, 53, 310
Boone, Daniel, 30
borer, 233
    European corn, 296
Bowdoin, James, 110
Brachycereus nesioticus (K. Scum.) Backeb., 271
brittlebush, 107
broomrape family, 51
Brower, Lincoln, 127
Brown, J. H., 101
Brugmansia Pers., 227
    arborea (L.) Lagerh., 227
Bryant, William Cullen, 161–162
buckbean, 24, 68, 83–86, 169, 245
buckbean family, 83
Buddha, 35
bug, 126
    ambush, 132–133, 185, 295
    assassin, 185
    large milkweed, 126
    seed, 126
    small milkweed, 126
bugleweed, 199
bumblebee(s), 9–12, 18, 40, 53, 71, 85, 102–103, 113–115, 122, 132–133, 145, 163, 166–167, 172–174, 176–177, 202–203, 213–216, 242–243, 249, 257, 261, 267, 281, 300–301, 310–312, 320, 328
    eastern, 53
    queen, 67, 70
bunchberry, 12, 88–95
    Alaska, 95
burdock, 37
burning bush, 195
Burroughs, John, 98, 231, 302
buttercup family, 100
butterfly(-ies)
    Anna blue, 326
    Arctic, Melissa, 11
        White Mountain, 11
    blue, karner, 326–329
        Melissa, 326
    duskywing, persius, 329
        wild indigo, 329
    elfin, frosted, 329
    fritillary, 253, 253
        great spangled, 131, 232
        gulf, 253
        White Mountain, 11
    heliconian, 253–255
    Juno long-wing, 254–255
    milkweed, 131
    monarch, 49, 59, 119–123, 125–135, 137–139, 236, 253–254, 295
    pearl crescent, 49
    skipper, 114, 172, 176, 301, 330
        little glasswing, 73
        Peck's, 113
        swarthy, 166
    sulphur(s), 172, 176
        orange, 49
    swallowtail, 172

    black, 295–296
    (eastern) tiger, 232, 236, 239, 299–300
    spicebush, 59, 102, 173
    viceroy, 121
butterfly-weed, 121–122, 128, 133–136, 139
buttonbush, 241

Cactaceae Juss., 265–266, 269
Cactoblastis cactorum, 272
cactus, lava, 271
cactus family, 100, 265, 269–270
Cade, J. J., 44
Callophrys irus, 329
Calluna Salisb., 17
Calopogon R. Br., 65, 68–70, 73
    tuberosus (L.) Britton, Sterns & Poggenb., 65–68
    fo. *albiflorus* Britton, 69
Cameron, Duncan, 81
Campanulaceae Juss., 97
Canada mayflower, 12
Candida albicans, 263
capitulum (capitula) (capitulescences), 45–46, 108, 111–112
Caprifoliaceae Juss., 25, 242
cardinal, northern, 97
cardinal flower, 96–105, 201
Cardinalis cardinalis, 97
Carex
    bigelowii Torr. ex Schwein., 11
    plantaginea Lam., 319
caribou, 94
Carnegiea gigantea (Engelm.) Britton & Rose, 275
carrot, cultivated, 287–288, 290
    wild, 287–290, 293
carrot family, 287
Cartier, Jacques, 285
Caryophyllaceae Juss., 3
Cassiopaia, 12
Cassiope D. Don, 12–13
caterpillar(s), asteroid owlet, 48
    asteroid paint owlet, 48–49
    brown-hooded owlet, 186–187
    calico paint owlet, 186
    golden-hooded owlet, 48
    saddleback, 314
Catesby, Mark, 285
cattail, narrow-leaved, 315
cattle, 226, 273
Cattleya violacea (Kunth) Rolfe, 171
cedar, Atlantic white, 67–68
Ceiba pentandra (L.) Gaertn., 133
Celastraceae R. Br., 193, 195
Celastrus orbiculatus Thunb., 195
celery, 290
Cephalanthus occidentalis L., 241
cerambycids, 295 (*See also* beetle, long-horned)
Chamaecyparis thyoides (L.) Britton, Stearns & Poppenb., 68
Chamaepericlymenum Hill, 95
Cheek, Martin, 279, 285
Chelone L., 141
chickadee(s), 189
chicorée, 109
chicory, 46, 106–115, 293
chicouryeh, 109
chiggers, 222
Chimaphila Pursh, 259–263
    maculata (L.) Pursh, 205, 259–260
    umbellata (L.) W. P. C. Barton, 258–260

chipmunk(s), 94, 233
chocolate family, 305
chrysalis, 119, 130–131, 133, 327–328
Cichan, Michael, 112
Cichorioideae Chevall., 110, 112
Cichorium intybus L., 107, 114–115, 293
    var. *sativum* Bisch., 109
Cicuta bulbifera L., 287
cilantro, 290
cinquefoil, Robbins', 3, 14–15
Cirsium pitcheri (Torr. ex Eaton) Torr. & A. Gray, 165
Clare, John, 153
claret-cup, 100
Clintonia borealis (Aiton) Raf., 11, 320
clover(s), 324
Clusius, Carolus, 285
coffee, 109, 241
coffeeweed, 109
Colchicaceae DC., 230
Coleosporium, 182
    solidaginis, 182
Colias spp., 172
    eurytheme, 133
Colocasia esculenta (L.) Schott, 37
columbine, 100, 161
Compositae Giseke, 107
composite family, 107
Congo cockatoo, 219
Conium maculatum L., 287
Conopholis americana (L.) Wallr., 142, 205
Conotrachelus fissinguis, 314
Convallaria majalis L., 230
copepods, 283
Coptis trifolia subsp. *groenlandica* (O.F. Müll.) Hultén, 11
Corchorus L., 305
cordgrass, 315
coriander, 290
Cornaceae Bercht. ex J. Presl, 89
cornel, Lapland, 94–95
    Swedish, 94
Cornus L., 90, 93, 95
    canadensis L., 12, 89–91, 94
    florida L., 89–90
    kousa Bürger ex Hance, 91
    suecica L., 90, 94–94
    unalaschkensis Ledeb., 95
cotton, 305, 308
cow, 226
cow-parsnip, 296
cranberry(ies), 5, 10, 16–25
    American, 17–18, 20, 24, 244
    high-bush, 25
    large, 18, 24
    mountain, 3, 4, 8, 10–11, 25, 244
    southern, 25
    small, 10, 18, 24–25
crane, 16–17
crane-berry, 16–17
Crematogaster, 53
cricket, 66
Crocosmia Planch., 57
Crocus L., 57
crowberry, 12
ctenucha, Virginia, 185
Ctenucha virginica, 185
Cucullia
    asteroides, 48
    convexipennis, 186

Index

cucumber family, 186, 252, 254
Cucurbitaceae Juss., 254
cucurbits, 254
Culver's root, 303
cumin, 290
Cuscuta gronovii Willd., 24
cushion plant(s), 5–6, 162
Cycnia
  inopinatus, 136–137
  tenera, 138, 184
cycnia, delicate, 138, 184
  unexpected, 136
cypress, bald, 212
Cypripedium L., 170, 299–300
  acaule Aiton, 299
  parviflorum Salisb., 303
    var. *pubescens* (Willd.) O.W. Knight, 76
  reginae Walter, 298–299, 301, 303
    fo. *albolabium* Fernald & B. G. Schub., 302
    fo. *album* (Aiton) Rolfe, 302

Dactylopius coccus, 273
daffodils, 319
daisy(ies), Michaelmas, 48
  ox-eye, 112–113, 293
daisy family, 44, 49, 129, 179
daisy trees, 107
Danaus plexippus, 49, 119, 121
dandelion, 46, 107–108, 110, 112
Darlington, Dr. H. T., 147
Darlingtonia Torr., 277
  californica Torr., 230
Darwin, Charles, 78, 170, 176, 216–217, 243, 245, 250, 294
Darwin, Erasmus, 216
date palm, 38
Datura L., 221–223, 225–227
  metel L., 226
  stramonium L., 221, 225
  wrightii Regel, 227
Daucus L., 288, 290
  carota L., 58, 287, 289–291, 293, 295–297
Davis, Chuck, 231
daylily(-ies), 58, 115, 230, 235, 238–239
deer, 19, 31, 33, 79–80, 94, 105, 135, 161, 218, 231, 233, 238, 268, 302
  white-tailed, 33, 79, 219
Dennstaedtia punctilobula (Michx.) T. Moore, 228
der wegewart, 108
devil's plague, 293
devil's trumpet, 221
devil's weed, 221
dewberries, 324
Diabrotica undecimpunctata, 186
diapensia, 2–8, 13
Diapensia, 4, 6
  lapponica L., 2–3, 6–8
    subsp. *lapponica*, 6
    subsp. *obvata* (F. Schmidt) Hultén, 6
Diapensiaceae Lindl., 3, 6
Dickinson, Emily, 98, 162, 201
Didierea family, 269
Didiereaceae Radlk., 269
dill, 290, 295
Dione juno, 254
Dioscorides, 193, 290
Distichlis spicata (L.) Greene, 315
dodder, 24

Doellingeria Nees, 43
  ptarmicoides Nees, 181
dogbane, 123, 134, 136–138, 184
dogbane family, 117
dogwood(s), 91, 95
  dwarf, 92, 94–95
  flowering, 89–40, 94–95
  gray-stemmed, 95
  red osier, 95
dogwood family, 89
Dolichovespula arenaria, 77–78
Donacia, 333–334
Douglas, David, 325
doves, 272, 303
dragon's mouth, 64–68, 72–73
Drosera L., 68
Dryas octopetala L., 162
duck(s) (ducklings), 37, 86, 162, 308, 334
duck acorns, 41
Dumosi, 44
Duvel, J. T., 222

Eaton's firecracker, 100
Echinacea purpurea (L.) Moench, 99
Echinocereus triglochidiatus Engelm., 100
Echium vulgare L., 58
edelweiss, 107
Edison, Thomas, 134, 182
elk, 94
Empetraceae Hook. & Lindl., 12, 17
Empetrum nigrum L., 12
Emphorini, 313
Empress Josephine, 199
Encelia farinosa A. Gray ex Torr., 107
Englemann, George, 44
Epiblema scudderiana, 188
Epidendrum domesticum L., 60, 63
Epifagus Nutt., 51, 53–55
  virginiana (L.) W.P.C. Barton, 50–51, 142, 205
Epigaea repens L., 245
Epilobium L., 156
Epipactis, 76–79, 81
  atrorubens (Hoffm. ex Bernh.) Besser, 76, 78–79
  gigantea Douglas ex Hook., 76, 78
  helleborine (L.) Cranz, 66, 75–81
  latifolia (L.) All., 78
Erechtites hieraciifolius (L.) Raf. ex DC. 49
Erica L., 17
Ericaceae Juss., 3, 5–6, 9, 12, 14, 17, 24, 201, 203, 205, 259, 324
Erythronium L., 230
Euchaetes egle, 123
euonymus, 195
Euonymus alatus Regel, 195
Eupeodes, 113
Euphorbia L., 269
Euphorbiaceae Juss., 268
Eurosta, 189
  solidaginis, 187–188
Eurybia (Cass.) Cass., 74
  divaricata (L.) G.L. Nesom, 318
  spectabilis, 46
Eurytoma, 189
  gigantean, 189
Euthamia, 190
  graminifolia (L.) Nutt., 183, 190
evening-primrose, 152–157, 164
evening-primrose family, 153
Exyra fax, 283

Fabaceae, 323–324
Fagus, 51
  grandifolia Ehrh., 51
    var. *grandifolia*, 55
    var. *mexicana* (Martinez) Little, 55
false hellebore, 12, 303
fenberry, 17
fennel, 290
fern(s), bracken, 324
  maidenhair, 26–27, 32
  marsh, 68
  rattlesnake, 26
figwort, 142
figwort family, 141
finch(es), 94, 272
  cactus, 272
  large cactus, 272
fir, 5, 129
  Douglas, 325
  Grecian, 193–194
fire-pink, 100
Firestone, Harvey, 182
flannel leaf, 143
fleabane, saltmarsh, 315
Fletcherimyia fletcherii, 281, 282
flower-of-the-forest, 260
fly(-ies), 7, 11, 13, 40, 59, 78, 85, 91, 112, 123, 126, 132–133, 145, 177, 187–189, 197, 213, 282, 214, 219, 281, 283, 294, 301, 320, 335
  bunch gall, 189
  flesh, 280–283 (*See also* sarcophagid fly)
  flower, 19, 113, 145, 155 (*See also* hover fly and syrphid fly)
  goldenrod ball gall, 188–189
  hover, 78, 113–114, 145, 262, 334 (*See also* flower fly and syrphid fly)
  leaf-mining, 126
  sarcophagid, 281 (*See also* flesh fly)
  syrphid, 19, 28, 49, 60, 78, 113, 145, 185, 214, 219, 301, 334 (*See also* flower fly and hover fly)
  tachinid, 92
  white, 314
foamflower, 303
Ford, Henry, 182
Freesia Eckl. ex Klatt, 57
Fritillaria L., 237
Frost, Robert, 70, 73, 161–162
Fry, Sir Edward, 44
fungus (pl. fungi), 17–18, 33, 80–81, 139, 146, 156, 190, 204–207, 263, 302
  ascomycetes, 81
  basidiomycetes, 81
  mycorrhizal 29, 80, 263
  relationship, 17, 81, 206, 263

Galax L., 7
Galerucella, 334
Galium L., 241
gardenia, 241
garlic, 317–320
garlic mustard, 320
Gaultheria procumbens L., 160, 260
Gaura L., 154, 156
geese, 86
Gemmingia Heist. ex Fabr., 60
gentian(s), 48, 108, 159–165
  bottle, 161, 166
  closed, 161, 166–167

367

Index

fringed, 158–159, 161–165, 169, 199
  narrow-leaved, 161, 166
  pine barrens, 165–166
  spotted, 160
  spring, 162
  stiff, 167
  yellow, 159–160
gentian family, 83, 159–160, 164
gentian violet (crystal violet), 160–161
Gentiana L., 159, 164–166
  acaulis L., 162
  autumnalis L., 165
  clusii Perr. & Songeon, 161–162, 161
  linearis Froel., 161, 166
  lutea L., 159
  punctata L., 160
  sedifolia Kunth, 161–162
  verna L., 161–162
Gentianaceae Juss., 83, 159
Gentianella Moench, 165, 167
  quinquefolia (L.) Small, 167
Gentianopsis Ma, 164–165
  crinita (Froel.) Ma, 158–159, 161, 163, 165, 199
  procera (Holm) Ma, 165
Gentius (Genthios), King, 159
Geospiza
  conirostris, 272
  scandens, 272
Gerbode, Sharon, 252
Geum peckii Pursh, 3, 13
gilia, scarlet, 100
Gillett, John M., 165
ginseng, 27–33, 321
  American, 26–28, 30–33
  dwarf, 30–31
ginseng family, 27
Gladiolus L., 57
gnats, 66
Gnorimoschema gallaesolidaginis, 188
goat's-beard, 115
gold thread, 11
Goldblatt, Peter, 60, 62–63
goldenrod(s), 44, 48, 101, 123, 129, 133, 156, 163, 179–187, 189–190
  alpine, 11
  blue-stemmed, 180
  Canada, 183, 187–190
  early, 179–180
  European, 182
  evergreen, 181
  flat-topped, 181, 190
  grass-leaved, 183, 190
  late, 183, 189
  rough-stemmed, 184, 190
  seaside, 181–182, 186
  showy, 178, 181, 184–186, 188, 190–191
  smooth, 189–190
  tall, 123, 183, 188, 190
  white, 180
  wrinkleleaf, 181
  zigzag, 180
goldfinches, 108, 134
Gomphocarpus R. Br., 131
Goodyera repens (L.) R. Br., 81
Gossypium L., 305
Governor Berkeley, 222
grape(s), Concord, 20
Graphosoma semipunctatum, 294

grass pink, 65–70, 72–73
  tuberous, 66
grass-of-Parnassus, 193–195, 197, 199
  eastern, 192, 194–195, 199
  marsh, 193–195, 197
Gray, Asa, 43–44, 170, 302
green comet, 137
grosbeak, 129
  black-headed, 129
Gross, Katherine, 146–147, 152
grouse, 94, 242
  ruffed, 242
  spruce, 10, 25
guayule, 183
Gurania (Schltdl.) Cogn., 254
  spinosa (Poepp. & Endl.) Cogn., 254

Habenaria Willd., 171
hares, snowshoe, 19
Harriman, Edward Henry, 12
Harrimanella Coville, 12–13
  hypnoides (L.) Coville, 3, 12
hawkmoth, five-spotted, 224
  pink-spotted, 224
hawkweed, 112–113
  orange, 113
Hawthorne, Nathaniel, 98–99
heath, 5–6, 8, 10, 13, 17, 259
  mountain, 3, 12
heath family, 5–6, 12, 17, 201, 259–260, 262, 324
heather, 17
  mountain, 12
Helenium autumnale L., 199
Heliamphora Benth., 277
hellebore, false, 12, 303
helleborine, 66, 75, 78–80, 204, 301
  broad-leaved, 66, 74–80
hell's bells, 221
Helophilus fasciatus, 49
Hemaris
  diffinis, 132
  thysbe, 177
Hemerocallidaceae R. Br., 230, 238
Hemerocallis L., 230, 238
  fulva (L.) L., 238–239
  lilioasphodelus L., 115
hemlock, 5, 29, 51, 201, 289
  poison, 287–290, 295
henna, 211
Hevea brasiliensis (Willd. ex A. Juss.) Müll. Arg., 182
Hibiscus L., 305, 308–310, 312–314
  bifurcatus Cav., 310
  laevis All., 310
  lasiocarpos Cav., 305
  moscheutos L., 304–307, 309–312
    moscheutos subsp. *lasiocarpos* (cav.) O.J. Blanch., 305
    moscheutos L. subsp. *moscheutos*, 305
  radiatus Cav., 310
  sabdariffa L., 310–311
  sororius L., 310
  syriacus L., 306–307
hickory, 27
Hicks, David, 243
Hieracium L., 112
  aurantiacum L., 113
Hippocrates, 296
hogweed, giant, 296

hollyhock, 306
  common, 307
honeybee(s), 18–19, 85, 145, 213–214, 257, 262, 300–301, 312, 328–329, 334
honeysuckle family, 242
Hoplitis, 301
hornworm, sweet potato, 224
  tomato, 224
horta, 109
hosta, 230
Hosta Tratt., 230
Hostaceae B. Mathew, 230
Houstonia, 14
  caerulea L., 3, 13–14
    var. *faxonorum* Pease & A.H. Moore, 3, 14
huckleberry, 67, 70
Huhndorf, Dr. Sabine, 146
hummingbird(s) (aka hummer), 98–102, 104, 139, 173, 212–214, 219, 222, 234, 254, 256–257
  ruby-throated, 99, 212, 232, 236
Hydrocera Blume ex Wight & Arn., 209
  triflora (L.) Wight & Arn., 209

ibimi, 17
Iltis, Hugh, 165
impatiens, 219
  'Elfin', 219
  'New Guinea', 219
Impatiens L., 209–212, 215
  capensis Meerb., 208–209, 214–215, 218
    fo. *albiflora* (E.L. Rand & Redfield) Fernald & B.G. Schub., 216
    fo. *immaculata* (Weath.) Fernald & B.G. Schub., 216
  glandulifera Royle, 219
  niamniamensis Gilg, 219
  pallida Nutt., 209, 215
Indian cucumber-root, 230
Indian pipe, 98, 200–207, 263
Indian tobacco, 105
Indian-pink, 100
indigo, wild, 329
Ionactis Greene, 43
  linariifolia (L.) Greene, 46
Ipomopsis aggregata (Pursh.) V.E. Grant, 100–101
Iridaceae Juss., 57, 60, 63
iris(es), 57–58, 61, 62
  crested, 58
  dwarf lake, 58
  dwarf violet, 58
Iris, 57–58, 60–63
  chinensis, 62
  domestica (L.) Goldblatt & Mabb., 56–57, 63, 230
  lacustris Nutt., 58, 165
  reichenbachi Heuff., 58, 61
  verna L., 58
  versicolor L., 58
  virginica L., 58
iris family, 57, 60
ironweed, 112
ixia, Chinese, 58–59
Ixia L., 60, 27
  chinensis L., 58

jack-in-the-pulpit, 27
James I, King of England, 294
Jasminocereus thouarsii (F.A.C. Weber) Backeb., 271

368

Index

Jefferson, Thomas, 58–59, 110
jewelweed, 52, 209–218
    orange, 208–209, 212, 215–216
    yellow, 209, 215
jimsonweed, 154, 220–226
junipers, 79
Jupiter, 45
jute, 305

Kalm, Peter, 30, 285
Kalmia L., 30
    procumbens (L.) Gift & Kron, 2–3, 9
kapok, 133–134, 308
kapok family, 305
kikhórion, 109
Kodric-Brown, A., 101
Kosteletzkya virginica (L.) C. Presl ex A. Gray, 305, 309, 314

Labdomera clivocollis, 124–125
Labrador tea, 12
Lactarius, 207
Lactuca L., 112
ladies-tresses, 199
lady-slipper, pink, 299
lady-slipper, showy, 298–303
lady-slipper, yellow, 76, 79, 303
lady's mantle, 37
lady's-earring, 209–210
Lasioglossum, 145
laurel, mountain, 30
Lawsonia inermis L., 211
Ledum groenlandicum Oeder, 12
leek, 320–321
    wild, 27, 316–321
legume(s), 84, 324–325, 327
Leontopodium R. Br. ex Cass., 107
leopard-flower, 61
leopard-lily, 61
Lepuropetalon Elliot, 195
lettuce, 112
Leucanthemum vulgare Lam., 112
Lewis and Clark, 58
Liatris Gaertn. ex Schreb., 112–113
lichen(s), 5–6
Liliaceae Juss., 229–230
Lilium L., 230, 237–238
    bulbiferum L., 234–235
    canadense L., 229–230, 232
        var. *canadense*, 232
        var. *editorum* Fernald, 232
    candidum L., 230
    catesbaei Kunth, 230, 235
    lancifolium Thunb., 233–235
    longiflorum Thunb., 230
    michauxii Poir., 233
    michiganense Farw., 232
    philadelphicum L., 229–230
    superbum L., 229–230, 232–233
    tigrinum Ker Gawl., 233
lily(ies), 58, 61, 229–233, 235–237–239
    Asian tiger, 234
    blue bead, 11, 320
    calla, 230
    Canada, 229–233, 236, 239
    Carolina, 233
    cobra, 230
    Easter, 230
    great yellow, 41

Madonna, 230
Michigan, 232–233
    orange, 234–235
    pine, 230, 235
    tiger, 233–234
    Turk's-cap, 229–234
    wood, 228–231, 234–236, 238–239
        western, 232
lily family, 57, 229–230, 237–238
lily-of-the-valley, 230
Limenitis archippus, 121
limes, 19
Limonium Mill., 285
    carolinianum (Walter) Britton, 285
    peregrinum (P.J. Bergius) R A. Dyer, 285
linden family, 305
Lindera benzoin Meisn., 173
lingonberry, 10–11, 25
Linnaea borealis L., 242
Linnaeus, 27, 30, 60, 68, 85, 97, 114–115, 119, 156, 164, 182, 193, 205, 221, 242, 285, 290, 332
Liriomyza asclepiadis, 126
L'Obel, Mathias de, 97
lobelia(s), 104, 199
    great blue, 99, 102, 105
    loose-flowered, 98
    red, 97, 199, 102
    scarlet, 97
Lobelia L., 97, 101–102
    cardinalis L., 97, 99, 101, 104–105
    inflata L., 105
    kalmii L., 199
    laxiflora Kunth, 97–98
    siphilitica, 99–102–103, 105, 199
Lobeliaceae Juss. ex Bonpl., 97
Lobelioideae, 97
Lobelius, 97
locoweed, 221
locust, black, 185
logania family, 100
Loiseleuria procumbens (L.) Desv., 2–3
lopseed family, 141
lotus, 35–40
    American, 34–37, 39–40
    sacred, 35–38
lotus family, 35
lupine, 324–329
    bigleaf, 325–326
    blue, 323
    Pacific, 325
    sundial, 323–324
    wild, 322–324, 326–329
Lupinus L., 324
    lepidus Douglas ex Lindl., 325
    perennis L., 322–323, 325–326, 328–329
        subsp. *perennis*, 323
        subsp. *gracilis* (Chapm.) D.B. Dunn, 323
    polyphyllus Lindl., 325
    ×*regalis*, 325
Lycaeides, 326–327
    idas, 326
    melissa, 326
    melissa subsp. *samuelis*, 326–327
    samuelis, 326
Lycopus L.
    americanus Muhl., 199
    virginicus L., 199
Lygaeus kalmia, 126

Ma, Yu-Chuan, 164–165
Mabberley, David, 60, 62
Macleania Hook., 17
madder, 241
madder family, 241
maggot, 188–189
Magnolia L., 335
Maianthemum canadense Desf., 12
mallow family, 305, 307
mallows, 307, 315
Malva, 307
    moschata, 307
    rotundifolia L. 147
Malvaceae Juss., 305–309, 312, 314
Manduca quinquemaculata, 224
mangroves, 212
mantid(s), 132
mantis, Chinese, 49
maple, sugar, 27, 51
maracuja, 255
marsh trefoil, 83
marsh-mallow, 307
marsh-marigold, 156
mayapple, 27
maypops, 247–248, 252–253, 255
McMahon, Bernard, 58
Medeola virginiana L., 230
Meerburgh, Nicolaas, 210
Megachile, 73
    concinna, 249
Megacyllene robiniae, 185
Melanthiaceae Batsch, 230
Menyanthaceae Dumort., 83, 86
Menyanthes L., 83, 86
    trifoliata L., 24, 68, 82–84
        var. *trifoliata*, 85
        var. *minor* Fernald, 85
Metriocnemus knabi, 282–283
Microstegium vimineum (Trin.) A. Camus, 29
milkweed(s), 117–135, 138–139, 156, 170, 217, 253
    blunt-leaved, 138, 139
    common, 116–120, 122, 124, 126, 131, 134–138, 217, 295
    few flowered, 137
    four-leaved, 138–139
    green, 137
    green comet, 137
    lance-leaved, 137
    Ozark, 128, 137
    poke, 135, 138
    purple, 126, 132, 137–139
    red, 139
    spider, 128, 137
    swamp, 122, 127, 132, 134–136, 139
    tall, 138
    tropical, 139
    whorled, 138
milkweed family, 117
milkweed subfamily, 131
Mimulus L., 141
mint, mountain, 199
mint family, 100, 149
Minuartia groenlandica (Retz.) Ostenf., 3, 11
Misumena vatia, 132–133
Mitchell, John, 242
Mitchella L., 242
    repens L., 241
    undulata Siebold & Zucc., 242
mite(s), 283, 296

Index

mockingbirds, 272
mold(s)s, 21, 33, 81
Monarda didyma L. 100
Moneses Salisb. ex Gray, 259, 261–263
    uniflora (L.) A. Gray, 259–261, 263
Monet, Claude, 35, 331
monkey-flower, 141
Monotropa L., 205
    hypopitys L., 205
    uniflora L., 201, 205
Monotropaceae Nutt., 17, 205
Monotropoideae Arn., 203, 205
Montgomery, Lucy Maud, 163
moose, 94
mosquito(es), 66, 185, 186, 282–283
    Aedes, 283
    Anopheles, 283
moss(es), 5, 12–13, 242, 244, 282
    sphagnum, 24, 67–68, 277, 281–282
moss plant, 3, 12–13
moth(s), 49, 66, 132–133, 154–155, 171–173, 177, 185, 187–189, 220, 222–225, 233, 272, 283–284, 329, 334
    ailanthus webworm, 185
    clearwing
        hummingbird, 177
        snowberry, 132
    delicate cycnia tiger, 138
    diurnal, 173, 184–185, 232, 236
    evening-primrose, 154–155
    goldenrod gall, 189
    gypsy, 135, 296
    hawk, 153–155, 171, 176–177, 223–224
        diurnal, 177
        five-spotted, 224
        pink-spotted, 224
    Io, 314
    milkweed tussock, 123
    nocturnal, 123, 153, 171, 176–177
    nocturnid, 177
    owlet, 48, 176
    snowberry clearwing, 132
    sphinx, 153–154, 223–224, 227
        diurnal, 176
        pink-spotted, 220
    tiger, 123
    tussock, 123
    unexpected cycnia, 136–137
Muir, John, 12
mullein, 141–146, 156
    black, 149–150
    clasping, 149–151
    common, 58, 140–151
    moth, 147–149
    purple, 149
    white, 149–150
mushroom(s), 81, 206–207, 320
    gilled, 207
muskrat(s), 36, 334
Myzocallis asclepiadis, 126

Nabokov, Vladimir, 326
Narthecium americanum Ker Gawl., 68
Nelumbo Adans., 35–36, 38–40, 331–332, 335
    lutea Willd. 35, 39, 41
    nucifera Gaertn., 35–36
Nelumbonaceae A. Rich., 35
nematodes, 124, 283
Newton, Mary Leslie, 293

nightshade, 221
nightshade family, 221
Nopales, 273
Nuphar Sm., 331–335
    advena (Aiton) W.T. Aiton, 332–334
    microphylla (Pers.) Fernald, 332
    variegata Engelm.ex Durand, 330–333
    ×*rubrodisca* Morong, 334–335
Nuttall, Thomas, 15
Nymphaea L., 38, 230, 331–332, 335
    odorata Aiton., 331–332
Nymphaeaceae Salisb., 38, 230, 331–332
Nymphaeales Salisb. ex Bercht. & J. Presl, 331
Nymphoides Ség.
    aquatica (J. F. Gmel.) Kuntze, 87
    cordata (Elliott) Fernald, 86–87
    peltata (S. G. Gmel.) Kuntze, 87

oak(s), 27, 242, 323, 326, 329
    black, 324
    red, 324
    scrub, 324, 326
    white, 324
Oakes, William, 15
Oenesis melissa semidea, 11
Oenothera L., 154–156
    biennis L., 153, 157
    guara W.L. Wagner & Hoch, 155
    parviflora L., 157
O'Keeffe, Georgia, 227
okra, 305, 307
Oligoneuron album (Nutt.) G.L. Nesom, 181
Onagraceae Juss., 153–154, 156
Oncopeltus fasciatus, 126
onion, 320
onion family, 317
Ophioglossum L., 70
Opuntia Mill., 265, 271–273, 275
    compressa J.F. Macbr., 265
    echios J.T. Howell, 271–272
        echios var. *barringtonensis* E.Y. Dawson, 271
    ficus-indica (L.) Mill., 265, 268, 272–273
    helleri K. Schum. ex B.L. Rob., 272
    humifusa (Raf.) Raf., 265–266, 268, 269
    macrorhiza Engelm., 265
orange(s), 20
orchid(s), 60, 65–67, 69–71, 73, 75–81, 170–171, 173, 176–177, 204, 231, 263, 299–303, 320
    bog, 66, 70–71
    crested yellow, 169, 174, 178
    fringed, 169–172, 176–178, 301
        orange, 169–170, 172–174, 176
        pale, 174
        purple, 175–176
            greater (large), 169, 175–177
            lesser (small), 169, 174–177
        ragged, 169, 176–177
        white, 168, 170–174, 176
        yellow, 169
    helleborine, 78, 80
    lady-slipper, 170
        pink, 299
        showy, 299
        yellow, 79, 303
    rein, 169
    slipper, 299
    tall northern white, 303
    weed, 79

orchid family, 60, 65–66, 75, 102, 169, 299
Orchidaceae Juss., 44, 65–66, 75, 107, 155, 169, 299
Orchis, 71, 73, 171
Oriental bittersweet, 195
oriole, black-backed, 129
Orobanchaceae Vent., 51, 141
Orthilia Raf., 259, 261, 263
    secunda (L.) House, 259–262
Osmunda cinnamomea L.
oyamel, 129

Padgett, Dr. Donald, 331–332
Panax L., 27, 30
    ginseng C.A. Mey., 27
    quinquefolium, 30
    quinquifolius L., 26–28, 30–31
    trifolius L., 30–31
Papaipema appassionata, 284
Papilio
    glaucus, 299–300
    polyxenes, 295
    troilus, 101–102, 173
Pardanthus Ker Gawl., 60
Parnassia L., 193–195, 197–199
    asarifolia Vent., 199
    glauca Raf., 192–194, 196, 199
    palustris L., 193–195
Parnassiaceae Martinov, 195
parrot flower, 219
parrot plant, 219
parrots, 219
parsley, 290
    fool's, 287, 290
parsnip(s), 157, 290, 295
    wild, 295–296
Parthenium argentatum A. Gray, 183
Parthenocissus quinquefolia (L.) Planch., 29
Passiflora L., 247, 250–255, 257
    biflora Lam., 255
    edulis Sims, 255–256
        fo. *edulis*, 255
        fo. *flavicarpa* O. Deg., 255
    foetida L., 253
    incarnata L., 247, 252–253
    lueta L., 246–248, 250–251, 253
    mollissima (Kunth) L.H. Bailey, 256
    tarminiana Coppens & V.E. Barney, 256–257
    tripartita var. *mollissima* (Kunth) Holm-Niels. & P. Jørg., 256
Passifloraceae Juss. ex Roussel, 247
passion-flower(s), 247–250, 252–257
    purple, 252
    yellow, 246–250
passion-flower family, 247
passion-fruit, 255
Peck, William D., 13
Penstemon Schmidel, 141
    eatonii A. Gray, 100
peppers, 226
Pereskia Mill., 269
    grandiflora Pfeiff., 269
Peterson, Roger Tory, 97
Philodendron Schott, 49
Phlomis L., 149
phlox, creeping, 99
phlox family, 100
Phlox stolonifera Sims, 99
Phoma Saccardo, 146

Index

Phrymaceae Schauer, 141
Phycoides tharos, 49
Phyllodoce, 12
Phyllodoce caerulea (L.) Bab., 3, 12
Phymata, 185
Phytolacca americana L., 138
Phytophora de Bary, 33
pilewort, 48
pine, 169, 205, 217, 323
    jack, 324, 326
    pitch, 174, 324
    white, 5, 51
pinesap, 205
pink family, 100
Pinus
    banksiana Lamb., 324
    rigida Mill., 174, 324
pipsissewa, 258–263
pitcher plant, 67–68, 276–285, 303
pitcher plant family, 277
Pitton de Tournefort, Joseph, 164, 285
Plantaginaceae Juss., 141
Plantago L., 141
plantain, 141
plantain family, 100, 141
Platanthera Rich., 169–170–171, 174, 176
    blephariglottis (Willd.) Lindl., 168–169, 171–173
    ciliaris (L.) Lindl., 169, 172–173
    cristata (Michx.) Lindl., 169, 174
    dilitata (Pursh) Lindl. ex L.C. Beck, 303
        var. *dilatata*, 303
    grandiflora (Bigel.) Lindl., 169, 175
    integra (Nutt.) A. Gray ex L. C. Beck, 174
    lacera (Michx.) D. Don, 169, 176–177
    pallida P.M. Brown, 174
    psycodes (L.) Lindl., 169, 174–177
    ×*andrewsii* (Niles) Luer, 177
    ×*bicolor* (Rafinesque) Luer, 172
    ×*canbyi* (Ames) Luer, 174
    ×*channellii* Folsom, 174
    ×*keenaii*, 177
Plebejus melissa subsp. *samueli*, 326
Pliny the Elder, 184
Pluchea odorata (L.) Cass., 315
Plumbaginaceae Juss., 285
pogonia(s), 71
    sweet, 36
Pogonia Juss., 65–66, 70
    ophioglossoides (L.) Ker Gawl., 65–66, 70
poison ivy, 211
poison-hemlock, 288–290, 295
pokeweed, 138
Popillia japonica, 314
potato(es), 33, 239, 318
Potentilla, 15
    robbinsiana (Lehm.) Oakes ex Rydb., 3, 14–15
Prenolepis impairs, 53
prickly burr, 221
prickly pear, 264–266, 268, 272–273
    eastern, 265–270, 272, 275
Primack, Richard, 231
Primulaceae Batsch ex Borkh., 156
prince's pine, 260
Protea L., 331
Proteales Juss. ex Brecht. J. Presl, 331
Pseudotsuga menziesii (Mirb.) Franco, 325
Psiguria Neck. ex Arn., 254
Pteridium aquilinum (L.) Kuhn, 324

Ptilothrix, 312–313
    bombiformis, 310–311
Puccinia Pers., 182
pupa (pupae), 49, 130, 146, 187, 313
purple coneflower, 99
Pycnanthemum Michx., 199
pyrola(s), 259–260, 262
    pink, 261–262
Pyrola L., 259–263
    asarifolia Michx., 262
    chlorantha Sw., 261
    elliptica Nutt., 205, 263
Pyrolaceae Lindl., 17, 205, 259
Pyrrharctia isabella, 49
Pyxidanthera Michx., 6–7

Queen Anne's lace, 58, 61, 107–108, 156, 164, 286–297
Quercus
    alba Loudon, 324
    ilicifolia Wangenh., 324
    rubra L., 324
    velutina Lam., 324
quick-in-the-hand, 210

rabbit(s), 94, 135, 218, 268
    cottontail, 268
Rafinesque, Constantine Samuel (Raf.), 285
ragged sailors, 108
ragweed, 184
ramp(s), 317–318, 320–321
ramson, 317, 320
Ranunculales Juss. ex Bercht. & J. Presl, 331
rats, 296
rattle nut, 41
rattlesnake-plantain, 81, 156
Redfield, John H., 44
Redouté, Pierre Joseph, 60
Rhabdopterus praetexus, 218
Rhingia nasica, 219
Rhinusa tetra, 146
Rhipsalis Gaertn., 270
Rhizanthella gardneri R. S. Rogers, 66
rhizobia, 324–325
Rhizocarpon geographicum (L.) DC., 6
rhododendron, 8–9, 17
Rhododendron L., 8–9, 17
    groenlandicum (Oeder) Kron & Judd, 12
    lapponicum (L.) Wahlenb., 6, 8–9
    ponticum L., 9
Rhopalomyia solidaginis, 188
Rhyssomatus lineaicollis, 125
rice, wild, 315
Robbins, James W., 15
Robinia pseudoacacia L., 185
Rosaceae Juss., 3
rose, pasture, 133
rose family, 13–14, 57, 162
rose pogonia, 65–66, 70–73
rosebay, Lapland, 2–3, 8–9
roselle, 310–311
rose-of-Sharon, 306–307
rotifers, 283
rubber, 134, 182–183
Rubia tinctorum L., 241, 273
Rubiaceae Juss., 3, 241
Rubus L., 324
Rudbeckia L., 101
rush(s), 3, 5

Rush, Benjamin, 242
Russell, George, 325
Russell hybrids, 325
Russula Pers., 206–207
Russulaceae Lotsy, 207
rust, 182
Rust, Richard, 214

sabra, 273
sage, scarlet, 100–101
St. Johnswort, marsh, 70
saltgrass, 315
Salvia coccinea Buc'hoz ex Etl., 100–101
sandwort, 11
    mountain, 3, 11
sang, 32
Sarcophaga sarraceniae, 281
Sarracenia L., 277–278, 284–285
    flava L., 278
    purpurea L., 68, 277–279, 281–282, 284–285
        subsp. *gibbosa* (Raf.) Wherry, 279
        subsp. *venosa* (Raf.) Wherry, 279
        var. *burkii* D.E. Schnell, 279
    rosea Naczi, Case & R.B. Case, 278–279
Sarraceniaceae, Dumort, 277
Sarrazin, Michel, 285
sarsaparilla, wild, 28–29
sassamenesh, 17
sawfly, 314
Saxifragaceae L., 195
scale insects, 273
Scalesia Arn. ex Lindl., 107
scallions, 320
Scelolyperus, 186
Schinia florida, 154
schinseng, 29
Schizachyrium scoparium (Michx.) Nash, 191
Schizomyia impatientis, 218
Scrophularia L., 142
Scrophulariaceae Juss., 141–142
seabirds, 94
sea-lavender, 285
seashore-mallow, 305, 309, 314–315
sedge(s), 3, 5, 65, 70, 199, 319
    Bigelow, 11
    seersucker, 319
Sedum L., 162
Semple, John, 181
Sericocarpus Nees, 43
sheperd's purse, 328
shinleaf, 205, 260, 262–263
shinleaf family, 259
Shortia Raf., 7
shy maiden, 260
sidebells, 259–263
Silene
    stenophylla Ledeb., 38
    virginica L., 100
silver-rod, 180
single delight, 259–263
Sisyrinchium L., 58
skunk cabbage, 40, 230
skunks, 268
slugs, 33, 126
snake mouth, 66
snap-weed, 210
sneezeweed, 199
Socrates, 289
Solanaceae, 221, 226

371

Index

Solidago L., 101, 179, 181–182, 190
   altissima L., 123, 183, 188, 190
   bicolor L., 180
   caesia L., 180
   canadensis L., 182–183, 187, 190
   cutleri Fernald, 11
   flexicaulis L., 180
   gigantea Aiton, 183
   juncea Aiton, 179–180
   ptarmicoides (Torr. & A. Gray) B. Boivin, 180–181
   rugosa Mill., 181, 184
   sempervirens L., 181, 186
   speciosa Nutt., 178, 181, 185–186, 188, 190–191
   vigaurea L., 182
Solidaster luteus M.L. Green ex Dress, 181
songbirds, 25, 28
sorrel, 310–311
Spartina Schreb., 315
spatterdock, 332–333
speedwell, 141
Speyeria cybele, 131
Sphagnum L., 24, 64–65, 67–68, 70–71, 82, 169, 172, 277, 281–282
   rubellum Wilson, 277
spicebush, 173
spider(s), 77, 121, 132–133, 183, 295, 327
   crab, 49, 113, 132–133, 177, 198, 283, 295
   sac, 283
Spigelia marilandica (L.) L., 100
Spiranthes Rich.
   cernua (L.) Rich., 199
   lucida (H.H. Eaton) Ames, 199
spruce, red, 5
spurge family, 268
squawroot, 205
Staphylococcus aureus, 263
starlings, 296
Sterculiaceae, 305
stiltgrass, Japanese, 29
stinkweed, 221
Streptopus Michx., 230
   lanceolatus (Aiton) Reveal, 237
sunbird(s), 219
sundews, 67–68
sunflower, 110–111
swamp pink, 66
swamp rose-mallow, 305–307, 309–315
swan(s), 162
   tundra, 25
   whistling, 25
sweet peas, 324
Swida Opiz, 95
Symphyotrichum Nees, 43
   foliaceum var. *apricum* (A. Gray) G.L. Nesom, 45
   lanceolatum (Willd.) G.L. Nesom, 46, 49
   novae-angliae (L.) G.L. Nesom, 42, 46–49, 129, 178
   novi-belgii (L.) G.L. Nesom, 43
   oblongifolium (Nutt.) G.L. Nesom, 46
   parviceps (E.S. Burgess) G.L. Nesom, 46
   puniceum (L.) Á. Löve & D. Löve, 45
Symplocarpus foetidus (L.) W. Salisb., 40

tansy, 46
Taraxicum, 112
   officinale L., 108
taro, 37
Taylor, George Lansing, 48

Tenodera aridifolia sinensis, 49
termites, 66
Tetraopes tetrophthalmus, 123–124
Theophratus, 156
Theseus, 45
thistle(s), 46, 134, 156
Thoreau, Henry David, 231
thorn apple, 221
thrush(es), 28
   hermit, 28
   Swainson's, 28
   wood, 28
Thuidium delicatulum (Hedw.) Schimp, 242
Tiliaceae Juss., 305
tobacco, 226
   Indian, 105
tomatoes, 226
tortoise(s), 271–272
   giant, 271–272
touch-me-not, 210, 216
touch-me-not family, 209
Tournefort, Joseph Pitton de, 164, 285
Toxomerus
   geminatus, 114, 214
   marginatus, 145
Tragopogon pratensis L., 115
trailing arbutus, 245
tree of heaven, 185
trillium, 230
Trillium L., 230
trout-lily, 156, 230
truffles, 81
Tuber P. Micheli ex F.H. Wigg, 81
Tulipa L., 230
tulips, 230, 319
tuna, 273
turkeys, 33
   wild, 268
turtles, bog, 303
turtlehead, 141
twisted stalk, 230, 237
Typha angustifolia L., 315
Typocerus
   lunulatus, 295
   veluntinus, 40

Umbelliferae Juss., 291
Uromyces, 182
Urquhart(s), 127
Utricularia cornuta Michx., 276
Uvularia L., 230

Vaccinium L., 10, 17, 24
   angustifolium Aiton, 10, 163
   cespitosum Michx., 10
   erythrocarpum Michx., 25
   macrocarpon Aiton, 10, 16–18, 20, 24–25, 68
   oxycoccus L., 10, 18, 24
   uliginosum L., 8, 10
   vitis-ideae L., 3–4, 10, 244
      subsp. *minus* (Lodd., G. Lodd. & W. Lodd.) Hultén, 10, 25
      subsp. *vitis-idaea*, 10–11, 25
Vanilla Mill., 60
Venus flytrap, 279
Veratrum viride Aiton, 12, 303
Verbascum L., 142–143, 146–147
   blattaria L., 147
   densiflorum Bertol., 151

   lychnitis L., 149
   nigrum L., 149–150
   phlomoides L., 149–150
   phoeniceum L., 149
   thapsus L., 58, 141, 143, 147–149, 151
Vernonia Schreb., 112
Veronica L., 141
Vespula
   consobrina, 77
   sylvestris, 78
vetches, 324
Viburnum opulus var. *americanum* Aiton, 25
Viburnum trilobum Marshall, 25
Victoria Lindl., 335
viper's bugloss, 58
Virgin Mary, 230
Virginia bluebells, 108
Virginia creeper, 29
Viscum L., 270
Vitaceae Juss., 29

Washington, George, 110
wasp(s), 10, 77–78, 112–113, 126, 132–133, 185–187, 189–190, 213–214, 249, 294
   ichneumon, 190
   parasitic, 189–190, 238, 327
   social, 77
water shamrock, 83
waterfowl, 334
water-hemlock, 287, 290
   bulbiferous, 287–288, 290
   common, 287–290
watering can, 41
water-lily, 38, 230, 332
   Amazon, 335
   white, 331–333
water-lily family, 38, 331
weevil(s), 123, 125–126, 146, 314
   milkweed, 125
   mullein, 146
Werner, Patricia, 147
Wherry, Edgar, 285
Williams, William Carlos, 293
Willmer, Pat, 226
willow(s), 5
   weeping, 6
wintergreen, 160, 260
   green-flowered, 261
   one-flowered, 260
   one-sided, 260
   pink, 262
   spotted, 205, 259–263
witlof, 109
wood nymph, 260
woodpecker, 189
   downy, 146, 189
woolly bear, 49
Wyeomyia smithii, 282–283

Xylocopa, 164, 252
   darwinii, 253, 267
   virginica, 184, 252

yeasts, 81
yellow pond-lily, 330–335
yockernut, 41

Zantedeschia aethiopica (L.) Spreng., 230
Zizania aquatica L., 315